INSECT PESTS OF STORED GRAIN

Biology, Behavior, and Management Strategies

Postharvest Biology and Technology

INSECT PESTS OF
STORED GRAIN

Biology, Behavior, and Management Strategies

Ranjeet Kumar, PhD

Apple Academic Press Inc.
3333 Mistwell Crescent
Oakville, ON L6L 0A2 Canada

Apple Academic Press Inc.
9 Spinnaker Way
Waretown, NJ 08758 USA

© 2017 by Apple Academic Press, Inc.

First issued in paperback 2021

Exclusive worldwide distribution by CRC Press, a member of Taylor & Francis Group

No claim to original U.S. Government works

ISBN 13: 978-1-77-463682-4 (pbk)
ISBN 13: 978-1-77-188503-4 (hbk)

Library and Archives Canada Cataloguing in Publication

Kumar, Ranjeet, 1982-, author
Insect pests of stored grain : biology, behavior, and management strategies / Ranjeet Kumar, PhD.
(Postharvest biology and technology book series)
Includes bibliographical references and index.
Issued in print and electronic formats.
ISBN 978-1-77188-503-4 (hardcover).--ISBN 978-1-31536-569-5 (PDF)
1. Grain--Storage--Diseases and injuries. 2. Grain--Diseases and pests. 3. Insect pests--Control. I. Title. II. Series: Postharvest biology and technology book series
SB608.G6K87 2017   633.1'0497   C2017-900751-3   C2017-900752-1

Library of Congress Cataloging-in-Publication Data

Names: Kumar, Ranjeet, 1982- author.
Title: Insect pests of stored grain : biology, behavior, and management strategies / author: Ranjeet Kumar, PhD.
Description: Waretown, NJ : Apple Academic Press, 2017. | Includes bibliographical references and index.
Identifiers: LCCN 2017005613 (print) | LCCN 2017009960 (ebook) | ISBN 9781771885034 (hardcover : alk. paper) | ISBN 9781315365695 (ebook)
Subjects: LCSH: Grain--Storage--Diseases and injuries. | Grain--Diseases and pests. | Insect pests--Control.
Classification: LCC SB608.G6 K86 2017 (print) | LCC SB608.G6 (ebook) | DDC 633.1/0468--dc23
LC record available at https://lccn.loc.gov/2017005613

Apple Academic Press also publishes its books in a variety of electronic formats. Some content that appears in print may not be available in electronic format. For information about Apple Academic Press products, visit our website at **www.appleacademicpress.com** and the CRC Press website at **www.crcpress.com**

Dedicated to my son ABHINAV with love

ABOUT THE AUTHOR

Dr. Ranjeet Kumar

Ranjeet Kumar, PhD, is an Assistant Professor cum Junior Scientist in the Post-Graduate Department of Entomology at Bihar Agriculture University, Sabour, Bhagalpur, Bihar, India. Dr. Kumar has worked on stored product entomology for the sustainable and herbal management of stored grain and seed insect pests, investigating several herbal fumigants for the sustainable management of stored grain and seed insect. He has published several books and a manual on the topic as well as authored over 20 research papers in national and international journals. He has also published 19 popular articles in magazines and periodicals.

Dr. Kumar is life member or fellow of several scientific committee and societies. He received an award from an international conference on entomology at Punjabi University in Punjab.

Dr. Ranjeet Kumar earned his PhD (Stored Product Entomology) in 2010 from the G. B. Pant University of Agriculture and Technology, Pantnagar, Uttarakhand, India.

ABOUT THE BOOK SERIES: POSTHARVEST BIOLOGY AND TECHNOLOGY

As we know, preserving the quality of fresh produce has long been a challenging task. In the past, several approaches were in use for the postharvest management of fresh produce, but due to continuous advancement in technology, the increased health consciousness of consumers, and environmental concerns, these approaches have been modified and enhanced to address these issues and concerns.

The **Postharvest Biology and Technology** series presents edited books that address many important aspects related to postharvest technology of fresh produce. The series presents existing and novel management systems that are in use today or that have great potential to maintain the postharvest quality of fresh produce in terms of microbiological safety, nutrition, and sensory quality.

The books are aimed at professionals, postharvest scientists, academicians researching postharvest problems, and graduate-level students. This series is intended to be a comprehensive venture that provides up-to-date scientific and technical information focusing on postharvest management for fresh produce.

Books in the series will address the following themes:

- Nutritional composition and antioxidant properties of fresh produce
- Postharvest physiology and biochemistry
- Biotic and abiotic factors affecting maturity and quality
- Preharvest treatments affecting postharvest quality
- Maturity and harvesting issues
- Nondestructive quality assessment
- Physiological and biochemical changes during ripening
- Postharvest treatments and their effects on shelf life and quality
- Postharvest operations such as sorting, grading, ripening, de-greening, curing etc
- Storage and shelf-life studies
- Packaging, transportation, and marketing
- Vase life improvement of flowers and foliage
- Postharvest management of spice, medicinal, and plantation crops
- Fruit and vegetable processing waste/byproducts: management and utilization
- Postharvest diseases and physiological disorders
- Minimal processing of fruits and vegetables

- Quarantine and phytosanitory treatments for fresh produce
- Conventional and modern breeding approaches to improve the postharvest quality
- Biotechnological approaches to improve postharvest quality of horticultural crops

We are seeking editors to edit volumes in different postharvest areas for the series. Interested editors may also propose other relevant subjects within their field of expertise, which may not be mentioned in the list above. We can only publish a limited number of volumes each year, so if you are interested, please email your proposal wasim@appleacademicpress.com at your earliest convenience.

We look forward to hearing from you soon.

Editor-in-Chief:
Mohammed Wasim Siddiqui, Ph.D.
Scientist-cum-Assistant Professor | Bihar Agricultural University
Department of Food Science and Technology | Sabour | Bhagalpur | Bihar | INDIA
AAP Acquisitions Editor, Horticultural Science
Founding/Managing Editor, *Journal of Postharvest Technology*
Email: wasim@appleacademicpress.com
wasim_serene@yahoo.com

Books in the Postharvest Biology and Technology Series:

Postharvest Biology and Technology of Horticultural Crops: Principles and Practices for Quality Maintenance
Editor: Mohammed Wasim Siddiqui, PhD

Postharvest Management of Horticultural Crops: Practices for Quality Preservation
Editor: Mohammed Wasim Siddiqui, PhD

Insect Pests of Stored Grain: Biology, Behavior, and Management Strategies
Ranjeet Kumar, PhD

CONTENTS

LIST OF ABBREVIATIONS

ACS	American Chemical Society
ADI	Acceptable daily intake
APC	Agricultural Price Commission
APEDA	Agricultural and Processed Food Products Export Development Authority
CA	controlled atmosphere
CACP	Commission on Agricultural Costs and Prices
CNS	central nervous system
CT	carbon tetrachloride
CWC	Central Warehousing Corporation
DIPA	destructive insect pest act
DPPQS	Directorate of Plant Protection and Storage
EC	economic community
EDB	ethylene dibromide
EPA	Environmental Protection Agency
FAO	Food and Agriculture Organization
FCI	Food Corporation of India
GC	gas chromatography
GCMS	gas chromatography and mass spectra
GIT	gastrointestinal tract effects
IGSI	Indian Grain Storage Institute
IGSRTI	Indian Grain Storage Research and Training Institute
IOBC	International Organization on Biological Control
IPPC	International Plant Protection Convention
IR	infrared
LC	lethal concentration
LD	lethal dose
MB	methyl bromide
NBARD	National Bank for Agriculture and Rural Development
NCI	National Cancer Institute
NCIPM	National Center for Integrated Pest Management
NE	natural enemies
NIR	near infrared
NRI	Natural Resource Institute

PFAA	Prevention of Food Adulteration Act
PPO	Plant Protection Organization
RCA	Royal Commission on Agriculture
SGC	Save Grain Campaign
SGRL	Stored Grain Research Laboratory
SWC	State Warehousing Corporation
TC	toxic concentration
TD	toxic dose
USDA	United Nation Department of Agriculture
WCA	Warehousing Corporation Act
WDRA	Warehouse Development and Regulation Act
WFD	World Food Day
WHO	World Health Organization

PREFACE

Stored products of agriculture and animal origin are attacked by more than 600 species of beetles, 70 species of moths, and about 355 species of mites, causing quantitative and qualitative losses. Insect contamination in food commodities is an important quality control problem of concern at different storage levels and for food and agricultural industries. According to recent estimates, about 70–75% of food grain is handled at farmer's level in India where we do not have adequate scientific storage facilities and grain-protection technologies; rest 25–30% of food grain is procured and stored by Food Corporation of India (FCI), Central Warehousing Corporation (CWC), and State Warehousing Corporation (SWC) who have scientific storage facility. The surplus food grain and seeds of the nation needs keeping facilities and care during storage.

Alternative options to traditional management of stored grain insects pests are being advocated in several countries since last four decades. Integrated pest management, secondary metabolites of plant and their products, use of bio agents, manipulation of abiotic factors, inert dusts, and application of microbial pathogens are recently utilized for sustainable management of stored grain insects all over the world under different storage conditions.

This book *Insect Pests of Stored Grains: Biology, Behavior, and Management Strategies* covers all the aspects of stored product entomology since the beginning to the modern era in detail. This book consists of 18 chapters. Chapter 1 provides an introduction of storage entomology and information related to one of the important branches of entomology. Chapter 2 describes the historical aspects of stored products since human civilization and important developments in the field of storage entomology for better discussion. The classification and identification of important stored grain insects are discussed in Chapter 3 with possible key of identification. Chapter 4 deals with the major stored product coleopteran and lepidopteran insects infesting our stored commodities. The storage infestation, either visible or hidden, and their detection techniques are elaborated in Chapter 5. The estimation of losses caused by stored grain insect pests on scientific lines is presented in Chapter 6. Chapter 7 deals with different sources and kinds of infestation. Chapter 8 discusses the factor responsible for infestability of stored grain insects. Chapter 9 deals with different structures used for storing commodities since the beginning to the modern era. Chapters 10 and 11 elaborate the alternative methods for management of stored grain insects by utilization of behavior modification techniques or utilization of secondary metabolites of plants. Chapter 12 discusses the integrated management of stored grain insects. Fumigation of stored grains for the protection of infestation is one of the important and age-old practices since the beginning to till date and it is elaborated in Chapter 13. Besides stored grain insects, mites, birds, rodent, and fungi also affect stored commodities which are thoroughly discussed in Chapters 14, 15, 16, and 17, respectively. Chapter 18 deals with some major stored grain insects of quarantine importance.

This book will provide comprehensive information in the field of stored product entomology in order to facilitate sustainable management of insects and other non-insect pests.

The author would appreciate receiving comments and suggestions from readers that will be helpful in subsequent editions.

—**Ranjeet Kumar**

ACKNOWLEDGMENTS

Due to the blessings of my god **HANUMAN JI,** I am able to complete this book. I express my heartiest gratitude to god **HANUMAN JI** for his continuous blessings for writing this book.

I express my sincere thanks to Bihar Agricultural University, Sabour, Bhagalpur, India, for providing an opportunity to execute this type of work.

I feel elated to owe my thanks to the founder and former honorable Vice Chancellor Dr. M. L. Chaudhary and former Dean Agriculture Dr. D. Roy of Bihar Agricultural University for their constant encouragement and guidance.

I am very much grateful to Dr. Washim Siddiqui, Assistant Professor cum Junior Scientist, Department of Food Science and Technology, Bihar Agricultural University, Sabour, for his inspiration and consideration of my project work in a very short time.

I am grateful to Mr. Ashish Kumar, President, Apple Academic Press, to complete my dream in the form of this book. I am also grateful to Ms. Sandra Jones Sickels and Mr. Rakesh Kumar of Apple Academic Press for their constant support and inspiration to publish my book.

I express my sincere thanks to the Food and Agriculture Organization for providing permission for reproducing some text. I am also thankful to all the workers associated with the field of stored product entomology. The information and resource generated by several publishers are duly acknowledged.

I feel immense pleasure to express my gratitude to my adored parents Sri Vijay Kumar Keshari and Late Durgawati Keshari for their continuous inspiration and blessings in my life.

I express my special thanks to my better half and heartiest benevolent Smt. Vineeta and my son Abhinav for silent prayers, moral support, and encouragement for the completion of this book.

CHAPTER 1

INTRODUCTION TO STORAGE ENTOMOLOGY

CONTENTS

ABSTRACT

The qualitative or quantitative damage and contamination of stored commodities by insects, mites, rodents, and fungi could be prevented by management practices with a minimal use of hazardous chemicals and avoiding harmful chemical residues. This could be termed as stored grain pest management or storage entomology or management of post harvest losses. Storage of food grain is indispensable to ensure continuous food supply, but it is not possible without the use of fumigants. None of the countries of world have any other viable alternative to protect food grain from infestation of insect pests during storage which are known to cause 5–20% damage in different countries. Most of the countries lack adequate scientific storage facilities which are also dependent on fumigation for protection of grain. Good food is the basic need to good health, thus demands are being placed on production system as there is increase in population. Any loss of food due to insect, mites, and fungi means decrease in the availability of food supply to support good health of people. Presently, problem of improving the quality of products is one of the important aspects of food commodities. It is necessary to improve in both terms of quality and quantity, but to renovate mechanically the range of commodities, keeping quality with scientific technical achievements and taking into account the increasing necessity of the country. The processing of food grains and then their conservation is of vital area of national and international importance. Fumigation plays a vital role in insect pest management in stored products. Currently, phosphine and methyl bromide are the two common fumigants used for stored-product protection worldwide. Insect resistance to phosphine is a global issue now and control failure has been reported from several countries. Methyl bromide, a broad spectrum fumigant, has been declared as an ozone depleting substance and therefore, is being phased out completely. Now a time has reached when we have to think whether it is a boon or bane. Scientists all over the world realize that presently it is impossible to survive without its use, but they also understand that it is necessary to minimize its use at least at farmers' level by searching other viable alternatives. The stored grain insect pests may be managed by biotechnological intervention being carried out to utilize the gene responsible for protease inhibition and production of secondary metabolites, and use of antisense RNA technology in combination with gene transfer. Transgenic plants may also be used in manipulating enzyme synthesis for protection of stored products.

1.1 INTRODUCTION TO STORAGE ENTOMOLOGY

India is a developing country with population of 125 crores, green revolution has increased our production about 230 million tones of food grains. The introduction of new high yielding variety and use in crop husbandry resulting increased production and food storage for long period have posed many problems in storage and increased the importance of prevention of loss in stored commodities. The directed efforts to preserve produce will solve problem of feeding of hungry men to a considerable extent. The damage and contamination by insects, mites, rodents, and fungi could be prevented by management practices with a minimal use of hazardous chemicals and avoiding harmful chemical residues. This could be termed as stored grain pest management or storage entomology or management of post harvest losses. Storage of food grain is indispensable to ensure continuous food supply, but it is not possible without the use of fumigants. None of the countries of world have any other viable alternative to protect food grain from infestation of insect pests during storage which are known to cause 5–20% damage in different countries. Most of the countries lack adequate scientific storage facilities which are also dependent on fumigation for protection of grain. The situation is very depressing at farmers' level who store about 70% grain but are not permitted or equipped to use these chemical fumigants due to legal restrictions, lack of knowledge, or non-availability of air-tight storage structures. Stored products of agriculture and animal origin are attacked by more than 600 species of beetles, 70 species of moths, 140 species of rodents, 15 species of fungi, and about 355 species of mites, causing quantitative and qualitative losses and insect contamination in food commodities. It is an important quality control problem of concern at different storage levels and for food industries (Rajendran, 2002).

There is a wide recognition of the fact that farmers either lack the proper means of handling, transportation to store the bumper crop. It is recognized by grain handlers that when grain is left unattended for even a short while, it is invaded by different types of insects, mites, and fungi. The cool and temperate climate and lower temperature sometimes incubate the development of pest populations. After harvesting, most of the grains contain high moisture that enhances the potency of infestations. In developing countries falling in sub-tropical and tropical areas, however, where sanitation standards are generally very poor, insect population often build up rapidly as insects are active here in most parts of the years. The development of insect infestation is a complex process followed by equal interactions of biotic and abiotic factors. The temperature, moisture, and relative humidity play a vital role in determining the storage ability of grain for a length of period. The introduction of grain market liberalization in many countries has resulted in a rapid decline or elimination of local marketing organizations. Farmers now have an opportunity to take advantage of seasonal price but the benefits can only be achieved if grain is held longer on the farm with no deterioration in quality. Thus the farmers are advised on the most cost effective type of storage system and methods of pest management to use.

Good food is the basic need to good health, thus demands are being placed on production system as there is increase in population. Any loss of food due to insect, mites, and fungi means decrease in the availability of food supply to support good health of people. There have been reports on records on intestinal diseases after eating the food with insect infestation (Palmer, 1970).

The industrial enterprises, commercial firms, bakeries, and agro based companies become aware of the need for specialized capable of evaluating the materials necessary for production. In this relation, a commercial education system arose in the country, conditions for the production of high quality commodities have been established in India. Only recently and state control of their quality introduced throughout, from producer, distributor, trader to consumers standardization of quality norms for commodities have been made the basis of system of standards. These along with advances in science led to development of science of food commodities as a specific discipline and turned into a major field of applied science. Further experts in the field of trade, supply, public relation, technology of different industries, economists, agronomists, food technology, and many others study commodities science to varying extents from the view of their consumer value. A research institute for this purpose first time established at Hapur in the year 1958 to carry out complete study in the field of storage quality control in India.

At present, problem of improving the quality of products is one of the important aspects of food commodities. It is necessary to improve in both terms of quality and quantity, but to renovate mechanically the range of commodities, keeping quality with scientific technical achievements and taking into account the increasing necessity of the country. The processing of food grains and then their conservation is of vital area of national and international importance. It is well established that it would provide better utilization of agricultural raw materials which are now either wasted or give a low return. Further, it would ensure a better nutrition to a mass of population, rural or urban, who suffer from malnutrition and protein deficiency. It has also interesting export possibilities of some of the agricultural produce of the country for which an export demand already exists.

1.2 STORAGE PESTS MANAGEMENT STRATEGIES

Research has been conducted on various aspect of stored grain pest management like storage structure of different capabilities, application of chemicals and non chemicals, and manipulation of environmental factors to prevent the infestation of insects, rodents, mites, and so many microorganisms. Fumigation plays a vital role in insect pest management in stored products. Currently, phosphine and methyl bromide are the two common fumigants used for stored-product protection worldwide. Insect resistance to phosphine is a global issue now and control failure has been reported from several countries (Taylor, 1989; Collins et al., 2002). Methyl bromide, a broad spectrum fumigant, has been declared as an ozone depleting substance and therefore, is being phased out completely. Now a time has reached when we have to think whether it is a boon or bane. Scientists all over the world realize that presently it is impossible to survive without its use. But they also understand that it is necessary to minimize its use at least at farmer's level by searching other viable alternatives. In view of the problems with the current fumigants, there is a global interest in alternative strategies including development of chemical substitutes, exploitation of controlled atmospheres and integration of physical methods (MBTOC, 2002). Presently, the use of sulfuryl fluoride, a structural fumigant for termite and woodborer control, has been expanded to food commodities and food handling establishment in USA, Canada, and European countries (Prabhakaran, 2006). New fumigants such as carbonyl sulfide (Desmarchelier, 1994) and ethane dinitrile (Ryan et al., 2006) and the old fumigant ethyl formate either alone or mixture with CO_2 (Damcevski et al., 2003) have also been investigated as alternatives for food and non food commodities. Furthermore, interest has been shown in plant products, that is, essential oils and their components, fumigant action since it is believed that natural compounds from plant sources may have the advantage over traditional fumigants in terms of low mammalian toxicity, rapid degradation, and local availability. For the last four decades serious attempts are being made to find its alternative in plant kingdom and hundreds of plants have been screened to search their fumigation potential. And most interesting so many plants have been identified which are equipped with fumigant properties against insect pests of stored grain. Now it has been proved beyond doubt that under laboratory condition the plant components can suppress feeding and breeding of stored grain insect or even kill them completely within a few hours just like chemical fumigants.

In controlling the stored grain insect pests the biotechnological intervention has got good potential. So an extensive research is being carried out to utilize the genes responsible for protease inhibition and production of secondary metabolites, and use of antisense RNA technology in combination with gene transfer is used for management of stored grain insect pests. Transgenic plants can thus be used in manipulating enzyme synthesis for protection of stored products.

With all advantages and disadvantages, essential that if not the mass population, at least grain handler and those interested should get the know how about grain preservation, problem, and management practices. As yet, however, no efforts have so far been made to prepare a comprehensive text covering, the essential aspect of grain and seed storage with special emphasis on several types of management strategies of food grain and seeds into single volumes. This book, therefore, aims to go someway toward filling the gaps, is based on experiments and practices of farmers with awareness and development of green pesticides for the sustainable management of store grain insects, rodents, birds, and microorganisms. I hope this book will arise an awareness in people involve in grain handling, transportation, and storage of food grain and seeds, and also helpful for students of post graduate and doctorate under agricultural university system in India and abroad.

KEYWORDS

- fumigants
- insect infestation
- protein deficiency
- food technology
- mammalian toxicity

REFERENCES

Collins, P. J.; Daglish, G. J.; Pavic, H.; Lambkin, T. M.; Kapittke, R. Combating Strong Resistance to Phosphine in Stored Grain Pests in Australia. In *Stored Grain in Australia 2000,* Proceeding of the Australian Postharvest Technical Conference, Adelaide, Aug 1–4, 2000; Wright, E. J., Banks, H. J., Highley, E., Eds.; CSIRIO Stored Grain Research Laboratory: Canberra, Australia, 2002; pp109–112.

Damcevski, K.; Dojchinov, G.; Haritos, V. S. VAPORMATE™, a Formulation of Ethy Formate with CO_2 for Disinfestations of Grain. In *Stored Grain in Australia 2003,* Proceedings of the Australian Postharvest Technical Conference, Canberra, Jun 25–27, 2003; Wright, E. J.; Webb, M. C.; Highley, E., Eds.; CSIRO Stored Grain Research Laboratory: Canberra, Australia, 2003;pp 199–204.

Desmarchelier, J. M. Carbonyl Sulphide as a Fumigant for Control of Insects and Mites. In *Stored Product Protection*, Proceeding of the Sixth International Working Conference on Stored product Protection, Canberra, Australia, Apr17–23, 1994, Highley, E.; Wright, E. J.; Banks, H. J.; Champ, B. R., Eds.; CAB International: Wallingford, UK, 1994; pp 75–81.

MBTOC. Reports of the Methyl Bromide Technical Options Committee Assessment: UNEP, Nairobi, Kenya, 2002.

Palmer, E. D. Entomology of Gastrointestinal Tract. A Brief Review. *Mil. Med.* **1970**, *13 (1)*, 41–42.

Prabhakaran, S. In *Commercial Performance and Global Development Status of Profume Gas Fumigant*, Proceeding of the 9th International Working Conference on Stored Product Protection, Oct 15–18, 2006; Sao Paulo, Brazil, Brazilian Postharvest Association: Campinas, Brazil, 2006; pp 635–641.

Rajendran, S. Postharvest Pest Losses. In *Encyclopedia of Pest Management;* Pimentel, D., Ed.; Marcel Dekker, Inc.: New York, 2002; pp 654–656.

Ryan, R.; Grant, N.; Nicolson, J.; Beven, D.; Harvey, A. *Sterigas^{TM} and Cosmic^{TM} Update on Proposed New Fumigants,* In Proceeding of the 9th International Working Conference on Stored Product Protection, Sao Paulo, Brazil, Oct 15–18, 2006; Brazilian Postharvest Association: Campinas, Brazil, 2006; pp 624–629.

Taylor, R. W. D. Phosphine-a Major Fumigant at Risk. *Int. Pest Control.* **1989**, *31,* 10–14.

CHAPTER 2

HISTORY OF GRAIN OR SEED STORAGE

CONTENTS

ABSTRACT

Since the civilization of agriculture in Mesolithic period, some indications are there for crop usage, for example mills stores, pestles, and mortars discovered during the Mount Carmel explorations of 1930. The history of grain and seed storage indicates that in ancient times, man practiced hunting wild animals. They further started collecting various foods including seeds of some wild plants to have a change in the way of life in a settled community. The discoveries to grow more crop and store made man to leave nomadic habit and live in a settled society. The settled life made man to have development of culture and agriculture as there would have been enough leisure time to think of storage of food grains and seeds. Cereals have been the most valued crop throughout all eras and were the first plant types cultivated and easy to store. The crop storage seems to have begun in 8000 B.C. as per radio carbon dating. The story of crop conservation and storage is interesting as it moves from the new Stone Age and Copper Age into Bronze Age and later Iron and Roman Ages. The grain storage was learnt indigenously at different times, but there has been a gradual transfer of knowledge from East to West and North to South. The storage structure constructed for wheat and rice has been observed in the excavations of Mohenjo-daro, Harappa, and Lothal of Indus Valley Civilization. These structures were dated back to about 2500–2000 B.C. and were made of sunbaked and fireclay materials. The reference of wheat grains is found in "Atharvaveda" an Indian holy book dated back to 1500–500 B.C. "Manu Smriti," an Indian book can also be quoted as an evidence for the knowledge of grain storage in India as there is a writing as good seeds in the good field yields abundant crops. In India, the establishment of warehousing came into being on the basis of recommendation by Royal Commission of Agriculture in the year 1928 and was brought to enact legislation for setting up of warehouses in 1944. Earlier to this, the storage of food grain seems to be the hand of farmers, grain handlers, and marketers. The structures, enclosures, or wrappers used for storing the food grains or seeds for the safety against any types of losses come under category of warehouse. The food habits of humans vary from tropical to sub-tropical; persons living by the costal side and mainly paddy growing area prefer rice and so it is a staple food for them.

2.1   HISTORY OF GRAIN OR SEED STORAGE

Storage of grains and seeds is a very important part for human health and better crop growth. Since the civilization of agriculture in Mesolithic period, some indications are there for crop usage, for example mills stores, pestles, and mortars discovered during the Mount Carmel explorations of 1930. Coles (1973) reported that the status of human groups, the organization of its economy, and the size of population, all depend to a certain extent on its sources in planned grain and seed utilization and its storage.

The history of grain and seed storage indicates that in ancient times, man practiced hunting wild animals. They further started collecting various foods including seeds of some wild plants to have a change in the way of life in a settled community. Nash (1978) published the knowledge on how to store crops probably developed before it was discovered, how to cultivate grain crops. Man could not hunt in rainy season and whatever he could have stored used during this season. The discoveries to grow more crop and store made man to leave nomadic habit and live in a settled society. The settled life made man to have development of culture and agriculture as there would have been enough leisure time to think of storage of food grains and seeds. Cereals have been the most valued crop throughout all eras and were the first plant types cultivated and easy to store. The crop storage seems to have begun in 8000 B.C. as per radio carbon dating. The story of crop conservation and storage is an interesting as it moves from the new Stone Age and Copper Age into Bronze Age and later Iron and Roman Ages. The grain storage was learnt indigenously at different times, but there has been a gradual transfer of knowledge from East to West. Nash (1978) gave an account of tome lag with the advances in the age, the time taken in spread in knowledge is clear when reference is paused, for example Neolithic (New Stone and Copper culture in East 7000 B.C., in West Europe 4000 B.C., in France and Britain 3500 B.C.). Bronze culture 3000 B.C. in East and 1000 B.C. in Europe, Iron age 100 B.C. and West Europe only 700 B.C. thus Iron age culture seems to have spread more rapidly requiring only 400 years. Storage pits have been known to be used dating back 4500 B.C. were used for storing of harvested wild crops in Syria and Israel. It is clear that in the beginning knowledge on how to store food grain and seeds was minimal and with the development of settled society knowledge and ability on hoe to store grains and seeds developed before it was discovered to cultivate and grow crops.

The storage structure constructed for wheat and rice has been observed in the excavations of Mohenjo-daro, Harappa, and Lothal of Indus Valley Civilization. These structures were dated back to about 2500–2000 B.C. and were made of sunbaked and fireclay materials. The reference of wheat grains is found in "Atharvaveda" an Indian holy book dated back to 1500–500 B.C. "Manu Smriti," an Indian book can also be quoted as an evidence for the knowledge of grain storage in India as there is a writing as good seeds in the good field yields abundant crops. The nomadic nature of man, on the other hand, does not testify the habit of man to store food. Anyhow, whatever may be the period of start of food grain storage, it is proved that man storing of food grain were known as used thousands of years ago. It may be that crop storage which was learnt indigenously in different parts of the world in different era by the way of survival. Mankind, in the starting storing grain for himself only, latter for the family and ultimately to a broader prospects. Due to pressure of huge population there are needs to commercial storage. The increase in population led to expansion and migration for space which in turn resulted in fighting for the daily food grain supply. The vagaries of climate coupled with war resulted in starvation at times. Edouard Saouma, former Director General, Food and Agriculture Organization, warned in September 1980, for the adverse sign of food supply chain under strain so long that it would reach to breaking point. In East Africa, drought and war resulted in a wide spread famine. Such types of warning have been issued from time to time by the persons on the whelm of affairs about the shortage of food supply. Each country has a chain or production system approach, where in surplus produce is brought to the market and still surplus is maintained as a buffer stock. In the beginning, there has been barter system of fulfilling daily requirements. Latter, a system of transaction with common standard money came into practice. The value of money fluctuates with the availability of supply. The broad expansion of production depends upon an increase in income and investment. Here, one constrained in its ability to undertake directly or indirectly to promote more investment because of the certain inflationary impact. A potential cost inflation in the production sector is being suppressed. Higher cost due to devaluation and the attempt to restore real income in the face of rising food costs not fully reflects in produce. The most important, though, is the potential for further demand which induces increase in food prices. If the deficits widen or credit to grower grows rapidly to pay for new investment, the associated prices will quickly be translated into demand for scarce food. Thus, without larger food supplies any general expansion of demand is bound to raise prices and thereby exacerbate an already precarious situation. Man, thus, learnt reserving grain for dearth periods. Thus the store houses were built of huge capacities even in earlier days too, to serve during starvation. The evidences of record on Egyptian monuments expressed that storage structures were made of clay, stones, and bricks. The Kautilya, famous minister during Maurya region had written about good store housekeeping and grading grain for quality standards. The Mauryans were named as promoters to save the grain and construct the storage structure at home and on ships or boat to developed trade with neighboring countries far and near. Chhatrapati Shivaji in Maharashtra, Tamil Nadu, Thanjavur are known for the architect comparing to updated silos. The construction of system of storage had been in practice as per the necessity and requirement depending upon the existing situation, for instance, to protect grain stored from ravages of bombing and theft. The underground storage was designed war. The tunnel type of storage structure could be seen in Argentina. It would be disgraceful to speak of knowledge of the person of time and place, if said, that as with advancement of science and technology, it would be better to spell out that with the need and requirement of the facility for storage of food grains the methods and structures were suited were designed. The production of steel bins were designed but because of heating and spoilage, concrete bins with inbuilt arrangements for temperature and humidity recording and treatment management were designed in some developed nations, whereas, in developing countries the traditional procedures are still in practice. The huge amount of surplus production of food grains has brought about revolution in marketing and international policies and newer mode of storage of food grains and seeds. The requirement of storage structures depends upon the time and place of production. At places emergent situations arise at and other places the food grain supply is surplus to be conserved for long term to come and distribution is desired at place of requirement of immediate supply. Thus the storage method employed is dependent upon the surrounding environment of places, socio-economic conditions of the mass population, state, and country.

The state or nation is responsible for the development of storage facility as per the production and needs. In India the establishment of warehousing came into being on the basis of recommendation by Royal Commission of Agriculture in the year 1928 and was brought to enact legislation for setting up of warehouses in 1944. Earlier to this, the storage of food grain seems to be the hand of farmers, grain handlers, and marketers. There have been several committees appointed and the recommendations of these committees led to the promotion and development of food grain warehouses such as Committee of Government of India, 1945 and 1949. And later conferences and symposiums were held to discuss the development of central and state level warehousing facilities. These committees were constituted in order to survey and study the problem of rural socioeconomic conditions and requirements of the country. The inadequacy of the storage facilities was considered to be the gravest economic backwardness of rural mass, particularly the mass of population directly involved in production.

The structures, enclosures, or wrappers used for storing the food grains or seeds for the safety against any types of losses come under category of warehouse. The food habits of humans varies from tropical to sub-tropical; persons living by the costal side and mainly paddy growing area prefer rice and so it is a staple food for them. Cultivators, on the other side where wheat is main crop, people prefer wheat and gram and less of rice. The pressure on supply of grain, therefore, differs from place to place. The demand of food grain due to natural calamities, coupled with growing population in state and center made government to create a central board. This board was named as Central Board of National Cooperative Development and Warehousing Board under the Agriculture Production Corporation Act, 1956. The state government was also developed including public financial partnership by development of state warehousing scheme. The national Co-operative Development and Warehousing Board was functionary of planning, directing, and co-ordinating the movement of grain of All India Warehousing Corporations and State Warehousing Companies. Thus there was a National Cooperative Development Board, All India Warehousing Corporation (Central Warehousing Corporation), State Warehousing Corporation. The Central Warehousing Corporation came into existence in March 1957. These warehouses were more or less licensed grain storage handler and their objective was to encourage, establish, and regulate independent storage facilities as warehouse to reduce storage losses assist orderly market. The state Warehousing Act and rules framed to provide the license to any person or organization on making an application satisfying the requirements. It is compulsory to become a licensed warehousing man for undertaking business of warehousing for the commodities specified in the act. In India, most of the states have laws to regulate business of warehousing of stored commodities. The first law that came into force was the Central Provinces and Berar Agricultural Warehouse Act, 1947. Letter other state warehouse acts came into being in different years to come like Madras, 1951; Bihar, 1953; Orissa, 1956; Punjab, 1957; Rajasthan and Hyderabad, 1958; Bombay and Assam, 1959; Kerala, 1960; Mysore, 1961; Uttar Pradesh and West Bengal, 1963. The first state warehousing corporation was formed in Bihar in 1957. Thus there are 26 state warehousing corporations in function of which some came into being on the acceptability of states. The review of the warehousing corporation was desired by the Government of India in 1974 in the contest of Warehousing Corporation Act 1962 and was aimed to suggest modifications needed. In the year 1976, the Warehousing Corporation Act 1962 was amended to streamline the working of warehouses in regard to reports, credits, and feed with information about their operations. The Food Corporation of India is serving as vital area of procurement, distribution, and movement of stored commodities. It also diversified its activities toward production of nutritious processed foods like establishment of rice milling, dal mill, solvent extraction plants, production of protein rich diet for children. The Food Corporation of India has established several rice mills in different part of the country to increase the availability of rice and extract oil from rice bran.

The reports of various Indian authorities and private entrepreneurs revealed that building up and maintaining of buffer stock of food grain have been important in Indian Food Policy which seems to be true with other nations and at international forum too. From time to time as per need of the nation at international level, recommendations have been made by the technical group on buffer stock of food grains. The buffer stock implies primarily to ensure stability in supplies and price over years. The Government of United Arab Republic has also set up mills, when it took over the rice export trade in 1961. All over the world, efforts are directed to increase output and reduce losses. In some part of the world where traditional methods are still in practice, ample improvement in storage, however, can be based on scientific principles. A simple method of fumigation was introduced in Ghana. It saved at least 20,000 tons one third of country whole production. In Tanzania improvement over existing clay structure by plastic sheeting was more effective. During the World War II, the need of buffer stock of food in Britain drew attention to import grain from tropics and consider the factors responsible for deterioration by biotic and abiotic means. Latter, India after independence faced an intrigue position of food scarcity and imported food grains from temperature and sub tropical nations under public law (PL 480) program. It was in 1941 and 1943 that in Britain system of inspectorate operating under infestation orders, was brought to attention. An enactment was framed, named as Pest Act 1949, under which grain handlers began to take steps to eliminate pests. The needs to consider the quality standards in import were emphasized and survey of British ports for pest infestation was made. In the year 1951 Pest Infestation Laboratory established liaison to promote improvement in food and seeds storage and handling. Several entomologists were studying various aspect of stored product in developing nations as per their interest. Though the Pest Infestation Laboratory was established in 1940 under Department of Scientific and Industrial Research, but with the transfer of this laboratory in 1970 to the Ministry of Agriculture, Fisheries, and Food, it had a wide coverage on the research, inspection, and advice concerned with the prevention and control of infestation by insects, mites, and fungi on stored grain and seeds. The importation of food stuffs into the tropics, on the other hand, accentuated storage, transportation, and handling problems. Thus need of a estimating losses of produce after harvest was felt since no systematic work was undertaken. The importance of postharvest losses was first focused by the Food and Agricultural Organization of United Nations conference held in London in 1947, as the loss figures were quite alarming. The conference recommended the development of training facilities on storage methods in by respective countries. The Department of Food, Government of India was established Grain Storage and Training Institute in the year 1958 at Hapur, Uttar Pradesh. This institute has been imparting training in scientific food grain handling, inspection, and storage. The training courses were organized by USSR and third fellowship training course in grain storage was furnished in the year 1964, organized at Timiryazev Agricultural Academy, Moscow; all Cereal Research Institute, Moscow, Institute for the Design of Structures and Facilities for Grain Industries, Moscow. Thus training and research on storage of food grains were started in most of the countries of the world. The pest control program including fumigation of all stored produce was introduced in the year 1957. A new motion picture "Grain Sampling" was introduced in Grain Division of Agricultural Marketing Service for use in training of grain inspectors.

The development of storage technique progressed rapidly and farm bins of different materials were developed. The design and structures made of varied materials came into being with the availability of local materials, details of storage bins were discussed in different storage structure chapter- Most of the farming communities continuing the bulk and bag storage even till date. Grain storage research continued to be a continuous process and the advancement in design of storage structure resulted in the development of cylindrical bins formed as elevators or silos. In the USA, there are two types of elevators, one known as country elevator and other as terminal elevator. The country elevators are principally to gather grain from nearby village to transport into the market. The size of elevators depends upon the size of farming communities. In India such elevator made of steel was established in Hapur and Pantnagar. The terminal elevators are located in larger markets and terminal points where rail transport and shipments are available and such elevators mostly popular in developed countries. Terminal elevators are equipped with high capacity equipments for cleaning, drying, and conditioning of grains and seeds.

All such developments aimed to protect food grain and seeds from losses of different sources, for example insects, mites, rodents, birds, and fungi. Due to advancement in research our pest management services are also became advanced toward starting from application of inert dust to several safer, economic and eco-friendly chemicals. The protection technology advancement has proven the application of system approaches with well-programmed protection based on ecological data. The improvement of storage environment with the prevailing atmosphere has been emphasized.

2.2   QUICK HISTORY

- Prevention of Food Adulteration Act 1954
- Prevention of Food Adulteration Act (revised) 1980
- Commission on Agricultural Costs and Prices 1985
- Agricultural and Processed Food Products Export Development Authority 1985
- National Bank for Agriculture and Rural Development 1982
- Prevention of Food Adulteration Rules 1955
- Agricultural Produce Corporation Act (APCA) 1956
- Warehousing Corporation Act (replaced APCA) 1956
- Central Warehousing Corporation 1957
- State Warehousing Corporation 1979
- Save Grain Campaign 1965
- Agricultural Prices Commission 1965
- Grain Storage Research & Training Centre (GOI, Dept. of Food) 1958
- Indian Grain Storage Institute Hapur 1968
- Royal Commission on Agriculture recommended that establishment of licensed warehouses 1928.
- Reserve Bank of India directed to state government for making the law of warehouse 1944.
- Government of India constituted Agriculture Finance sub committee under chairmanship of Pro. D. R. Gadgil for the suggestion on warehousing 1944.
- Pro. D. R. Gadgil recommended the establishment of warehouse in which place where produce must be store, and receipt can be generated for farmers 1945.
- Government of India constituted a committee for village Banking Inquire 1949.
- RBI constituted All India Village Credit Survey Committee for survey of farm family 1951.
- On the recommendation of All India Village Credit Survey Committee, Government of India constituted the Agricultural Producer Development and warehousing Act 1956.
- Warehousing Corporation Act 1962

- International storage conference 1982 on the completion of 25 years of Central Warehouse Corporation.
- International storage conference 2007 on the completion of 50 years of Central Warehouse Corporation.
- Central Railside Warehouse Limited 2007
- Warehousing development and regulatory law 2007
- National conference on warehousing 2008
- Upgradation of Central Warehousing Corporation from article B to article A 2009, Field Station Ludhiana, Baptla
- Indian Grain Storage Management and Research Institute Hapur, Field Station Hyderabad, Ludhiana, Jabalpur, Jorhat, Udaipur
- World Food Day (FAO) (16th October)
- Food and Agriculture Organization, Rome, Italy 1943

KEYWORDS

- **storage structure**
- **buffer stock**
- **warehousing**
- **terminal elevators**
- **crop storage**

REFERENCES

Coles, J. *Archeology by Experiment;* Hutchinson and Company Limited Publication: London, 1973; p 205.

Nash, M. J. *Crop Conservation and Storage in Cool Temperature Climates;* Pergamon Press Publication: New York, NY, 1978; Vol. 12, p 393.

CHAPTER 3

CLASSIFICATION AND IDENTIFICATION OF IMPORTANT STORED GRAIN/SEED INSECTS

CONTENTS

ABSTRACT

The correct identification of any insect is the base of successful pest management. The protection against insect infesting in food grain and seeds, going into storage, must be started with the raw commodity and be continued through processing, packaging, handling, transportation, and storage. The population of stored grain insect pests directly depends upon climatic conditions like high or low temperature, poor sanitation, relative humidity, and moisture content of the produce. Insect pest species that rarely damage grain except by the contamination due to their presence are termed as incidental pests. Besides insect pests and mite some insects act as predators and parasitoids on the insect pests in the stored grain. On the basis of their feeding behavior these insects may be classified as internal feeders and external feeders. Internal feeders are those insects which enter into the grain by making a hole and passing the immature stage inside them, coming out as adults. The external feeders start the feeding through the tender part of the grain and seeds like germs while others web the grain kernels together and larvae feeds inside them. To identify the stored grain insect on the basis of their marks of identification, taxonomic keys have been prepared and with the help of these most dominant species may be identified under Indian situation. The stored commodity deteriorated by insects belongs to order Lepidoptera (moth) and Coleoptera (beetles and weevils). There are several keys are available for correct identification of insect pests up to family and species level.

3.1   CLASSIFICATION AND IDENTIFICATION OF IMPORTANT STORED GRAIN/ SEED INSECTS

The correct identification of any insect is the base of successes full pest management. The protection against insect infesting in food grain and seeds, going into storage, must be started with the raw commodity and be continued through processing, packaging, handling, transportation, and storage. The main objective of such protection is prevention for insect, dead or alive, present in stored commodities. The first step to achieve the goal of protection is to identify the insect causing contamination and/or damage. The population of stored grain insect pests directly depends upon climatic conditions like high or low temperature, poor sanitation, relative humidity, and moisture content of the produce. Insect pest species that rarely damage grain except by the contamination due to their presence are termed as incidental pests. Besides insect pests and mite some insects act as predators and parasitoids on the insect pests in the stored grain. On the basis of their feeding behavior these insects may be classified as internal feeders and external feeders. Internal feeders are those insects which enter into the grain by making a hole and passing the immature stage inside them, coming out as adults. The external feeders start the feeding through the tender part of the grain and seeds like germs while others web the grain kernels together and larvae feeds inside them. To identify the stored grain insect on the basis of their marks of identification, taxonomic keys have been prepared and with the help of these most dominant species may be identified under Indian situation.

3.2   STORED-PRODUCT COLEOPTERA

The stored-product Coleoptera consists of beetles and weevils and responsible for the major damage or contamination of stored produces. Some most dominant species of stored-product Coleoptera with the family are described here.

3.2.1   FAMILY ANOBIIDAE

These are small oval or cylindrical beetles with head strongly deflexed below the thorax.

3.2.1.1   CIGARETTE BEETLE (LASIODERMA SERRICORNE) (FABRICIUS)

The cigarette beetle is also known as tobacco beetle, is a most destructive insect of stored tobacco. It is cosmopolitan in nature and found worldwide. It was reported to found in Egypt more than 3500 years ago (Alfeiri, 1931). Cigarette beetles infest the cacao, spice, dried fruits, ginger, turmeric, and processed tobacco. Bry and Lang (1966) reported that cigarette beetle served the viscose rayon filling yarn in sized mattress ticking and caused considerable damage to mattresses in storage. Their larvae may cause 100% damage to rayon threads used in cotton textile bags in which produced are stored. It also feeds on wheat flour, cotton bolls, and books. Mossop (1932) reported that in Southern Rhodesia, the adults live five to six weeks in summer but longer in winter. The average life span of female is 18 days while male requires 20–22 days to complete their life cycle. The adult beetles are small, stoutly oval, 2–2.5 mm long, light brown yellow in color. The size and color vary as per food and stores. Elytra smooth, with very short hairs, but without striae, antennae about half as long as body, 11 segment, 4–10 segment serrate, generally females are larger than males. The females lay eggs in folds and crack or crevices of host. The newly hatched larvae are small and measured about less than 1 mm in length, covered with fine hairs. The body of larvae is white and scarabaeiform. After four to five molting in 40–45 days the full grown larvae are measured about 1.5–2.0 mm, and then they construct a smooth cell for pupation. The newly formed pupa is glossy white in color, then changes into reddish brown in color.

3.2.1.2   DRUG STORE BEETLE (STEGOBIUM PANICEUM) (LINNAEUS)

The drug store beetles are similar to L. serricorne and they only can be distinguished by their length. It is 2.5–3.5 mm longer than cigarette beetle. It generally infests the drug and pharmaceutical, and is found in European countries. Cittenden (1896); Zacher (1953); Laibach (1960) reported that the drug store beetles infest almost any dry plant and animal product; however, it has made spectrum of food stuffs like cured and uncured tobacco, black and red pepper, drugs, spice, leather, books, and textiles. The life cycle of this pest is similar to cigarette beetle. The eggs followed by five to six larval instars, then pupation in cocoon. The total life cycle is completed in 73 days at 25 °C and 70% relative humidity and 231 days at 17 °C and 50% relative humidity; 71 days at 30 °C and 90% relative humidity (Lefkovitch, 1967). Last three segments of antennae form a large loosely segmented club. Elytra have longitudinal striae.

3.2.2   FAMILY BOSTRICHIDAE

The body is cylindrical, head ventral to prothorax. Pronotums have rasp-like teeth or hooks, antennae straight and have a loose three or four segmented club. Apically elytra are flattened and sloped ventrally more or less steeply (declivity). Tarsi 5 segmented.

3.2.2.1   LESSER GRAIN BORER (RHYZOPERTHA DOMINICA) (FABRICIUS)

R. dominica is first time reported in the year 1972 from the stored speci-mens from South America and it is also known as Australian wheat weevil. It is a native of India (Pruthi & Singh, 1950). Besides India it is reported from different parts of the world like Argentina, USA, South Wales, Okla-homa, and Kansas. The adult beetles are strong flier and spreads with great rapidity and often infesting wheat in field (Cotton, 1938; Dean, 1947). These insects having strong teeth with which it damages not only the sound grain but also any parts of the storage receptacles and it also bores into leather and dry fruits. The first stage larvae being straight in shape may bore into sound grain in latter stage they become curved shape and measured about 3.2 mm in length and 0.8 mm in breadth. It is dark brown to black with rough body surface. Its head is inverted into a hood like triangular structure under the thorax, prothorax hood shaped covered with tubercles which are rather coarse especially in the front. The male and female may be identified by examining the segment of ventral side of the body. The "V" segment of sternum is light yellow in female while in male it is brown in color. The third and fourth sternites are light doted in female but in male they are green in color. Generally the female lays eggs on the grain near the embryo but sometimes on surface of bags, walls, and cracks. It can lay eggs even at 8% moisture content in grain. The adults are 2–3 mm long typical bostrichid. Pronotum rounded at front with transverse rows of teeth. Tubercles are flat-tened centrally and posteriorly. Scutellum between pronotum and elytra is almost square shaped. Elytra have regular row of punctures and short setae that curve posteriorly. Apically elytra are gentle convex structures. Antennae have 10 segments with a loose three segment club.

3.2.2.2   LARGER GRAIN BORER (PROSTEPHANUS TRUNCATUS) (HORN)

The adults having typical cylindrical bostrichid shape and the body is 3–4.5 mm long. Declivity is flattened and steep, many small tubercles over surface. The limit of declivity is apically and laterally marked by carina. Antennae are 10 segmented with a loose three segmented club. Stem of antennae is slender and clothed with long hairs and the apical club segment is as wide as, or wider than, preceding segments. Larvae are white and parallel-sided with short legs and small head capsule. Thoracic segments are considerably larger than those of abdomen.

3.2.3 FAMILY BRUCHIDAE

On the basis of food habits the insects of Bruchidae family may be divided into three groups, one remains in the store for breeding and feeding, second breeds and feeds in both field and store, and third only causes infestation in the field only. Raina (1970) reported five species of *Callosobruchus* from India, but only three species occur during storage of pulses.

3.2.3.1 PULSE BEETLE (CALLOSOBRUCHUS CHINENSIS) (LINNAEUS)

Adults have a pair of distinct ridges (inner and outer) on the ventral side of each hind femur, and each ridge has a tooth near the apical end. The inner tooth is slender, rather parallel-sided, and equal to (or slightly longer than) the outer tooth. Antennae pectinate are present in male and serrate in female. Elytra are pale brown in color with small median dark marks and larger posterior patches which may merge to make entire posterior part of elytra dark in color. The side margins of abdomen have distinct patches of coarse white setae. The host specificity of *C. chinensis* is restricted mainly, but not exclusively, to seeds of the legume tree Viciaea. As its common name indicates, these beetles prefer cowpea but it has been reported to feed on several types of pulses. *C. chinensis* is primarily a pest of seeds of moong, red gram, and lentil but occurs in almost all pulses in India. The strains of *C. chinensis* from Israel and Japan were reported to have some morphological differences (Applebaum et al., 1968). The eggs hatch in about 4–5 days, hatching of eggs may be easily seen under microscope. Raina (1970) reported average incubation period at 30 °C and 70% relative humidity 3–5 days with 98% hatching. The first instar grub bears a large spine on either side of the first abdominal segment and two groups of smaller spines dorsally on the tergal plate of the pronotum (Van Emden, 1946). The grub undergoes four molts before pupation it chews a circular hole near the seed till only a thin layer of seed covering is left intact. The grub and pupal periods last for about 18–20 days, generally it remains for four days in pupal stage during summer. The total life cycle is completed in 22–25 days in winter season while 45–48 days during summer season. Halstead (1973, 1980) reported that insects develop in mature and dry seeds and do not require for survival or reproduction. Males are strong flier than female. Male lives for seven to ten days while female live for five to ten days.

3.2.3.2 PULSE BEETLE (C. MACULATUS) (FABRICIUS)

Adults have a pair of distinct ridges (inner and outer) on the ventral side of each hind femur, and each ridge has a tooth near the apical end. The inner tooth is triangular, and equal to (or slightly longer than) the outer tooth. The antennae of both sexes are slightly serrate. Females often have strong markings at elytra consisting of two large lateral dark patches mid-way along the elytra and smaller patches at anterior and posterior ends, leaving a paler brown crossed shaped area covering the rest. Males are much less distinctly marked. It infests so many pulses and also develops in those pulses which inhibit the development of *C. chinensis* like soybean and French beans. Howe and Curri (1964) gave an account of the development period of *C. maculatus* on more than six foods. The eggs are laid by the female with a range of 110–160. The highest number of eggs laid on chickpeas was 75 (Sawaf, 1956), where Howe and Curri (1964) noted 116 average number of eggs. The eggs hatch in about 3–5 days and average grub and pupal period is about 20–22 days.

3.2.3.3 PULSE BEETLE C. ANALIS (FABRICIUS)

The antennae of these beetles are filiform type. Coloration and patterning of both sexes are similar to that of an average female *C. maculatus*, except that the pronotum of *C. analis* is uniformly red-brown (usually black or dark brown in *C. maculatus*). In freshly emerged specimens, white setae at the center of posterior half of each elytron are very conspicuous in *C. analis* but inconspicuous in *C. maculatus*. As in all *Callosobruchus* species, each hind femur has a pair of ridges on the ventral edge; in *C. analis* the outer ridge bears the usual noticeable blunt tooth, but the inner tooth is usually much smaller or even absent. The life cycle of this pest has been studied by Fletcher (1930), mating of female is found to be similar to that of *C. chinensis*. The single female lays an average of 75–80 eggs over a period of nine days. Raina (1970) reported that grub and pupal period was completed in 25–28 days at 70% relative humidity and 30 °C temperature.

3.2.4   FAMILY CURCULIONIDAE

All adults are characterized by a rostrum, which is a forward snout-like extension of the head and carries the mouth part. The antennae are elbowed in shape while at rest.

3.2.4.1   RICE WEEVIL (SITOPHILUS ORYZAE) (LINNAEUS)

S. oryzae is the one of the destructive pests of stored cereals, throughout the world. A small weevil is measured about 3 mm in length with head protruded into a snout-like structure or a distinct beak or proboscis. At the end of this structure, there is a pair of mandibles. It is generally radish brown in color, and some time dark brown to black. The both sexes look apparently alike but when carefully examined, the male can be distinguished from the female by the rostrum which is shorter and broader in male than in female, well developed hind-wings are present beneath elytra punctures on pronotum rounded, the dorsal surface is much more closely punctured in male than that of female. It has been reported to fly toward wind direction and infest standing crops up to 2.5 km distance from storage (Khare & Agrawal, 1964). Both grubs and adults feed voraciously so much so that grain is rendered unfit for mankind consumption. In India *S. oryzae* has been reported to remain in old stores and grain carried by rodents and birds. The females prefer to lay eggs in the hairy end of the wheat grain, and lay about 300–400 eggs during life time at 14% moisture content and between 25 and 29 °C temperature and 70% relative humidity. The eggs are white, translucent, and measure 0.7 mm long and 0.3 mm broad, grub hatch in four days (Khare & Agrawal, 1963) in summer season and 6–9 days in winter season. Grubs are small, white, fleshy, and legless have brown head and jaws, burrow inside the seeds or grain and whole developmental stage passed inside them. The grub passes through several molt at 27–30 °C (Howe, 1952) then goes into pupation. Soderston & Wilbur (1966) reported that at 76% relative humidity and 27 °C temperature weevils complete life cycle in 42–45 days. In India this insect passes 5–6 generation in a year. Weevils emerged in summer live for 3–6 weeks, but winter generations live longer. In USA it is reported to live for five months (Cotton, 1960). One pair of adult weevil may produce one million adults within three months. The largest damage by weevil is done between August and November, comparatively less in February and March, and least in December and January.

3.2.4.2 MAIZE WEEVIL (S. ZEAMAIS) (MOTSCHULSKY)

S. zeamais is most serious species of stored seeds and grain worldwide. It can readily fly from stocks of stored grain to growing maize (Giles, 1968) and at harvesting time all stages of development of this insect are present. Several scientists have shown that amongst all the factors affecting the distribution and number of *S. zeamais* in maize ears on the growing crop, the proximity and intensity of a source of infestation, usually cereals in stores, and the extent of kernel coverage by the husk in the variety grown, appears to be the most important factor (Mossop, 1940; Floyd et al., 1959; Davies, 1960; Eden, 1952; Giles & Ashman, 1971). So many fruits and seeds are infested by this insect and these insects are also collected from habitate of *Sitophilus* sp. (Mayne & Donis, 1962), but their most preferred food is maize, sorghum, and other cereals. *S. zeamais* and *S. oryzae* may be distinguished by aedeagal characters; in *S. oryzae* the curve of the aedeagus is more or less uniform to the top whereas in *S. zeamais* the tip of the aedeagus is distinctly hooked and tip varies from slight to most pronounce in *S. zeamais*. It is absent in *S. oryzae* and no gradation between the species has been observed (Proctor, 1971; Kuschel, 1961). The life cycle of *S. zeamais* is nearly same as *S. oryzae*. However, Baker and Mabie (1973) after rearing both species on Meridio diet reported that the grub period of *S. zeamais* on an average was 13–15 days.

3.2.4.3 GRANARY WEEVIL (S. GRANARIUS) (LINNAEUS)

S. granarius is similar in appearance to *S. oryzae* and *S. zeamais*, but can be distinguished by the absence of metathoracic flight wings (only small vestigial wings are present) and by the oval shape of punctures on the prothorax. *S. granaries* were earlier known as *Calandra granaria* (L). The archeological evidences in barley in the Icheti's tomb under the step pyramid at Saqqara indicate that it was known during sixth dynasty, about 2300 BC. It was also recorded from sites of Roman occupation. It was a cereal pest some 5000 years ago and observed with barley seeds in Egyption tombs (Solomon, 1965). It is a pest of cereals and has also been reported to infest shelled acorns (Howe, 1965). *S. granaries* are one of the oldest known pests of stored grain and grain products and are almost cosmopolitan in nature, have been carried by transportation to all parts of the world and in particular in temperate zone. It does not fly as it has lost the power of flight and is depend upon the domestic animals for transportation. The elytra are fused and reduced to short narrow straps. It is only found in places where either grain is stored or grain residues are left over as sweeping of stores. *S. granaries* are most prominent pests of temperate zone. *S. granaries* live on an average for seven to eight months. The females lay eggs in a slit like opening made in grain and covered with a gelatinous mass like rice weevil, on an average a single female can lay about 200–300 eggs. Richards (1947) reported that female weevils lay eggs almost indiscriminately except that they avoid grain containing large grub so that more than one egg may frequently be found in a single grain. The eggs hatch within 6–28 days depending upon temperature and relative humidity, grub passes through several molt than goes for pupation and only one adult develops inside one grain but occasionally two have also been reported.

3.2.5   FAMILY DERMESTIDAE

Beetles are generally ovoid but sometimes stoutly oval in shape and vary in length from 1.5 to 12 mm, usually clothed in hairs or colored scales. Antennae are relatively short with 10 or 11 segments. A fairly distinct three segmented antennal club is common. The grubs are characteristically dense hairy.

3.2.5.1   KHAPRA BEETLE (TROGODERMA GRANARIUM) (EVERTS)

Adults are reddish-brown, with or without darker vague marking, but the thorax is darker brown. They are oval in shape and vary in size from 2–3 mm, females being somewhat larger than the males. The dorsal surface is moderately clothed in fine hairs. A median ocellus present between compound eyes. The number of antennal segment is usually 11 but some fusion of segments may take place so that there can be as few as nine. The fairly distinct antennal club consists of 3–5 segments. In the male the apical segment of the club is much elongated in comparison with that of female. The antennae fit into ventral grooves in the prothorax. The larvae are typically very hairy. Spicisetae of various lengths are arranged over the dorsal surface and a "brush" of long spicisetae on the ninth abdominal segments project posteriorly like a tail. *T. granarium* is a major pest of stored grain in India. The first records were made by Coles in the year 1894. It has several times been recorded as destructive insect in Europe. It has been introduced a number of times during transportation of grain (Cotton, 1960). Similarly, within the confines of the USA, there are well defined areas in which certain species floursh in temperature prevailing between 32 and 49 °C, whereas they are of little importance in other insects. *T. granarium* occurs in India in Rajasthan, Bundelkhand, and Punjab. It is reported to be a pest of hot arid regions of our country as well as in world (Burges, 1962). *T. granarium* is a native of Ceylon and Malaya. It is one of very few storage insects that can breed in a hot dry environment, in addition it can survive normal winter of cool country (Solomon & Adamson, 1955). This beetles slipped into the USA in 1946 or earlier during World War II.

3.2.6   FAMILY SILVANIDAE

These beetles are generally parallel-sided and rather short (2–4 mm). Most species have projections on the prothorax which are in the form of teeth or swelling at the anterior angle or several teeth along the lateral margin.

3.2.6.1   SAW GRAIN TOOTH BEETLE (ORYZAEPHILUS MERCATOR (FAUVEL) AND O. SURINAMENSIS (LINNAEUS))

Adults of both species are slender dark brown beetle, 2.5–3.5 mm long. The antennae are relatively short and clubbed. The prothorax has six distinct teeth like projections along each side. The length of the temple (the side of head behind eye) is much shorter in *O. mercator* than that in *O. surinamensis*. The larvae are white, elongate, somewhat flattened, and about 4–5 mm long when fully grown. It is associated with starchy food and generally found in warm areas throughout the world. Linnaeus came across specimens of this insect in Surinam and for that reason he named it as surinamensis in 1967 (Cotton, 1960). The insect bears six saw like toothed on the each side of thorax. The merchant beetle, *O. mercator* was described in the year 1889. The both insects are cosmopolitan pests of stored grain and grain products. It is reported to occur in flour mills and warehouses. It is able to multiply in cool condition but its ability to overwinter in the fabric of warehouses enables quite large populations to persist until heating produce provides conditions needed for rapid increases (Howe, 1956). It never infests the whole grain, as per report of (Freeman 1973) both species occurs in 10% in the farm of England and Wales and 3% in Scotland. Both the species have wings but they use them rarely or not at all, spread throughout the tropics and are continually introduced by trade into temperate zone. *O. mercator* is more hygrophilic than its congener *O. surinamensis* because former is more vulnerable to low humidities. Its behavioral avoidance of dryness is more highly developed (Arbogast & Carlhon, 1972). Generally both the species live from six to ten months, but they have been found to live three years. The female lays 6–10 eggs per day and 50–300 eggs in her life span. Grub molts three times then goes for pupation (Howe, 1956). The total life cycle is completed in 100–150 days depends upon the environmental conditions.

3.2.7   FAMILY TENEBRIONIDAE

The adults are 3–10 mm long and usually rather parallel-sided. Antennae are 11 segmented, are of moderate length, and either are simple in form or have a more or less distinct club. Tarsi of hind legs have four segments while those of the front and middle legs have five. The larvae have a characteristic shape and are generally well sclerotized, often having a distinctly banded appearance and one or two pointed projections (urogomphi) at the end of body.

3.2.7.1   RUST-RED FLOUR BEETLE (TRIBOLIUM CASTANEUM) (HERBST)

In the year 1797, it was named as *T. castaneum* Herbst is placed next to *T. confusum* on the basis of its extents of damage (Cotton, 1960). All the species are secondary pests of stored grain, but able to feed on broken grain. Adults are typically tenebrionid in shape, 2.3–4.4 mm in length and red-brown in color. The males possess a hairy puncture on the ventral surface of anterior femur; this puncture is absent in female. The grubs are typically tenebrionid and possess two upwardly curved urogomphi on the ninth abdominal segment. The antennae are well developed, the last few segments being abruptly much larger than the preceding ones, and generally females are heavier than males. The five species of *Tribolium* are found fairly frequently in stored produce and all are easy to culture but two species are most dominant. The grub is small, worm like, slender, cylindrical, and wiry in appearance. The body is segmented with fine hairs. The exoskeleton is soft and whitish. The fully developed grub is 48 cm long and light yellow in color.

3.3   IDENTIFICATION OF STORED-PRODUCT COLEOPTERA WITH THE HELP OF TAXONOMIC KEY

3.3.1   KEYS TO ADULT OF COLEOPTERA

The keys include the adult Coleoptera commonly found in tropical storage and also some of those less commonly encountered or only of local importance. Identification should normally start at couplet 1 below. Where three or more species of a family are regularly found in tropical storage, a separate family key is given (Halstead, 1986).

3.3.2   GENERAL KEY TO FAMILIES AND DISTINCTIVE SPECIES

1. - Head (in front of eyes) with a long narrow "snout"; antennae arising from the sides of this distinctive "snout" which is between 2 and 7 times as long as wide. ………... …………. ………… ……….. ……….**2**

    - Head never with a distinctive long "snout". …………………….……**3**

2. - Antennae is elbowed (i.e. bent sharply near the middle) and the first (basal) antennal segment is very long (usually about half as long as rest of antenna). Central area of elytra is not strongly convex, often rather flat, and elytra are usually not much wider than proth orax...Curculionidae
       (Follow the key of family)

    - Antennae are straight or evenly curved (never bent sharply near the middle) and the first (basal) antennal segment is not very long. Elytra are entirely convex and often much wider than prothorax ……………………………………………………………… Apionidae

       And if: body is almost black, sometimes with paler greyish-brown striae; elytra are very strongly convex, twice as wide as base of prothorax, and with swollen "shoulders" near the front angles; length 3–4 mm (infesting pulses, but not common). ………………....……

       ….....………………………………………….Piezotrachelus sp.

3. - Elytra are extremely short, leaving five, six, or seven abdominal segments visible from above. (Predatory beetles, usually only present in very small numbers in stores.)

       ……………………………………………….…...Staphylinidae

- Elytra either completely cover abdomen or, if shortened, only leave one, two, or three abdominal segments exposed. **4**

4. - Antennae are elbowed, with a distinct angle between the long first segment (which is about a third of the length of the antenna) and the remainder of the antenna. Head contains a deep groove on each side, into which the first antennal segment may be folded. Surface of body is strongly shining and usually black. Elytra are slightly shortened and exposing one or two abdominal segments, which are usually shining and black like the rest of the body (predatory beetles, not common). ....Histeridae

- Antennae are rarely elbowed, usually more or less straight or evenly curved and first antennal segment is usually of normal length (not more than a quarter of the length of the antenna); if the antennae are elbowed, with a long first segment, then the body is cylindrical and the elytra completely cover the abdomen. If one or two abdominal segments are exposed behind the elytra, then these are not shining black. Without the above characters combined. **5**

5. - With two or three large abdominal segments visible from above. Antenna with a compact round club is formed of three short wide segments. (Infesting various commodities, but usually only if the moisture content is high; more common in initial phase of storage.)............... .... Nitidulidae, in part

- Elytra either completely cover abdomen or expose only one abdominal segment (the pygidium). Antenna is of various forms, rarely with such a compact round three-segmented club.....................
.... **6**

6. - Last abdominal segment (the pygidium) is comple1ely exposed behind the elytra; this segment is either rather small, narrow, triangular, and more easily seen from slightly behind than from above, or large, rather wide, steeply (almost vertically) inclined, and sometimes only visible from behind. Stout, robust beetles that are never dorso-ventrally flattened...7

- Elytra completely cover the abdomen* (including the pygidium) **8**

7. - Antenna with last (apical) three segments is distinctly thicker than the other segments (thus forming a long loose antennal club). Body surface is grayish-brown with small alternating patches of light and dark setae, giving a mottled appearance. Elytra contain a number of shallow rounded bumps (not easily visible but giving an uneven appearance to the surface). Exposed segment of abdomen is small, narrowly triangular, and often sloping downward, but rarely almost vertical. It is having length of 3–4.5 mm. (Mainly infesting coffee and cocoa, but also found on other commodities, especially cassava and maize *Araecerus fasciculatus,*

- Antennal form is various but never with the last three segments noticeably larger and forming a club. Body colo is various, not often greyish-brown but if so then other characters are not as described above. Exposed segment of abdomen is almost vertical; this segment is usually large (width and length not much less than the width of the elytra), sometimes smaller, but if smaller then the body length is more than 5 mm (common major pests either of pulses or of groundnuts). ....Bruchidae

8. - Prothorax contains a short, narrow apical neck; the back of the head is thus completely exposed. General body shape shows a slight resemblance to an ant...Anthicidae

- Prothorax does not contain an apical neck; the back of the head thus fits partly into the front of the prothorax (except in abnormally distended specimens) **9**

9. - Coxae of hind legs are large and set at an oblique angle forming a V-shape which completely divides the first ventral abdominal segment. Very active fast-moving beetles have long legs; mandibles are strong and often relatively large; antennae are rather long and filiform -all segments are oval/oblong shape and of similar size; all tarsi contain five segments (predatory species, only found in small numbers). ....Carabidae

- Coxae of hind legs are either small or large, but if large then never set at an oblique angle and never completely divide the first abdominal segment. (General appearance is not usually as above, but if similar then the antennae are not filiform or some tarsi have less than five visible segments.) **10**

10. - With a fringe of conspicuous bristly setae along the margins of the elytra and (usually) at the sides of the prothorax; these setae are distinctly stiff and blunt, and are conspicuously erect (more or less perpendicular to the surface). Upper surface of body usually completely or partly has a metallic shining of bluish or blue- green

coloration (though other colors may also be present); but if body is entirely mid-brown then shape is rather cylindrical and the prothorax has a short (but distinct and narrow) constricted "neck" just in front of the elytra. Antennae are 11-segmented with the last segments forming a more or less distinct club.

These are mostly predators, but two species are pests of high-protein commodities. Cleridae

- Without a conspicuous fringe of bristly erect setae (as described above). Rarely metallic coloration. If color is mid-brown and shape cylindrical, then prothorax does not have a narrowly constricted "neck" just in front of the elytra……………………………………………….. **11**

11. - Prothorax contains six or more teeth on each lateral margin (but never with teeth along the anterior margin). If with six teeth each side these are large and conspicuous; if with more than six teeth (usually 9 or 10) then these are rather small and indistinct …………………...…..**12**

- Prothorax is either without teeth, or with only one or two teeth on each side (including, often, a tooth or projection on the front angle of the prothorax), but occasionally with teeth along the anterior margin of the prothorax……………………………………………………………....**13**

12. - Prothorax has six large teeth on each side; three are conspicuous longitudinal ridges on the prothorax. Body is dark brown in color, rather elongate, and flattened (common pests, especially of cereals and oilseeds). …………………...…………..*Oryzaephilus* sp.Silvanidae

- Prothorax contains more than six teeth (usually 9 or 10) which are rather small; without longitudinal ridges on the prothorax and have length of 1.7–2.1 mm (associated with moldy commodities; uncommon)…………………………*Henoticus* sp., Cryptophagidae

13. - Prothorax has a deep cavity (visible from above) at each front angle. Antennae are 10-segmented with the last segment expanded as a large one-segmented club which, at rest, fits into the deep cavity on the dorsal front angle of the prothorax. Body is oval, strongly convex, and brightly shining. These are very small beetles (1.5 mm or less) (associated with moldy commodities). *Murmidius* sp., Cerylonidae

- Prothorax does not have deep cavities on the dorsum of the front angles. If antennal cavities are present, these are on the ventral surface and are never visible from above. Never with all the above characters in combination……………………………………………………………..…….**14**

14. - Prothorax (from above) has a complete longitudinal carina (sharp ridge) near each side, more or less parallel to the lateral margin of the prothorax, thus dividing the prothorax into three areas (one large median area and two narrow lateral areas) **15**

   - Prothorax (from above) does not have a complete distinct carina on each side. **16**

15. - Carinae (ridges) and lateral margins of the prothorax are evenly and rather strongly curved (prothorax narrower at the front). Body is convex and conspicuously hairy and having length of 1.5–1.8 mm. Antennae are between a quarter and a third of the length of the body, and with a distinct three-segmented club (occasionally found on moldy commodities; not common)..................*Mycetaea hirta*, Endomychidae,

   - Carinae (ridges) and lateral margins of prothorax are rather wider apart at the front, and carinae are almost straight. Body is very distinctly flattened and without conspicuous hairs, usually light brown in color. Prothorax is about as long as wide; elytra are rounded oblong in shape; head and prothorax are together rather large (often nearly half the length of the body) and having length less than 3 mm. Antennae are long (often more than half the length of the body, sometimes nearly as long as body), and without a club (all segments similar in size and shape). (Several closely related species, most of which cannot be distinguished by external examination, including three common and important pests, and four minor pests, of various commodities.) ....*Cryptolestes* sp., Cucujidae

16. - Front part of prothorax usually contains a number of triangular rasp-like teeth (sometimes also contains similar smaller teeth at the sides of the prothorax). Body is cylindrical and usually dark-brown. Head is strongly deflexed (positioned below the front of the prothorax so that it is almost completely hidden from above)........................ **17**

   - Front part of the prothorax never contains rasp-like teeth on its surface (occasionally with one or two teeth on the side margin of the prothorax). Body is usually not completely cylindrical, head is rarely strongly deflexed; never with above characters in combination **18**

17. - Antenna has long first segment (about one-third length of antenna) and a compact three segmented club (segments of club fitting very closely together); antenna is often rather elbowed (bent at the outer end of the first segment) and often with setae of elytra distinctly flattened, broad, and scale-like. (Mostly wood-boring species occasionally found in stores, but one species is a pest of maize in and another attacks coffee cherries.)..Scolytidae

- Antenna does not have conspicuously long first segment; antennae are usually straight and with a loose three- or four-segmented club (segments of club are separated by distinct narrow constrictions at the base of each segment). Setae of elytra are normal, never broad, and scale-like. (Two major pests of cereals, and several minor pests of various commodities.) ,...................................... Bostrichidae

18. - Prothorax at front angles is slightly or distinctly produced laterally to form a tooth or a flattened projection or a rounded swelling. Occasionally this tooth or swelling projects slightly forward as well as laterally (i.e. obliquely antero-laterally), but it never projects directly forward. Antennae are always with a three-segmented club..........**19**

- Prothorax at front angles is almost always smoothly rounded or bluntly angled; if a tooth is present on the front angles, it is clearly pointing directly forward; never with a laterally or antero-laterally projecting tooth or swelling. Antennal form is various (sometimes with a three-segmented club)... **20**

19. - Prothorax has a tooth near the middle of each side in addition to the swelling or projection at the front angle. These have length of 1.5–3.5 mm. (Several species, which feed on mold in damp produce; not common.) *Cryptophagus* sp., Cryptophagidae

- Prothorax does not have a tooth near the middle of each side, only with a small or large lateral or antero-lateral tooth at the front angle; lateral margin of prothorax is either smooth or with minute serrations. ....Silvanidae, in part

20. - Antennae are never thickened toward the apex; all antennal segments are similar in width, sometimes with apical segments slightly narrower; never with an antennal club. Antennae are long (often more than half the length of the body). Body is either hairy or shining, but elytra are always strongly or extremely convex (elytra are curved underneath at the sides so that their lateral margins are not visible from above). Prothorax is often slightly constricted posteriorly and legs are rather long, so that the beetles resemble spiders . (Several species are usually associated with unhygienic storage, but not common in hot climates.).. Ptinidae

- Antennae are always with apical segments thickened, often forming a distinct club. Antennae are of normal length (less than half length of body). These are not strongly resembling a spider...................**21**

21. - Antennae have a distinct compact two-segmented club. Body is parallel-sided, narrow, and elongate (sometimes very elongate). Eyes are large and strongly convex. General appearance is distinctive. (Occasionally found in stores, but usually boring in wood.)... Lyctidae

- Antennae have either apical segments gradually thickened or a club, which is very rarely two-segmented, but if with a two-segmented club then body is not narrow and parallel-sided. Eyes are usually normal or small and never very strongly convex. ....…...……….**22**

22. - Upper surface of body is completely covered with conspicuous colored scales (white, yellowish, brown, or black) forming a distinct color pattern; body shape is almost round from above. (Several species, which feed on materials of animal origin; occasionally found in unhygienic store) *Anthrenus* sp., Dermestidae

- Body does not have conspicuous colored scales: occasionally with colored setae but these are normal (hair-like); body shape is seldom almost round.. **23**

23. - Antenna has an abrupt and very compact three-segmented club which is very nearly round (never more than 1.5 times as long as wide). Either with eyes extremely small and inconspicuous, or with middle two-thirds of antenna very thin (about a quarter of the width of the club)... **24**

- Antennal form is various, but if antennal club is three-segmented then this club is never nearly round but is always more than twice as long as wide…...................................…...........................………**25**

24. - Eyes are extremely small and inconspicuous (embedded in a small depression at side of the head); head is small and concealed from above by the prothorax; prothorax has base constricted slightly (forming a short wide "neck") and front three-quarters of prothorax are almost round in shape. Body is light brown and shining, with a few short setae (minor pest). ……….*Thorictodes heydeni*, Dermestidae

- Eyes are normal and conspicuous (never embedded in a small depression); head is clearly visible from above, never completely hidden beneath prothorax; shape of prothorax is various but never almost round with a basal "neck." Antenna with middle two-thirds very thin, approximately a quarter of the width of the club (on damp or moldy commodities; uncommon)...........................Nitidulidae, in part

25. - Upper surface is distinctly hairy: setae are often quite short but always rather dense (even if some setae have been rubbed off, patches of dense setae are still usually visible on some parts of the prothorax or elytra). **31**

   - Upper surface is not distinctly hairy, sometimes rather shining: usually with a few short setae which are, however, inconspicuous except at high magnification. **26**

26. - Lateral margins of prothorax and elytra are conspicuously projecting horizontally, forming a wide flange; elytra have very conspicuous longitudinal carinae (sharp ridges) evenly spaced across each elytron. Body is distinctly flattened, and having length of 2.7–3.2 mm (common pest of cereals, especially on paddy rice in Southeast Asia). ....……….. *Lophocateres pusillus*, Lophocateridae

   - Lateral margins of prothorax and elytra do not have a wide horizontal flange (but sometimes with a narrow marginal strip); elytra do not have very conspicuous, evenly spaced carinae (sometimes with shallow indistinct blunt carinae)....................................... **27**

27. - Base of prothorax contains a short longitudinal carina on each side, halfway between the mid-line, and the lateral margins; each carina is about one-sixth the length of the prothorax and is set in a broad depression. Body is shining dark brown to reddish-black; these are 4–4.5 mm long (rare and unimportant pest of damaged cereals in North and South America) ....*Pharaxonotha kirschi*, Languriidae

   - Base of prothorax does not contain longitudinal carinae of the type described above. **28**

28. - Prothorax is distinctly separated from abdomen by a narrow "neck" (less than half width of the prothorax) and has distinctive forwardly projecting front angles. Body is distinctly flattened, shining, dark brown, or black; length is variable, but usually 7–10 mm. General appearance is distinctive. Tarsi of all legs are appearing four-segmented, but actually five-segmented with the first segment extremely small. (Infesting various commodities, especially cereals and cereal products; also predatory on other pests; widespread, but especially common in Oriental region.) .....*Tenebroides mauri-tanicus*, Trogossitid

- Prothorax and general appearance are not as above. Tarsi are either three-segmented on all legs or visibly five-segmented on the front and middle legs and four-segmented on the hind legs....................**29**

29. - Tarsi of front and middle legs are five-segmented, and of hind legs are four-segmented; antennae are either with a distinct club of three or more segments or gradually broadened toward the apex, but if with a distinct three-segmented club then the elytra are parallel-sided. Side margin of head is usually expanded as a lateral flange, often partly dividing each eye .... Tenebrionidae

- Tarsi of all legs are three-segmented; antenna has a two- or three-segmented distinct club; elytra are not parallel-sided....................**30**

30. - Dorsal surface is smooth and shining; elytra have only one stria (close to the suture). (Fungus-feeders; one species occurs frequently on damp produce in the tropics.) ....*Holoparamecus* sp., Merophysiidae

- Dorsal surface is sculptured, or strongly punctured, or densely pubescent; never smooth and shining. (Only found on moldy commodities; uncommon in hot climates.)................................Lathridiidae

31. - These are large beetles (5.5–10 mm long). Upper surface of body is mainly very dark brown, sometimes with distinctly pale areas either at the sides of the prothorax or at the base of the elytra. (Either infesting animal products and high-protein plant commodities, or scavenging in stores of other commodities.) *Dermestes* sp., Dermestidae

- These are smaller beetles (less than 5.5 mm long)....................**32**

32. - Head is dorsally with a small but distinct ocellus positioned on the mid-line between the eyes. Elytra are never with striae. Dermestidae, in part

- Head is never with an ocellus. Elytra are either with or without striae **33**

33. - Head is in normal position, visible from above, at least in part. Antennal club is rather compact (usually three-segmented but sometimes two-, four-, or five-segmented) and distinctly shorter than the rest of the antenna **34**

- Head is held almost completely under prothorax and not visible from above. Antennal club is longer than remainder of antenna: club is either serrate, with several triangular segments, or composed of three large elongate segments **35**

34. - Tarsi are all five-segmented (recorded in small numbers from various commodities). ....*Cryptophilus* sp., Languriidae

- Tarsi are four-segmented, except front tarsi of males which are three-segmented. (Usually only found on damp commodities, especially when moldy.)..….............Mycetophagidae

And if: Antennal club is three-segmented. Elytra have punctures and striae usually arranged in distinct longitudinal rows. Without a distinct deep oval pit on each side of the posterior of the prothorax (sometimes with a very shallow depression in this position). Body is evenly colored (usually golden-brown with yellowish-golden hairs), never with distinct pale patches on elytra, and 2.2–3 mm long.

(Infesting various commodities, usually when these are damp and infested with other pests.) ....*Typhaea stercorea*

35. - Antenna has segments 4–10 serrate. Elytra do not have striae. (Infesting a wide variety of commodities, including tobacco.) *L. serricorne*, Anobiidae

- Antenna has a large, elongate, three-segmented club. Elytra have striae. Central area of prothorax is evenly and only slightly convex (never with a strong central "hump"). (Infesting a wide variety of commodities in most climates, but not very common in the tropics.): ....…........................*S. paniceum*, Anobiidae

3.3.3 KEY OF BOSTRICHIDAE

The body is cylindrical and the head ventral to the prothorax so that it is not visible from above. Characteristically, the pronotum has rasp-like teeth, hooks, or horns, the antennae are straight, that is not elbowed, with a loose three- or four-segmented club. At the posterior of the beetle the elytra are usually somewhat flattened and sloped more or less steeply downward: this sloping region is called the declivity. The elytral declivity is frequently decorated with ridges, tubercles, or hooks, all of which are useful recognition features. The tarsi are all five-segmented.

1. - Posterior tarsus is always shorter than tibia. Anterior region of pronotum contains several transverse rows of teeth. Never with large hooks or horns on the pronotum or elytra. These have length of 2.5–4.5 mm. (Principally storage species.............Dinoderinae ...2

- Posterior tarsus is never shorter than tibia. Pronotum or elytra are frequently decorated with large hooks or horns. The length exceeds 4.5 mm. (Mainly wood-boring species, but some pests of dried root crops.). ....…...................................…..........Bostrichinae

2. - Posterior half of pronotum is with flattened tubercles; elytral are declivity gently convex and all elytral hairs are distinctly curved. (Common pest throughout warm and tropical regions; infesting various commodities, especially whole cereals.)..........*R. dominica*

- Posterior half of pronotum is without flattened tubercles but with punctures; declivity strongly convex or flattened and steep and elytral hairs are all erect or only erect on apical one-third of erytra. **3**

3. - Elytral declivity is steeply sloping but without ridges or ornamentation. Posterior region of pronotum is often with two more-or-less distinct shallow depressions. Punctures of elytra are irregular, not arranged in rows. (Common in tropics; frequent pests of felled bamboo, but occasionally infest maize and dried cassava.).........*Oinoderus* sp.

- Elytral declivity is flattened and with pronounced curved ventral-lateral ridges forming a semicircle when viewed from behind. Posterior of pronotum does not have shallow depressions. Punctures of elytra are arranged in longitudinal rows. (Found in South and Central America, East Africa and West Africa as an important or serious pest of maize and cassava.).......................…...*Prostephanus truncatus*

3.3.4 KEY OF BRUCHIDAE

The most valuable feature in the recognition of bruchid species is the arrangement of teeth and ridges on the hind femur. Many adult bruchids have a colorful pattern on the elytra; unfortunately, the form of this is rather variable so that it is usually unreliable for specific identification. However, certain color characteristics are useful and when examining a specimen it is essential that it is completely dry as the colors of damp specimens are distorted or indistinct. It is frequently useful to confirm an identification by examination of the male genitalia. To do this, first remove the pygidium (tergum of the last abdominal segment), then carefully lift out the genitalia and remove adherent tissue and the "hood-shaped" sclerite that may still be covering the posterior tip of the genitalia. If cleaning the genitalia proves difficult treat them in hot 10% aqueous KOH (this is a dangerous caustic reagent—eyes should be protected) for a few minutes. Finally, place the genitalia on a slide with a drop of glycerol or water and examine with, preferably, a compound microscope.

1. - Prothorax contains a carina (ridge) and an impressed line (groove) around much or all of the dorsal margin; hind femur is much broader than coxa and with a ventral comb of teeth............Pachymerinae

    And if: Carina and impressed line of prothorax are present around the base and the posterior of the lateral margins but absent in the anterior third; hind femur has a comb of one long tooth and 8–12 smaller teeth; elytra have vague uneven pattern of darker coloration. These have length of 3.5–6.8 mm. (Widespread in the tropics and, in some areas, a major pest of groundnuts in shell; commonly found infesting tamarind pods; a closely related species with pale evenly colored elytra is associated with senna pods.) ....*Caryedon serratus*

   - Prothorax contains a surrounding carina and impressed line; hind femur is not as above...**2**

2. - Hind tibia has two long movable spurs (as in Plate 8d) at apex; hind coxa is twice as broad as femur.........................Amblycerinae

    And if: Hind tibial spurs are approximately equal in length and red in color; elytra are only slightly longer than width. Females have black cutlde and white pubescence; males are uniformly grey-brown. These have length of 2.0–2.5 mm. (Widespread in tropical and subtropical regions; major pest of pulses, especially common beans.) ...*Zabrotes subfasciatus*

   - Hind tibia has fixed spines or teeth at apex, or without teeth; hind coxa is less than twice as broad as femur. ....Bruchinae
    ....**3**

3. - Pronotum is broadly transverse (distinctly wider than long) and with a tooth, sometimes very small, near middle or near front of each lateral margin. (Usually temperate and subtropical species, but several species are widely spread by introduction into tropical areas; field pests of a range of pulses and rarely found on stored pulses).........
    ....*Bruchus*

   - Pronotum does not have lateral teeth and not broadly transverse........**4**

4.   -   Ventral edge of hind femur is concave between two distinct sharp parallel ridges; These always have a large tooth near the apex of the outer ridge and usually have a similar tooth on the inner ridge also. (Infesting pulses, especially cow peas and grams.)...*Callosobruchus* sp. **6**

   -   Ventral edge of hind femur is rather flat between two weak indistinct ridges; never with a tooth on the outer ridge, often contains one or more teeth near the apex of the inner ridge..............................**5**

5.   -   Inner ridge of hind femur consists of three or four teeth (one large tooth and two or three small sharp teeth) near apex; usually mainly grey and reddish in color. *Acanthoscelides* sp.

   And if: Last (apical) antennal segment is reddish (never dark grey), first five antennal segments are also reddish, remainder of antenna is dark grey; abdomen is reddish in color; legs are also reddish except for ventral part of the hind (and, often, middle) femur, which is black; pronotum consists of yellowish-brown hairs; elytra have dark brown and yellowish hairs. These have length of 3–4.5 mm. (distributed in tropical and subtropical regions; major pest of pulses, especially common beans........*Acanthoscelides obtectus*

   -   Inner ridge of hind femur either has a single tooth or does not have a tooth. (Several species; field-to-store pests of pulses.) ...*Bruchidius* sp.

   And if: Inner ridge of hind femur has a single tooth; pronotum is conical with distinct gibbosities (raised bumps) on the basal margin; antennae are strongly pectinate in male, strongly serrate in female. (Field-to-store pest of cow peas in West Africa and Sudan; does not persist in storage.) ....*Bruchidius atrolineatus*

6.   -   In males whole body is uniformly colored (dark reddish-brown to black cuticle with pale setae); females are similar but have a vague whitish pattern of setae on the elytra. These have length of 4.5–5.5 mm. (mainly infesting bambara groundnuts but able to breed on some other pulses.................................... *C. subinnotatus*

   -   Body has some red or yellowish-red markings (sometimes also with whitish or yellowish marks); elytra of mature specimens usually consist of a distinct and contrasting color pattern. These have length less than 4 mm. **7**

7. -   Tooth on inner ridge of hind femur is small (sometimes absent), usually smaller than tooth on outer ridge; central section of inner ridge has minute irregular serrations; pronotum is uniformly red-brown. These have length of 2.5–3.5 mm. (Found in Asia, but not usually very common; infesting pulses, mostly *Vigna* spp.).........*Callosobruchus* analis

    -   Inner tooth of hind femur is about as long as or longer than outer tooth; central section of inner ridge is smooth; pronotum usually has a distinct pattern of coloration. (Some species are major pests of pulses.) ;.. **8**

8. -   Side margins of most abdominal segments (just below edges of elytra) consist of small but dense patches of coarse white hairs: these hairs are usually much coarser and whiter than the normal hairs on the rest of the ventral surface of the abdomen....................................**9**

    -   Side margins of abdominal segments do not have dense patches of coarse white hairs: instead with the same fine yellowish or whitish hairs found over the whole of the ventral surface.....................**11**

9. -   Antenna of males is pectinate and dark in color; pygidium (tergum of last abdominal segment) has predominantly white or silver hairs. Male genitalia has parameres rounded at apex and median lobe longer than parameres; basally with two small sclerotized plates. These have length of 2.5–3.5 mm. (Major pest of pulses especially cow peas and grams; originally an Oriental species now widely distributed in tropical and subtropical regions.) ....*C.* chinensis

    -   Antenna of males is serrate (Plate 8e) or strdngly serrate and usually rather yellowish in color. Pygidium consists of white or yellowish hairs...**10**

10. -   Dorsally with prominent eyes, width of one eye has at least twice width of head between eyes; pygidium consists of whitish hairs. There is pale brown coloration with complex pattern of dark brown and whitish spots on elytra and prothorax, occasionally females have black cuticle. Male genitalia is very long and narrow, median lobe with an acutely pointed apex and two "hand" shaped plates situated mid-way down the median lobe. (Mainly known as a field or field-to-store pest of pigeon peas in South Asia.) ....*Callosobruchus theobromae*

    -   Dorsally with less prominent eyes, width of one eye is about equal to width of head between eyes; pygidium has yellowish hairs. Male genitalia is relatively short, parameres are very thin and pointed at apex, and median lobe is shorter than parameres; median lobe basally has several (six) strongly toothed sclerotized plates. (Found in Africa, especially in southern and eastern parts; infesting cow peas and grams.), ...*Callosobruchus rhodesianus*

11. -   Pronotum is dark red to black, usually with a pattern of paler hairs which often form two white oval spots close together at the middle of base of the pronotum. Male genitalia may take one of the two forms illustrated; in both there are two clearly defined longitudinal "streaks" in the median lobe, composed of small spines. (Major pest in tropical and subtropical regions; infesting cowpeas, grams, and sometimes soybeans.)..............................*Callosobruchus maculatus*

-   Pronotum is pale yellowish or reddish brown with dark brown central spots or lines. Male genitalia contains longitudinal "streaks" in the median lobe. (Widespread in the tropics; most often found on dolichos beans but also infests other pulses.)..........*Callosobruchus phaseoli*

3.3.5   KEY OF CURCULIONIDAE

The family Curculionidae is the largest in the animal kingdom, comprising about 50,000 species, but only a few species are encountered in stores and most are typical, being easily recognized by their long snouts (rostra) and elbowed antennae. In the tropics, only two species are frequently encountered: *S. oryzae* and *S. zeamais*. These can only be identified with certainty by examination of their genitalia, which must be dissected out; techniques for this dissection are described at the end of the key below.

1. -   Antennae consist of eight segments; elytra are very slightly shorter than abdomen so that tip of abdomen is visible from above; snout (rostrum) elongates about four times as long as wide. These have length of 2.4–4.5 mm. ....*Sitophilus* sp. **2**

-   Antennae consist of nine segments; elytra completely cover the tip of abdomen; snout (rostrum) is not very long (only twice as long as wide). These have length of 2.5–3.5 mm. (Recorded in tropical America and West Indies; minor pest of cereals and ginger.) ....Caulophilus oryzae

2. -   Pronotum has elongate oval punctures which are relatively large and widely spaced; elytra are evenly colored (never with distinct paler spots) and with narrow longitudinal grooves (striae) containing fine punctures. These do not have hind-wings (with very small vestigial lobes only). (Found in temperate and subtropical areas and in cool regions in the tropics; major pest of cereals, especially wheat and barley.) ....*Sitophilus granarius*

- Pronotum has round punctures which are relatively small and not widely spaced (Plate 6b); longitudinal grooves (striae) of elytra have coarse punctures; ridges (interstices) between striae are usually narrower than striae. These have fully developed membranous hind-wings. (Found in tropical and subtropical regions; infesting cereal grains, rarely spices or pulses.) **3**

3. - Pronotum has fine punctures, separated from each other by a distance at least as great as the diameter of a puncture, area between punctures is smooth and shining. Elytra are usually uniform dark brown, but paler areas may be present at both ends of each elytron and these are sometimes connected by a pale longitudinal band. (Widespread in tropics; infesting spices but not cereals.) ....*S. linearis*

- Pronotum has coarse punctures which are very close together (often nearly touching), area between punctures is dull and slightly roughened. Elytra often consist of four reddish-yellow spots but these are not normally joined by longitudinal bands. (Common major pests of cereals.)..**4**

4. - In males the surface of the aedeagus is completely smooth and convex. In females the "prongs" of the V-shaped sclerite are rounded and the gap between them is narrower than their combined width. (Major primary pest of cereals in the tropics and subtropics; rarely, strains have been found, infesting peas and grams.) ....*S. oryzae*

- In males the aedeagus has a central ridge between two longitudinal depressions. In females the "prongs" of the V-shaped sclerite (Plate 6f) are pointed at the ends and the gap between them is wider than their combined width. (Major primary pest of cereals in the tropics.) *S. zeamais* Motschulsky

3.3.6 KEY OF DERMESTIDAE

1. - Prothorax has base constricted slightly (forming a short wide "neck") and anterior three-quarters almost round in shape; eyes are extremely small and inconspicuous (embedded in a small depression at the side of the head); antennal club is three-segmented, very compact, and almost round in shape; head is small and concealed from above by the large prothorax; body is light brown and shining, with a few short setae. (Minor pest of various commodities; not often very common.) ....*Thorictodes heydeni*

- Prothorax is widest across base; eyes are normal; antennal club is never so compact or round; head is normal and usually at least partly visible from above; upper surface of body is either distinctly hairy or covered with colored scales..**2**

2. - Head has a small ocellus positioned centrally between the compound eyes. These have length of 2–5.9 mm.....................................**7**

- Head never has a median ocellus. These have length of 5.5–10 mm. *Dermestes* sp. **3**

3. - Pronotum at sides (and also sometimes in front) has distinct patches of white or yellowish hairs. Ventral side of abdomen is mainly clothed with dense white hairs, usually with distinct small patches of black hairs.. **4**

- Pronptum everywhere has dark hairs, never has distinct patches of pale hairs at the sides. Ventral side of abdomen contains black or brownish hairs, sometimes has darker patches, but never has dense white hairs.. **6**

4. - Each elytron has a small but distinct sharp tooth at apex. (Cosmopolitan; commonly infesting animal products and scavenging in stores of plant commodities.) ....*Dermestes maculatus*

- Posterior apex of each elytron is rather rounded, never with a sharp tooth. 5

5. - Abdomen is ventrally with black patches both laterally and at the posterior tip. Males have a distinct brush of hairs on only the fourth abdominal sternite (Cosmopolitan; frequently infesting animal products, sometimes found scavenging in stores of plant commodities.) ....*Dermestes frischii*

- Abdomen is ventrally with black patches laterally but without a black patch at the posterior tip. Males have a distinct brush of hairs on both the third and fourth abdominal sternites. (Widespread distribution, but less common in stored products than preceding species of this genus; attacks animal products, sometimes found scavenging in stores of plant products.) ....*Dermestes carnivorus*

6. - Basal half of each elytron has a large area of pale greyish-brown or yellowish hairs enclosing three small patches of black hairs; apical half of elytra is dark brown or black. (Not common in tropics; usually infesting animal products, sometimes found as a scavenger in stores of plant commodities.) ....*Dermestes lardarius*

- Elytra are covered by dark hairs, never with basal half conspicuously paler than apical half. Ventral side of abdomen is golden-brown and clearly patterned with darker patches. (Found throughout tropics; often found on high protein plant commodities (e.g. copra), also on animal products and scavenging in stores.)............Dermestes ater

7. - Body is covered with colored scales. Antennae fit into sockets at front of prothorax. (Several species which feed on material of animal origin; occasionally found in unhygienic food stores.) *Anthrenus* sp.

- Body is covered with hairs. **8**

8. - Elytra are evenly colored or with a transverse pale band in the basal third or with two small central white spots, but if evenly colored or transversely banded then body length is usually 3–6 mm. First segment of hind tarsus is only half (or less than half) as long as second segment; antennal club composed of three segments. (Several species that infest animal products and scavenge in stores of plant commodities.)...*Attagenus* sp.

- Elytra are either evenly colored or with a complex (sometimes delicate) pattern of coloration (never with two central white spots), but if evenly colored then body length is less than 3 mm. First segment of hind tarsus is as long as, or longer than, second segment; antennal club commonly has four or more segments. (Several species, including one major and serious pest, that infest various commodities, especially cereals, cereal products, oilseeds, and oilseed cake.) *Trogoderma* sp.

3.3.7   KEY OF FAMILY SILVANIDAE

1. - Pronotum has six large teeth at each side. **2**

- Pronotum never has six large teeth, but has front angles produced laterally and somewhat anteriorly to form a rounded or indistinct lateral swelling...**3**

2. - Temple (side of head immediately behind eye) is relatively long and distinct, but eyes are rather small; the length of the temple is thus about two-thirds the length of the eye. These have length of 2.1–3.2 mm. (Cosmopolitan; infesting a wide range of commodities, especially cereal grains and cereal products.) .... *O. surinamensis*

- Temple is short and inconspicuous, but eyes are rather large; the length of the temple is thus less than a third of the length of eye. These have length of 2.2–3.1 mm. (Cosmopolitan; commonly infesting dried fruit, oilseeds, and oilseed cake; less common on cereal grains and cereal products.) .... *O. mercator*

3. - Pronotum is wider than long, and with side margins distinctly curved and with minute serrations; front angle of pronotum has a distinct rounded tooth. These have length of 2–3 mm. (Cosmopolitan; common on damp produce and in stores where the ambient relative humidity is generally high.) .... *Ahasverus advena*

- Pronotum is either slightly longer than wide (females) or much longer than wide (males), and with side margins almost straight and smooth; front angle of prothorax has a very indistinct lateral swelling. These have length of 2.5–4 mm. (Widespread; sometimes infesting cereals, especially maize cobs before harvest and during the initial phase of storage.) .... *Cathartus quadricollis*

3.3.8   KEY OF FAMILY TENEBRIONIDAE

About 100 species of tenebrionid have been recorded in association with stored products. The adults vary considerably in size but those found on stored produce are generally 3–10 mm long and more or less parallel-sided. *Alphitobius* is exceptional in being more oval in shape. The antennae of the majority of Tenebrionidae have 11 segments, are of moderate length, and either are simple in form or have a weak club. In most genera, but notably not in *Coelopalorus* or *Palorus*, the compound eyes are partly divided horizontally by a backward projection of the side of the head. In all Tenebrionidae the tarsi of the hind legs have four segments whereas those of the front and middle legs have five segments.

1. - Large parallel-sided beetles (length 12–18 mm). (Usually found on cereal products; not common in the tropics.) .... *Tenebrio* sp.

- Smaller beetles (length less than 7 mm). **2**

2. - Head has a conspicuous pair of mandibular horns and a pair of prominent tubercles between the eyes. .... *Gnatocerus* sp. males .. **8**

- Head does not have horns on mandibles. 3

3. - Rather broad beetles: oval in shape with sides of elytra distinctly rounded. Dark brown or black in color. These have length 4.5–7 mm. (Usually found on damp or moldy produce.) 4

- Rather narrow beetles: more or less oblong in shape with parallel-sided elytra. Color is usually mid-brown, reddish-brown, or yellowish-brown, rarely dark brown. Length is almost always less than 4.5 mm, rarely 5–6 mm long...5

4. - Eye is only partly divided by side margin of head which extends across a half to two-thirds of the eye, leaving a wide band at the back of the eye (as wide as three or four eye facets); tibia of front leg is strongly broadened toward its apex. These have length of 5.5–7 mm. ....*Alphitobius diaperinus*

- Eye is nearly completely divided by side margin of head, leaving a very narrow band at the back of the eye (only as wide as one facet); tibia of front leg is normal (expanding gradually and evenly toward its apex). These have length of 4.5–6 mm. ....*A. laevigatus*

5. - Eyes are not round but with vertical diameter distinctly greater than horizontal, and always at least partly divided by expanded side margins of head. 6

- Eyes are more or less round, and never even partly divided by side margins of head...11

6. - Antennae are shorter than head and with a distinctive compact five-segmented club, with the final segment conspicuously narrower than the others. Side margin of head does not extend more than a quarter of the distance across the eye. These have length of 2.7–3 mm. (Tropical and subtropical but especially found in the Oriental region; an important pest of cereals and cereal products, especially in hot conditions, 30–40 °C.)...*Latheticus oryzae*

- Antennae are distinctly longer than head, and with a compact three-segmented club, or with a loose four-segmented club, or without a distinct club but gradually thickened toward apex. Side margin of head extends a half or more than a half of the distance across the eye...7

7. - Prosternal process is distinctly broader at apex than elsewhere. Lateral regions of elytra consist of shallow longitudinal carinae (ridges) on

the intervals between the striae; sometimes with similar carinae on the central area of the elytra...........................*Tribolium* sp. ...**9**

- Prosternal process is almost parallel-sided, not distinctly broadened at apex. Elytra have intervals entirely flat, without carinae. ....*Gnatocerus* sp. ...**8**

8. - Males have strongly triangular mandibular horns. Females (without horns) have widest part of head across side margins in front of eyes. These have length of 3.5–4.9 mm. (Cosmopolitan; minor pest of cereals, cereal products, oilseeds, and various other commodities.) ....*Gnatocerus cornutus*

- Males have curved narrow horns. Females (without horns) have head across back of eyes as wide as or wider than in front of eyes. These have length of 2.5–4.0 mm. (Minor pest of cereals and cereal products.) ....*G. maxillosus*

9. - Eyes are ventrally closed together, with the distance between them about equal to the transverse ventral diameter of an eye. Eye is three or four facets wide at its narrowest point. Antenna have a distinct three-segmented club. Body is usually reddish-brown in color. These have length of 2.3–4.4 mm. (Cosmopolitan, but particularly common in tropical and subtropical regions; major pest of many commodities, but especially cereals and cereal products.)..............*T. castaneum*

- Eyes are ventrally more widely separated, with the distance between them about two or three times the transverse ventral diameter of an eye. Eye is one or two facets wide at its narrowest point. Body is usually brown or dark brown in color. **10**

10. - Small species (2.6–4.4 mm). Body is usually mid-brown, occasionally darker. General appearance is as in. (Cosmopolitan, but particularly common in temperate climates and much less common than *T. castaneum* in most parts of the tropics; important pest of many commodities, but especially cereals and cereal products.).......*T. confusum*

- Larger species, usually more than 5 mm long. Body is very dark brown or black. (Restricted to cool areas in the tropics, more often found in warm temperate and subtropical regions, also found in heated premises in cool temperate regions; minor pest of cereals and cereal products.) ....*T. destructor*

11. - Each elytron has a prominent lateral ridge and almost vertical flattened sides, and pronotum has two longitudinal lateral depressions (foveae). Body is strongly shining. These have length of 3.6–4.3 mm. (Found in India and the Orient, minor pest, infesting a wide range of commodities especially those with a high moisture content.) ....*Coelopalorus foveicollis*

- Elytra do not have lateral ridges, pronotum with or without foveae. Body is not strongly shining. The length is not more than 3 mm... *Palorus* sp.

12. - Pronotum has a deep longitudinal fovea on each side. These have length of 2.2–2.9 mm. (Common in cool highland areas in East Africa; minor pest in farm stores.)...................*Palorus laesicollis*

- Pronotum may not have foveae, if present, they will be shallow and ill-defined...**13**

13. - Side of front of head reaches dorsal margin of eye and conceals anterior part of eye from above. These have length of 2.7–3.0 mm. (Found in tropical and subtropical regions; minor pest of various commodities, especially cereals and cereal products.)............*P. subdepressus*

- Side of front of head does not conceal anterior part of eye from above..**14**

14. - Side of front of head is produced to form a somewhat triangular lobe on each side. These have length of 2.1–2.6 mm. (Oriental, but imported into Africa and West Indies; minor pest of various commodities.) ....*P. genalis* males

- Side of front of head is different. **15**

15. - Eyes are very small, length is equal to about half breadth of clypeus, and sides of pronotum do not have a small projection on basal third. These have length of 2.4–3.0 mm. (Cosmopolitan; minor pest especially of cereals and cereal products.)...................*P. ratzeburgii*

- Eyes are larger, length is almost equal to breadth of clypeus.........**16**

16. - Lateral margins of pronotum have a small projection (sometimes not very distinct) near basal third; pronotum is broadest toward apex; sides of front of head is not thickened. These have length of 2.2–2.8 mm. (Widespread in Africa, also recorded from South and Southeast Asia; minor pest of a wide commodities.)............................*P. ficicol*

3.4 STORED-PRODUCT LEPIDOPTERA

3.4.1 FAMILY GELECHIIDAE

3.4.1.1 ANGOUMOIS GRAIN MOTH (SITOTROGA CEREALELLA) (OLIVIER)

S. cerealella is an important and destructive insect of stored grain belonging to order Lepidoptera. This insect infests our stored produce as well as field crops. A review of literature revealed, however, that its life cycle has apparently never been worked out in details (Barnes & Grove, 1916; Simmons & Ellongton, 1933; Crombie, 1943; Back, 1922; Candura, 1926; Back, 1922; Cotton, 1960). The genus *Sitotroga* was formerly known as genus *Alucito*. The common name of insect obtained because it was found depredating wheat in the province of Angoumois, France, where it is known to have been imported into from the USA by early settlers in supplies of wheat brought in from the old country. In states it was reported as early as1728. The insect was not described and given a scientific name until 1789 (Cotton, 1941). *S. cerealella* is cosmopolitan in distribution and most abundant in throughout country. It is a strong flier and infests the produce from field to storage. The sizes of grain play a vital role in infestation of this insect. The bigger size grain may provide more space for its infestation. The adults having fore-wings, pale ocherous brown color and often have a small black spot in the distal half; they have a span of 10–18 mm. The hind-wings have a long fringe of hairs, longer than half the width of the wing and are sharply pointed at the tip. The labial palps are long, slender, and sharply pointed. The larva possesses true legs but the prolegs are greatly reduced and have only two crochets each. The life cycle of *S. cerealella* on different hosts has been studied by the so many workers (King, 1918; Simmons &Ellington, 1933). The moths feed and breed in threshing yard granaries under waste grain and near the straw stacks. After mating female starts egg laying in the grove of grain or nearby area of godown.

3.4.2 FAMILY PYRALIDAE

3.4.2.1 ALMOND MOTH (EPHESTIA (CADRA) CAUTELLA) (WALKER)

E. cautella is known to be one of the destructive insect of stored wheat and other cereals and fruits and nuts in our country since old age (Cotton, 1960).

It is imported into Britain on both cargo and ship during transportation of dried fruits (Prevett, 1968). The common name of insect is the tropical warehouse moth, the dried currant moth as almond moth. In India recorded from stored vegetables, rasins, dried fruits, lac, almond, and walnut. Before 1950 the infestation of this insect was not serious but latter it posed a serious problem to the country. The fore-wings of adult are greyish-brown with an indistinct pattern. The wing span is 11–20 mm. And both fore- and hind-wings have broadly rounded tips and only short fringes of hairs. The labial palps curve upward in the front of head and are rather blunt at the tip.

3.4.2.2   INDIAN MEAL MOTH (PLODIA INTERPUNCTELLA) (HUBNER)

P. interpunctella is a cosmopolitan pest of stored grain and developed successfully on the several stored produced (Williams, 1964; Reyes, 1969; Spitler & Clark, 1970). This insect first described in the year 1827 by Hubner, but first referred to as the Indian meal moth by Fitch because he found the larvae feeding on corn meal. *P. interpunctella* occurs in India, Burma, Srilanka, Pakistan, Bangladesh, Belgium, Canada, USA, and African countries. The larvae of this insect infest the stored grain and foodstuffs, generally all grain, seeds, and cereal products. The adult of *P. interpunctella* is small moth 5 to 10 mm long with the fore-wing is cream colored in the basal two-fifth, while rest of the wing is copper colored (dark reddish brown) with some dark grey markings. The wing span is 14–18 mm. And the labial palps point directly forward. The larvae are yellowish white, pinkish, and sometimes greenish white in color, depend upon color of host. "Fluttering" during which the wings beat rapidly as the insect walks and dance is a typical characteristic of this species of moth during matting. Single female lay eggs about 200 on the food materials, after hatching young larvae feed on kernels near the surface. The larvae completely web over the surface by making the surface grains together with silken threads they may seen webbing over the bag surface or on grain surface. Bell (1975) studied its developmental period at different temperatures and advocate that at 20 °C it took 55 days to emerge as adult moth.

3.4.2.3   RICE MOTH (CORCYRA CEPHALONICA) (STAINTON)

C. cephalonica is important pest of stored grain with special reference to stored rice and other grain products (Hodges, 1979). This insect rarely found in flour mills but reported to be infesting rice, wheat, maize, and other cereals (Cotton, 1960). It is reported that this insect originated from India then spread worldwide through transportation of food materials and finally became the cosmopolitan insect. The adult moth is nocturnal being most active during night (Pajni & Gill, 1974), moth rests on shady places in store with their anterior ends raised and their bodies, therefore, inclined than other moths of stored grains.

In the adults the hind-wings are pale buff, and the fore-wings are mid-brown or greyish-brown with thin vague lines of a darker brown along the wing veins. The fringes of hairs along the wing margins are relatively short and wing span is usually 15–25 mm. The labial palps point forward or downward; in the female they are long and pointed and in the male they are very short, blunt, and inconspicuous.

3.5   IDENTIFICATION OF STORED-PRODUCT LEPIDOPTERA WITH THE HELP OF TAXONOMIC KEY

3.5.1   KEY TO ADULTS LEPIDOPTERA

Important identification features for stored-product moths include the shape of the wings, the wing venation, the shape of the labial palps, and occasionally the coloration. The last feature relates to the presence of colored scales on the wings. The scales start to falloff soon after emergence and more are rubbed off when the moth's wing touches sack fibers, etc: within 2 or 3 days few, if any, may be left so that a once colorful moth may become dull brown. Thus, the color characteristics can usually only be judged in fresh specimens that still have most of their scales in place.

1. - Hind-wings narrows strongly at the tip to form a sharp point, and with a posterior fringe of long hairs that are more than half as long as the breadth of the wing. Labial palps are curved upward in front of head, elongate, and pointed at tips. Coloration is golden or light brown with a few dark scales. (Common in tropical and subtropical areas; primary pest of cereals, especially in farm stores and other small-scale storage, attacking whole or sometimes broken grain.) .....*S. cerealella*

    - Hind-wings are never strongly pointed at tips. If labial palps curve upward in front of head they are blunted at tips...........................2

2. - Labial palps are curved strongly upward in front of head, rather stout and blunt at tips. Fore-wings of fresh specimens are covered

with brownish-grey scales and with some darker scales forming and indistinct pattern. (Common in tropical, subtropical, and warm temperate areas; several species including three important pests. Infesting whole and broken cereals, cereal flours, and many other commodities.)..*Ephestia* spp.

- Labial palps are either minute (and thus apparently absent) or, if present, pointing forward or slightly curved downward, never curving sharply upward in front of head. Fore-wing coloration is different; if with brownish-grey scales, then without the pattern of darker scales. 3

3. - Labial palps are distinct in both male and female. Basal segment of antenna is similar in shape and size to other antennal segments. Fore-wings of fresh specimens with basal (inner) third are very pale yellow and apical (outer) two-thirds are dark reddish brown. (Common in tropical and subtropical areas; infests whole and broken cereals, cereal flours, dried fruit, nuts, etc.) ,.................. *P. interpunctella*

- Labial palps of male are very short and inconspicuous, apparently absent; palps of Female are very long and conspicuous, being more than 1.5 times diameter of eye. Basal segment of antenna is broad and flattened, more than two times breadth of other antennal segments. Fore-wings of fresh specimens are covered with grayish-brown or dull brown scales, and sometimes with very thin lines of darker scales following the pattern of the wing veins. Fresh specimens also have a prominent forward pointing tuft of long pale scales on the front of the head. (Common in tropical and subtropical areas; infests cereals especially if broken or milled, and a wide range of other commodities.) ....*C. cephalonica*

3.5.2 KEY TO LARVAE

The larvae of *S. cerealella* are distinctive and rarely seen (because they are usually inside cereal grains). The larvae of stored-product Pyralidae have a pigmented ring encircling the base of the main seta above the spiracle on the eighth abdominal segment. The difference between the two stored-product subfamilies of Pyralidae concerns the position of a similar pigmented ring on the anterior part of the body: in the Galleriinae there is a pigmented ring on the first abdominal segment, whereas in the Phycitinae a similar ring is found on the mesothorax (second thoracic segment).

1. - Abdominal false legs are very short, narrow, and inconspicuous. Body is distinctly fatter in the middle (Common in tropical and subtropical areas; primary pest of cereals, infesting whole (or sometimes broken) grain. Larva tunnels inside individual grain and is rarely seen.) ....*S. cerealella*

   - False legs are well developed and conspicuous (Plate 4a). Body is never distinctly fatter in the middle...**2**

2. - All abdominal spiracles with posterior part of rim are thicker than anterior part. First abdominal segment with the base of the long seta above the spiracle is surrounded by a complete or incomplete brown ring that encloses a pale membranous area. Such a ring is absent on the mesothorax. (Common in tropical and subtropical areas; infests broken cereals, flours, and many other commodities.) *C. cephalonica*

   - Rim of spiracle is more or less evenly thickened. First abdominal segment has the long seta above the spiracle arising from either a brown spot or an unpigmented area. Mesothorax, however, has a brown or yellowish ring around the base of the long dorso-lateral seta **3**

3. - Most setae on the first seven abdominal segments arise from dark brown spots on the cuticle. Mesothorax and eighth abdominal segments have brown or black ring around long seta. (Common in tropical and subtropical areas, and some species also in temperate zones; infest whole and broken cereals, cereal flours, and many other commodities. Several closely related species including three important pests.) ....*Ephestia* spp.

   - All setae on the first seven abdominal segments entirely do not have pigmented spots at their bases. Body color is yellowish. Mesothorax and eighth abdominal segment have pale yellow ring around long seta. (Cosmopolitan in warm climates; infests whole and broken cereals, cereal flours, dried fruit, nuts, etc.)............*P. interpunctella*

KEYWORDS

- **grub**
- **larvae**

- **weevil**
- **mesothorax**
- **moth**
- **elytron**

REFERENCES

Alfeiri, A. Les Insect de la Tombe de Tautan Khamon. *Bull. Soc. Ent.* **1931,** *115*, 188–189.

Applebaum, S. W.; Southgate, B. J.; Podolev, H. The Comparative Morphology Specific Status and Host Compatative of Geographical Strains of *Callosobruchus chinensis* L (Coleoptera: Bruchidae). *J. Stored Prod. Res.* **1968,** *4,* 135–146.

Arbogast, R. T.; Carlhon, M. Humidity and Temperature Effect on Development and Adult of *Oryzaephilus surinamensis* (Coleoptera:Salvanidae). *Environ. Entomol.* **1972,** *1,* 221–222.

Back, E. A. *Insect Control in Stores;* Farmers Bulletin No. 1275; U. S Department of Agriculture (USDA): US, **1922**.

Baker, J. E.; Marbie, J. M. Growth Responses of Larvae of Rice Weevil, Maize Weevil and Grainary Weevil of Meridic Diet. *J. Eco. Ent.* **1973,** *66* (3), 681–683.

Barnes, J. H.; Grove, A. J.; The Insect Attaching Stored Wheat in the Punjab and Method of Combating them Including a Chapter on the Chemistry of Respiration. *Mem. Deptt. Agric. (Chem. Series).* **1916,** *4*, 262–270.

Bell, C. H. Effect of Temperature and Humidity on Development of Pyralid Moth Pests of Stored Products. *J. Stored Prod. Res.* **1975,** *11*, 167–175.

Bry, R. E.; Lang, J. H. Damage to Mattress Licking by Cigarette Beetle Larvae. *J. Ga. Entomol.* **1966,** *11*(4), 21–23.

Burges, H. D. Studies on the Dermestid Beetle *Trogoderma granarium* Everts Π Reactions of Diapauses Larvae to Temperature. *Bull. Ent. Res.* **1962,** *53* (1), 193–213.

Candura, G. S. Contributo alla Conoscenza della Vera Tignola del Grano *Sitotroga cerealella* Oliv. (A Contribution to Knowledge of the True Grain Moth *Sitotroga cerealella* Oliv.). *Bell. Lab. Zool. Gen.* **1926,** *19*, 102.

Cittenden, F. H. Insect Affecting Stored Cereal and other Products in Mexico. *U. S. Deptt. Agric. Div. Ento. Tech. Sr.* **1896,** *4*, 27–30.

Cotton, R. T. *Control of Insects Attacking Grain in Farm Storage;* Farmers Bulletin No. 1811; U.S. Department of Agriculture (USDA): US, **1938**.

Cotton, R. T. *Insect Pests of Stored Grain and Grain Products, Identification, Habits and Method of Control;* Burges Pub. Co.: Minneapolis, MN, **1941**; p 242.

Cotton, R. T. *Pests of Stored Grain and Grain Products;* Burges Pub. Co.: Minneapolis, MN, **1960**; p 289.

Crombie, A. C. The Development of Angoumois Grain Moth, *Sitotroga cerealella* Oliv. *Nature.* **1943,** *152* (3852), 246.

Davies, J. C. Experiments on the Crip Storage of Maize in Uganda. *E. Agric. J.* **1960,** *26*, 71–75.

Dean, G. A. Lesser Grain Borer in Wheat in the Field. *J. Eco. Ento.* **1947,** *40* (5), 751.

Eden, W. G. Effect of Husk Cover on Corn on Rice Weevil Damage in Alabama. *J. Eco. Ento.* **1952,** *5,* 543.

Fletcher, T. B. *Report on Imperical Entomologist*; Science Report, Agriculture and Research Institute: Pusa 1928-1929, **1930**; p 42.

Freeman, J. A. Problems of Infestation by Insects and Mites of Cereals Stored in Western Europe. *Ann. Technl. Agric.* **1973,** *22 (3),* 509–530.

Floyd, E. H.; Oliver, A. D.; Powell, J. D. Damage to Corn in Lousiana Caused by Stored Grain Insects. *J. Eco. Ento.* **1959,** *52* (4), 612–615.

Giles, P. H. Observation in Kenya on the Flight Activity of Stored Product Insects Particularly *Sitophilus zeamais* Motsch. *J. Stored Prod. Res.* **1968,** *4,* 317–329.

Giles, P. H.; Ashman, F. A Study of Post Harvest Infestation of Maize by *Sitophilus zeamais* Motsch. (Coleoptera : Curculionidae) in Kenya, Highlands. *J. Stored Prod. Res.* **1971,** *7,* 69–83.

Halstead, D. G. H. A Revision of the Genus *Oryzaephilus ganglbauer*, Including Descriptions of Related Genera (Coleoptera: Silvanidae). *Zool. J. Linnean Soc.* **1980,** *69,* 271–374.

Halstead, D. G. H. A Revision of the *Genus Silvanus latreille (s.l.) (Coleoptera: Silvanidae);* Bulletin of the British Museum (Natural History). *Entomology.* **1973,** *29 (2),* 37–112.

Halstead, D. G. H. Keys for the Identification of Beetles Associated with Stored Products. I -Introduction and Key to Families. *J. Stored Prod. Res.* **1986,** *22* (4), 163–203.

Hodges, R. J. A Review of the Biology and Control of Rice Moth Corcyra Cephalonica Stant. (Lepidoptera : Pyralidae). *Trop. Stored Prod. Inst. Pub. G.* **1979,** *20,* 125.

Howe, R. W. The Biology of Rice Weevil *Caladra oryzae. Ann. Appl. Biol.* **1952,** *39* (2), 168–180.

Howe, R. W. The Effect of Temperature and Humidity on the Rate of Increase of an Insect Population. *Ann. Appl. Biol.* **1956,** *40,* 134–151.

Howe, R. W. *Sitophilus granaries* L. (Coleoptera: Curculionidae) Breeding in Acorns. *J. Stored Prod. Res.* **1965,** *1,* 99–100.

Howe, R. W.; Curri, J. E. Some Laboratory Observations on the Rate of Development, Mortality and Oviposition of Several Species of Bruchidae Breeding in Stored Pulses. *Bull. Ent. Res.* **1964,** *55,* 437–477.

Khare, B. P.; Agrawal, N. S. Effect of Emperature, Relative Humidity, Food Material and Density of Insect Population of *Sitophilus oryzae* L. and *Rhyzopertha dominica* F. *Bull. Grain Tech.* **1963,** *1,* 48–60.

Khare, B. P.; Agrawal, N. S. Rodent and Ant Burrows as a Sources of Insect Innoculum in the Threshing Floors. *Indian J. Ent.* **1964,** *26,* 97–102.

King, J. L. Notes on the Biology of the Angoumois Grain Moth. *J. Econ. Ent.* **1918,** *11* (1), 87–93.

Kuschel, G. On Problems of Synonymy in the *Sitophilus oryzae* L. Complex (Colcoptera: Curculionidae). *Ann. Mag. Nat. Hist.* **1961,** *13* (4), 241–244.

Laibach, E. Insecten Schiillin on Textile. *Angiv. Ent.* **1960,** *12,* 142–149.

Lefkovitch. A laboratory Study of *Stegobium paniecualla* (Coleoptera : Anobiidae). *J. Stored Prod. Res.* **1967,** *3,* 235–249.

Mayne, R.; Donis, C. *Notes Entomologiques du Bois 2nd . Distribution Congo an Rivanda, et. An Burindi Observations Ehtologiques;* Publication d Institute National Etude Agronomique de Congo: Belgium, 1962; p 100.

Mossop, M. C. Pests of Stored Tobacco in Southern Rhodesia. *Rhodesia Agric. J.* **1932,** *29,* 245–265.

Mossop, M. C. *Control of Maize Weevil*; Bulletin No. 1161; Ministry of Agriculture: Laul, Salisbury, S. Rhodesia, **1940**; p 27.

Pajni, H. R.; Gill, K. M. Effect of Light on Stored Product Pests. *Bull. Grain Tech.* **1974,** *12* (2), 151–153.

Prevett, P. F. Some Laboratory Observations on the Life Cycle of *Cadra cautella* G. (Lepidoptera : Pyralidae). *J. Stored Prod. Res.* **1968,** *4*, 233–238.

Proctor, D. L. An Additional Aedeagal Character for Distinguishing *Sitophilus zeamais* M. from *Sitophilus oryzae* L. (Coleoptera : Curculionidae). *J. Stored Prod. Res.* **1971,** *6*, 351–352.

Pruthi, H. S.; Singh, M. Pests of Stored Grain and their Control. *Indian J. Agric. Sci.* **1950,** *18*(4), 1–87.

Raina, A. K. *Callosobruchus* spp. Infesting Stored Pulses in India and Comparative Study of their Biology. *Indian J. Ent.* **1970,** *32* (4), 303–310.

Reyes, A. V. Biology and Host Range of *Plodia interpunctella* H. (Lepidoptera : Pyralidae). *Philip. Entomol.* **1969,** *59*, 124.

Richards, O. W. Observations on Grain Weevils Calandra General Biology and Oviposition. *Proc. Zool. Soc.* **1947,** *117*, 1–43.

Sawaf, S. K. Some Factors Affecting the Longevity, Oviposition and Population. *Ecol. Kyoto Univ.* **1956,** *7*, 43.

Simmons, P.; Ellington, G. W. Life History of the Angoumois Grain Moth in Marryland. *USDA Tech. Bull.* **1933,** 351–355.

Soderston, E. L.; Wilbur, D. A. Biological Variations in Three Geographical Populations of Rice Weevil Complex. *J. Kansas Ent. Soc.* **1966,** *39* (1), 32–41.

Solomon, M. E. Archeological Records of Storage Pests *Sitophilus granaries* L. (Coleoptera : Curculionidae) from an Egyptian Pyramid. *J. Stored Prod. Res.* **1965,** *1*, 105–107.

Solomon, M. E.; Adamson, B. E. The Power of Survival of Storage and Domestic Pests Under Winter Condition in Britain. *Bull. Ent. Res.* **1955,** *46*, 311–355.

Spitler, G. H.; Clark, J. D. Laboratory Evaluation of Malathion as a Protectant for Prunes during Storage. *J. Econ. Entomol.* **1970,** *63*, 1668–1669.

Van Emden, F. I. Egg Bursters in Some Families of Polyphagus Beetles and Some General Remarks on Egg Bursters. *Proc. Royl. Ent. Soc.* **1946,** *21*, 89–97.

Williams, G. C. The Life History of the Indian Meal Moth *Plodia interpunctella* H. (Lepidoptera : Pyralidae) in a Warehouse. *Ann. Appl. Bio.* **1964,** *53*, 459–475.

Zacher, F. Schadlinge in Pflanzlichen Dogen Pharm. *Ztg. Berl.* **1953,** *26*, 3.

CHAPTER 4

IMPORTANT STORED GRAIN AND SEED INSECT PESTS

CONTENTS

ABSTRACT

Stored-product insects are serious pests of dried, stored, durable agricultural commodities and of many value-added food products and nonfood derivatives of agricultural products worldwide. Stored-product insects can cause serious post-harvest losses, estimated from up to 10% in developing countries, but they also contribute to contamination of food products through the presence of live insects, insect products such as chemical excretions or silk, dead insects and insect body fragments, general infestation of buildings and other storage structures, and accumulation of chemical insecticide residues in food, as well as human exposure to dangerous chemicals as a result of pest control efforts against them. On the basis of their feeding behavior the stored-product insects may be categories as "Internal feeders" and "External feeders." The internal feeders are those insects that bore into grain by making a hole and passing the immature stage inside the grain and emerged as adult. The external feeders start feeding on tender part of the grain or processed stored produce, while some insect feeding by making web on grain.

4.1   IMPORTANT STORED GRAIN AND SEED INSECT PESTS

Stored commodities of agricultural and animal origin are infested by more than 600 species of beetles, 70 species of moths, and about 355 species of mites (Rajendran, 2002). Stored grain insect causing contamination in grain and seeds is also matter of great concern for food and seed industries. In developed countries there is zero tolerance for insects in food grain (White, 1995; Pheloung & Macbeth, 2000). According to recent estimates about 70–75% of food grain is handled at farmer's level in India where we do not have adequate scientific storage facilities and grain protection technologies. Rest 25–30% of food grain is procured and stored by Food Corporation of India (FCI), Central Warehousing Corporation (CWC), and State Warehousing Corporation (SWC) who have scientific storage facility (Commodity Online, 2007). In developing countries, food grain production often fall below demand as a result of post-harvest losses, especially in the course of storage caused by pests and other factors. In India, major insect pests of stored grains and seeds are described in Table 4.1. Stored-product insects are serious pests of dried, stored, durable agricultural commodities and of many value-added food products and nonfood derivatives of agricultural products worldwide. Stored-product insects can cause serious post-harvest losses, estimated from up to 10% in developing countries, but they also contribute to contamination of food products through the presence of live insects, insect products such as chemical excretions or silk, dead insects and insect body fragments, general infestation of buildings and other storage structures, and accumulation of chemical insecticide residues in food, as well as human exposure to dangerous chemicals as a result of pest control efforts against them.

On the basis of their feeding behavior the stored-product insects may be categories as "Internal feeders" and "External feeders." The internal feeders are those insects that bore into grain by making a hole and passing the immature stage inside the grain and emerged as adult. The external feeders start feeding on tender part of the grain or processed stored produce, while some insect feeding by making web on grain. The most common stored grain insects (Table 4.1) are described here.

4.1.1 STORED PRODUCT COLEOPTERA

4.1.1.1 RICE WEEVIL, SITOPHILUS ORYZAE (LINNAEUS) (COLEOPTERA: CURCULIONIDAE)

Distribution

Sitophilus oryzae are probably the most destructive stored grain insects and distributed throughout the world. It was first seen breeding on rice and hence named as rice weevil. India is considered to be the native of rice weevil (Fletcher, 1911; Zacher, 1937). It is a common insect of warm climate and infests paddy, wheat, maize, and almost all cereals and their products. It was not known in Europe but it seems that it must have established itself at least in Southern Europe many years ago. In colonial days this species was not common in North America and it is until recently become the predominant stored grain insect in the USA (Cotton, 1960).

Host range

Rice weevil is able to develop on a wide range of cereals and cereal products. Although *S. oryzae* is primarily a pest of stored products, but it can also attack on crop at ripening stage under field condition.

TABLE 4.1 List of Common Stored Grain Insects.

Common name	Scientific name	Order	Family	Damaging stage	Preferred host
Rice weevil	*Sitophilus oryzae* L.	Coleoptera	Curculionidae	Grub & adult	Wheat, rice, maize
Maize weevil	*Sitophilus zeamais* M.	Coleoptera	Curculionidae	Grub & adult	Maize, wheat
Granary weevil	*Sitophilus granaries* L.	Coleoptera	Curculionidae	Grub & adult	Wheat, rice, maize
Lesser grain borer	*Rhyzopertha dominica* F.	Coleoptera	Bostrichidae	Grub & adult	Wheat, rice
Larger grain borer	*Prostepohanus truncates* H.	Coleoptera	Bostrichidae	Grub & adult	Wheat, rice
Cigarette beetle	*Lasioderma serricorne* F.	Coleoptera	Anobiidae	Grub & adult	Stored tobacco
Drug stored beetle	*Stegobium paniceum* L.	Coleoptera	Anobiidae	Grub & adult	
Pulse beetle	*Callasobruchus chinensis* L.	Coleoptera	Bruchidae	Grub & adult	All pulses
	Callasobruchus maculates F.	Coleoptera	Bruchidae	Grub & adult	All pulses
	Callasobruchus analis F.	Coleoptera	Bruchidae	Grub & adult	All pulses
Khapra beetle	*Trogoderma granarium* E.	Coleoptera	Dermestidae	Grub & adult	Wheat, rice, maize
Rust red flour beetle	*Tribolium castaneum* H.	Coleoptera	Tenebrionidae	Grub & adult	Wheat flour
Grain moth	*Sitotroga cerealella* O.	Lepidoptera	Gelechiidae	Larva	Paddy, maize
Rice moth	*Corcyra cephalonica* S.	Lepidoptera	Pyralidae	Larva	Rice, wheat, maize
Almond moth	*Ephestia cautella* W.	Lepidoptera	Pyralidae	Larva	Almond, cashew nut
Indian meal moth	*Plodia interpunctella* H	Lepidoptera	Pyralidae	Larva	Cereals and oilseeds

Primary hosts: *Oryza sativa* (rice), *Manihot esculenta* (cassava), Sorghum, stored products (dried stored products), *Triticum aestivum* (wheat), *Triticum spelta* (spelt), *Zea mays* (maize).

Secondary hosts: *Cicer arietinum* (chickpea), *Hordeum vulgare* (barley), *Lens culinaris* ssp. *culinaris* (lentil), *Panicum* (millets), *Pennisetum glaucum* (pearl millet), *Vigna angularis* (adzuki bean), *Vigna radiata* (bean, mung), *Pisum sativum* (pea), Secale cereale (rye), *Sorghum bicolor* (common sorghum), Triticale, *Vicia faba* (broad bean), *Vigna unguiculata* (cowpea).

Nature of damage

The adult and grub feed on grain voraciously so that grain is rendered unfit for human consumption. In case of severe infestation only pericarp of the kernel is left behind, while rest of the mass is eaten up. The eggs grub and pupae are not normally seen because they develop inside the grains. The grub chews makes irregular holes in the germ and endosperm of the kernel. Adult emergence holes (about 1.5 mm diameter) with irregular edges are apparent some weeks after the initial attack. Adults can be found wandering over the surface of grain.

Life cycle

The adult of *S. oryzae* (Fig. 4.1) makes small excavation in the soft part of the grain where they lay eggs (Pruthi & Singh, 1950). The female weevil makes a slit like opening with her mandibles and rostrum, lays an egg in the whole and plugs it with secreted material, when egg laying has finished only a small stopper is visible externally (Richards, 1947). A single female can lay eggs between 300 and 400 during entire lifetime. Female normally lays only one egg in one kernel but two or more eggs are also commonly seen. Birch (Birch, 1945) reported that at 14% moisture content, temperature between 25 and 29 °C, and 70–75% relative humidity are most favorable for optimum egg laying. The eggs are white, translucent in egg plug and measure about 0.7 mm long and 0.3 mm wide, legless grub hatches in four days in summer and 6–9 days in winter (Khare & Agrawal, 1963). The grub is fleshy and legless has brown head and jaws, burrow inside the kernels. The grub period has been reported to 6–25 days (Harein & Soderstron, 1966). The pre-pupal stage last for one day and pupal stage for 20–25 days; pre-emerged adults 2–3 days and adult lives for 4–5 months. Development from egg to adult requires 28–31 days (Soderstron & Wilbur, 1966). This insect passes into five generations in a year but sometimes may have 7–8 generations depend upon food and environmental condition. Cotton (Cotton, 1960) reported that the weevil emerged in summer lives for 3–6 weeks but winter generations live longer. Adults are usually red-brown, dull with coarse microsculpture. Scutellum usually with lateral elevations closer together than their length and evidently more than half as long as scutellum. They have the characteristic rostrum and elbowed antennae of the family Curculionidae. The antennae have eight segments. There are usually four pale reddish-brown or orange-brown oval markings on the elytra, but these are often indistinct. Males with median lobe of edeagus evenly convex dorsally in cross section. Females with lateral lobes of internal, Y-shaped sclerite broader and rounded apically, more narrowly separated. The one pair of weevil can produce one million adults within three months. The severe infestation caused by the weevil between August and November.

FIGURE 4.1   Adult and grub of *Sitophilus oryzae* and infested grain by *Sitophilus oryzae*.

4.1.1.2   LESSER GRAIN BORER, RHYZOPERTHA DOMINICA (FABRICIUS) (COLEOPTERA: BOSTRICHIDAE)

Distribution

Rhyzopertha dominica is strong flier and also known as Australian wheat weevil. It was originally described from South America but their native place is to be India (Pruthi & Singh, 1950). Besides India *R. dominica* also reported from Australia, Argentina, and so many countries. It is also found in temperate countries, either because of its ability for prolonged flight or as a result of the international trade in food grain and their products.

Host range

The beetle often found infesting wheat in field, adults and grub of *R. dominica* feed primarily on stored cereal seed including wheat, maize, rice, oats, barley, sorghum, and millet. They are also found on a wide variety of foodstuffs including beans, dried fruits, turmeric, coriander, ginger, cassava chips, biscuits, and wheat flour. There are several reports of the lesser grain borer being found in or attacking wood, as is typical of other member of family Bostrichidae (Surtees, 1964).

Nature of damage

The adult and grub of *R. dominica* feed on grains from outsides and in haphazard manner. It is also causing the damage to mature crops in field. It possesses powerful jaws which used in causing serious damage to the grain. The grub feed on starchy contents of grain leaving the outer husk only while beetles destroy whole grain which are reduced to frass and waste flour left out by adults.

Life cycle

The female of *R. dominica* (Fig. 4.2) lays eggs on grain near the embryo and where young grub can easily bore into the grain, sometimes eggs may laid on the bags, walls and in cracks and crevices. On an average a single female lays about 300–500 eggs in their entire life- times on wheat in optimum environmental conditions that is, 34 °C and 14% moisture content and 70% relative humidity (Birch, 1945). The eggs are pear shaped, glistening white when freshly laid, but they become pinkishly opaque as the grub develop inside the eggshell. The eggs hatch in about 5–6 days under normal conditions, but during spring it take about 7–10 days and in winter it takes longer period to come out as a first instar grub. Bains (Bains, 1971) reported that temperature had a linear relationship with the rate of development and at egg stage the shortest period at 39 °C being 5.55 days and grub developmental period along with pupal period was shortest 28.31days at 33 °C. The first instar grub is straight and undergoes five molts, the grub presumably has greater difficulty in feeding on whole grain than in damaged wheat grains (Howe, 1950). The full-grown grub is dirty white in color with a slight brown head and cured abdomen and occupied 27–28 days. The pupal stage last for 5–6 days and eggs to adult it takes 40–41 days (Pruthi & Singh, 1950). Adults are 2–3 mm in length, reddish-brown and cylindrical. The elytra are parallel-sided, the head is not visible from above, and the pronotum has rasp-like teeth at the front.

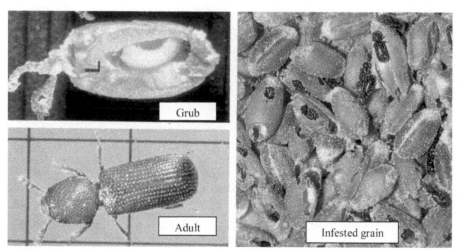

FIGURE 4.2   Adult and grub of *Rhyzopertha dominica* and infested grain by *Rhyzopertha dominica*.

There are generally 5–7 generations have been reported (Bains, 1971). In Northern India this insect breeds actively from April to October. The incidence of these pests is expected to be higher in area where the temperature ranging 27–36 °C and relative humidity is not too low 40% (Bains, 1971).

4.1.1.3   RED RUST FLOUR BEETLE, TRIBOLIUM CASTANEUM HERBST (COLEOPTERA: TENEBRIONIDAE)

Distribution

Tribolium castaneum is the most wide spread insect of stored grain and it is cosmopolitan in nature.

Host range

The major hosts of *T. castaneum* include several stored product and processed stored product but they commonly feed on cereals, millet, wheat bran, flour, grain spillage, broken grains, grain products, mixed feeds, beans, peas, lentils, butter beans, alfalfa seed, groundnut seed, flax, flax seed, rubber seed, cotton seed, cottonseed meal, safflower seed, sunflower seed, soybean meal, ginger, mustard, chillies, cinnamon, nutmeg, cocoa, copra, copra meal, yams, tapioca, raisins (including sultanas), dried figs, dried fruit, arecanuts, brazil nuts, walnuts, almonds, snuff, derris root, and oilseed cakes.

Nature of damage

T. castaneum does not cause damage to whole grain but mainly feeds on processed stored products, broken or infested grain by other insects. In case of severe infestation processed stored products turn grayish yellow and become moldy which produce foul smell and become unfit for human consumption. Both adult and worm like grub are responsible for damage and contamination of the produce.

Life cycle

T. castaneum (Fig. 4.3) start egg laying after 2–3 days of matting. The eggs are laid singly and freely in flour or among grains, eggs are most sticky when freshly laid, small particle of dust and flour wrap around it. A single female can lay on an average 450–500 eggs at 20 °C and 70% relative humidity. Eggs are small, slender, and cylindrical in shape rounded at both ends and whitish in color. The egg developmental period is 4–12 days at normal condition (Howe, 1956). The grub is small, worm like, slender, cylindrical, and wiry in appearance; body bears a number of fine hairs. The well-developed grub is measured 48 mm long and pale yellow in color, it molt 6–7 times and larval period is 22–25 days at 30–35 °C and 70% relative humidity. Usually pupation takes place on surface of food is naked white in color but gradually turns into yellowish color, the dorsal surface having hairs and resembling the grub. The pupal period lasts for about 4–24 days at 35 °C and 70% relative humidity (Howe, 1956). Adults emerge from the pupa within 3–7 days. The adult is 2.3–4.4 mm long, rather flat, oblong, and reddish-brown in color. The head and upper part of the thorax are covered with minute punctures and the wing covers are ridged lengthwise. The life cycle completed in

40–110 days depend upon environmental conditions and many generations completed in a year.

A. Adult of *Tribolium castaneum*

B. Grub of *Tribolium castaneum*

C. Grain infested by of *Tribolium castaneum*

FIGURE 4.3   Adult (A) and grub (B) of *Tribolium castaneum* and infested grain (C) by *Tribolium castaneum*.

4.1.1.4   KHAPRA BEETLE, TROGODERMA GRANARIUM EVERTS (COLEOPTERA: DERMESTIDAE)

Distribution

Trogoderma granarium is a major pest of stored grain in India. The first time record was made by Coles in 1894 and it has been recorded as destructive insect in Europe. It has been introduced a number of times in shipments of grain (Cotton, 1960). Similarly, within the confines of the USA, there are well defined areas in which certain species flourish in temperature prevailing between 32 and 49 °C, where as they are of little importance in others. It is reported to be pest of hot and arid regions of India and the world (Burges, 1962). *T. granarium* is a native of Ceylon and Malaya, and able to feed and breed in hot dry environment in addition it can survive normal winter of cool countries (Solomon & Adamson, 1955). It has been transported and established in Africa, Asia, Australia, and Europe and widely distributed throughout the world.

Host range

Primary hosts: *Arachis hypogaea* (groundnut), *Gossypium* (cotton), *H vulgare* (barley), *O. sativa* (rice), *Panicum miliaceum* (millet), *Sesamum indicum* (sesame), Sorghum, stored products (dried stored products), *Triticum* (wheat), *Z. mays* (maize).

Secondary hosts: *V. unguiculata* (cowpea), *C. arietinum* (chickpea), *Helianthus annuus* (sunflower), *P. glaucum* (pearl millet), *S. bicolor* (common sorghum), *T. aestivum* (wheat), *V. faba* (broad bean).

Nature of damage

Both adult and grub feed on grain but grub is most destructive. The grub bore into the stored product usually hollowing out the grain. It is polyphagous in nature and can survive in facultative diapauses for a year or longer in the absence of food.

Life cycle

The adult female of *T. granarium* (Fig. 4.4) lays eggs singly on grain, a single female can lay about 125 eggs in her lifetime. In the humid condition the incubation period is 5–9 days. Eggs generally hatch in 5–26 days (Laudani 1961) depending upon the prevailing temperature and humidity. The grub passed minimum 5–6 instars, or more instars depend up on duration of development. The duration of various instars was minimum at 35 °C and was prolonged with the decrease and increase of temperature (Atwal & Bains, 1974). The grub is brownish white in color, the body covered with bundles of long reddish broken hair. The hair are movable erectate on the posterior segments and forming a sort of tail at the posterior end. Dorsal surface uniformly creamy white to light brownish yellow or rarely light brown with some grayish pigmentation on sides of target. The total grub period for female is 26 days and for male it was 25 days. It has two set of life cycle one known as short while the other is long life cycle. Laudani (Laudani, 1961) reported that the eggs of Khapra beetle normally hatch between 5 and 26 days and grub period takes 27–87 days, pupal period 2–23 days and complete life cycle from egg to adult stage completed in 33–136 days. In long life cycle grubs enters into diapauses. The cause of this phenomenon is fully understood but believed to be due to unfavorable conditions or substances in the frass. After hatching the grub burrow in their food, feeding until they reach apparent maturity. Thus, if they have entered diapauses, their behavior pattern changes. They leave the food in search of a refuge, such as a crevice in a wall, where they cluster in large numbers in dormant state. In this state their rate of respiration is greatly reduced. At 30 °C and 70% relative humidity, which are good conditions for grub development, the average grub period of individuals not in diapauses is 33 days for female and 28 days for males (Burges, 1957). Generally this species of insect has 12 generations in a year (Khare, 1962).

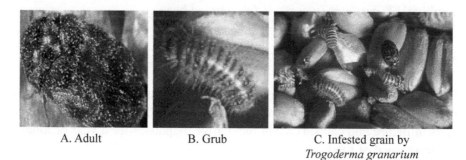

A. Adult   B. Grub   C. Infested grain by
                                    Trogoderma granarium

FIGURE 4.4   Adult (A) and grub (B) of *Trogoderma granarium* and infested grain (C) by *Trogoderma granarium*.

4.1.1.5   SAW TOOTHED GRAIN BEETLE, ORYZAEPHILUS SURINAMENSIS (COLEOPTERA: SILVANIDAE)

Distribution

Oryzaephilus surinamensis is a cosmopolitan pest of stored grain and grain products. It is reported to occur in flour mills and warehouses. It is able to multiply rapidly in cool conditions but its ability to overwinter in the fabric of warehouses enables quite large populations to persist until heating produce provides conditions needed for rapid increase (Howe, 1956).

Host range

O. surinamensis are typical secondary pests, and does not attack on whole grain. They infest cereals in stores like oatmeal flour or damage to broken kernels. Generally the cereals and oilseeds are preferred host of these insects.

Nature of damage

The grub could feed and breed on endosperm part of grain. They frequently found in processed cereals products and oilseed.

Life cycle

The adult female of *O. surinamensis* (Fig. 4.5) lays 6–10 eggs per day and 50–300 eggs in entire lifetime. Howe (Howe, 1956) reported the average number of eggs laid by these insects is 375. The eggs are either laid loosely in the food medium or deposited in crevices in the grain. Eggs are small white and slender in shape. The eggs hatch in about 3–5 days at 35–40 °C and 70% relative humidity (Howe, 1956). The grub molt three times but lower temperature enhance the molting number.

A   **B**

FIGURE 4.5   Grub (A) and adult (B) of *Oryzaephilus surinamensis*.

4.1.1.6   PULSE BEETLE, CALLOSOBRUCHUS CHINENSIS (LINNAEUS) (COLEOPTERA: BRUCHIDAE)

Distribution

Callosobruchus chinensis are important pests of pulses in India but also occur in tropic and sub tropics. It is known as cowpea weevil in America. It was first described in China 1758. It is said that from Asian countries it spread to west to Africa and Mediterranean basin (Caldoron, 1958) and east to America (Goncalves, 1939).

Host range

The host range of *C. chinensis* is restricted mainly, but not exclusively, to seeds of the legume tree *Viciaea*. As their common name indicates these beetles prefer cowpea but it has been reported to feed on pea, gram, pigeon pea, lentil, mung, urd, soybean, and other so many pulses.

Nature of damage

C. chinensis infestation commonly seen in the field, where eggs are laid on maturing pods of pulses. As the pods dry, the pest's ability to infest them decreases (Rahman, 1945), thus dry seeds stored in their pods are quite resistant to attack, whereas the threshed seeds are susceptible to attack throughout storage. In the early stages of attack the only symptoms are the presence of eggs cemented to the surface of the seeds or grain. The development occurs entirely within the seed; the immature stages are not normally seen. The adults emerge through windows in the grain, leaving round holes that are the main evidence of damage.

Life cycle

The female of *C. chinensis* (Fig. 4.6) prefers smooth whole seeds for oviposition, after the eggs are laid female covers it for about few seconds for dry of eggs. Normally eggs are singly laid on surface, on an average of 63–90 eggs laid by female in entire lifetime. In the field the eggs are laid on green pods. Freshly laid eggs are translucent, smooth, and shining but become light yellow or grayish white in color. Howe and Currie (Howe & Currie, 1964) reported that average number of eggs laid by a female is 45 with a range of 20–64 depend upon environmental conditions. The average incubation period at 30 °C temperature and 70% relative humidity is 3–5 days and% hatch ranged from 94 to 99 (Raina, 1970). The first instar grub bears a large spine on either side of the first abdominal segment and two groups of smaller spines dorsally on the tergal plate of pronotum (Van Emden, 1946). The grub turns at right angle after boring into seed, and forward horizontally feeding continuously on the cotyledons. The grub passes through four molts before pupation, grub and pupal period last for about 18–20 days. The complete development from eggs to adult takes an average of 22–23 days depend upon food source and environmental condition. The developmental stage is fast in cowpea and chickpea but slow in garden pea (Srivastava & Bhatia, 1958). The adults are 2.0–3.5 mm long bearing antennae pectinate in the male and serrate in the female, elytra are light brown, with small median dark marks and larger posterior dark patches, which may merge to make the entire posterior part of the elytra dark in color. The side margins of the abdomen have distinct patches of coarse white setae. *C. chinensis* has a pair of distinct ridges on the ventral side of each hind femur, and each ridge has a tooth near the apical end. The inner tooth is slender, rather parallel-sided, and equal to outer tooth.

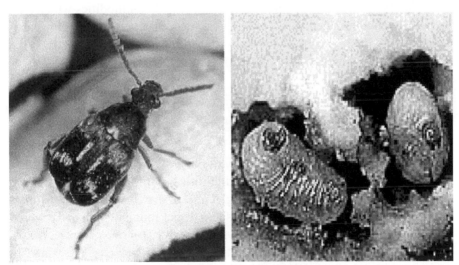

FIGURE 4.6   Adult (left) and grub of *Callosobruchus chinensis.*

4.1.1.7   CIGARETTE BEETLE, LASIODERMA SERRICORNE FABRICIUS (COLEOPTERA: ANOBIIDAE)

Distribution

The *Lasioderma serricorne* is commonly kwon as tobacco beetle or cigarette beetle and is most destructive insect of stored tobacco and other products. It is cosmopolitan in nature and is normally present at all times in one or the other stage in stores. Alfeiri (Alfeiri, 1931) reported that the *L. serricorne* is known to be present in Egypt more than 3500 years ago and was first recorded as a pest in North Carolina 70 years ago.

Host range

Besides infesting tobacco *L. serricorne* also infest the several food materials like ginger, oilseeds, cacao, spice, dry fruits, and other products (Le Cato, 1978). It also feed on wheat flour, cotton bolls, and books.

Nature of damage

The adult bores small round holes through infested commodities which may be heavily contaminated with dead bodies, frass, and pupal cells. Fine particles of food, dust, and frass adhere to the minute hairs of the body, often changing the appearance as per color of food.

Life cycle

The female of *L. serricorne* (Fig. 4.7) lays eggs in the folds and crevices of food materials. The egg laying capacity of the female ranging from 18 to 112 (Kurup et al., 1961). The egg is ovoid elliptical, whitish becoming opaque and dull in color just before hatching. The egg hatch in about 6–10 days (Bare et al., 1947). Newly hatch grub is less than 1 mm in length and covered with fine hairs, head is yellowish brown and body semitransparent. It passes 4–5 molt within 40 days to form a cell for resting. They pass the winter in grub stage in a stage of quiescence. The newly formed pupa is glossy white but gradually turned to a reddish brown after some times and average pupal period is about five days. Kurup et al. (Kurup et al., 1961) reported pupal period varying 3–13 days. The total life cycle of this insect is shortest in the months of May and June about 41 days and largest in the months of January to February about 58 days. They complete six generations in a year under tropical conditions (Kurup et al., 1961).

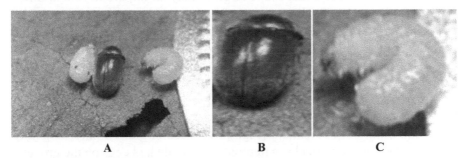

| A | B | C |

FIGURE 4.7   Adult and grub (A), adult (B), and grub (C) of *Lasioderma serricorne.*

4.1.2   STORED PRODUCT LEPIDOPTERA

4.1.2.1   ANGOUMOIS GRAIN MOTH, SITOTROGA CEREALELLA (OLIVIER) (LEPIDOPTERA: GELECHIIDAE)

Distribution

Sitotroga cerealella is an important pest of stored and field grain since ancient times. This insect obtained its name because it was found depredating wheat in the province of Angoumois, France, where it is known to have been imported into from the USA by the early settlers in supplies of wheat brought in from the old country (Cotton, 1941). It is cosmopolitan in nature, and widely distributed in all over the country.

Host range

S. cerealella is a pest of stored products particularly to cereals but it has also been found to infest stored spices, bell pepper (*Capsicum annuum*), coriander (*Coriandrum sativum*), black pepper (*Piper nigrum*), ginger (*Zingiber officinale*), and turmeric (*Curcuma longa*).

Nature of damage

It is strong flier and flies to the field at ripening stage of grains and start egg laying on them. They initially infest the grain at milking stage of crops than to stored grain. The size and shape of grain play an important role in infestation of this insect. The bigger size of grain with more air space provides better penetration and infestation at depth. Mahihu (1984) reported that maize permitted more infestation as compared to wheat and sorghum.

Life cycle

The adult moth of *S. cerealella* (Fig. 4.8) breeds in grain in threshing yard granaries near to straw stacks. After matting female laid eggs on the grove of grain or any place in store, a single female can lay on an average 40 eggs in range of 100–150 eggs. The eggs are creamy white cylindrical cigar shaped and hatch within three days at 30 °C temperature and 70% relative humidity. The first instar larvae feed on endosperm and enter into the seed after 5–7 days. The larvae pass through four molts than they spin cocoon inside cut space by making webbing. Khare and Mills (1968) reported the larval and pupal period of this insect is 20 and 22 days in wheat respectively. The total life cycle completed in 30–35 days depend upon environmental conditions. The adult moth is small, pale brown, 5–7 mm long with wings folded, wingspan 10–16 mm. The head, thorax and filiform antennae are pale brown in color; labial palpi are long, slender, sharply pointed and up curved, pale brown with dark tips, terminal segment longer than second segment. The forewing is elongate, light brown, or ochreous-brown, with a few black scales at the base of the dorsum and a concentration of black scales toward the apex.

A   B   C

FIGURE 4.8   Adult (A), adult with wing span (B), and larva (C) of *Sitotroga cerealella.*

4.1.2.2   RICE MOTH, CORCYRA CEPHALONICA (STAINTON) (LEPIDOPTERA: PYRALIDAE)

Distribution

Corcyra cephalonica is a major insect of stored grain and grain products (Hodges, 1979). It is common insect occurs in the humid tropics, especially in South and South-East Asia. It is a major storage pest in India, Thailand, Brazil, Ghana, Myanmar, Sri Lanka, and Indonesia where major cereal crops are rice, maize. It has been reported to breeding on groundnut also (Freeman, 1977).

Host range

C. cephalonica is a major insect of cereals, legumes, and their products but they also infest the stored oil cakes, dry fruits, cocoa, flex seeds, and processed products.

Nature of damage

The larvae of *C. cephalonica* cause damage to stored commodities by moving, feeding, and leaving silken threads. These threads are left over on the food which latter form dense and tough webbings, larvae may enter into grain and formed silken galleries, cocoons, frass, frothy mass, and excreta and feed on starchy content of grain (Pruthi & Singh, 1950).

Life cycle

The adult female of *C. cephalonica* (Fig. 4.9) starts egg laying just after matting. A single female lays about 150–160 eggs on walls, on rough surface, flour stacks in store at 24–32 °C and 60% relative humidity (Kamel & Hassanein, 1967). The eggs hatch in 3–6 days during summer and longer period during winter. The larva passes through eight molt but up to 16 molt occurs in both sex of rice moth (Kamel & Hassanein, 1968). They prefer warm and moderately moist climate, higher humidity and higher temperature provide faster development of larvae. They are able to develop more rapidly on wholegrain as compared to flour (Uberoi, 1991). The pupal period lasts for 7–10 days and life span of male is larger than female (Ayyar, 1954). The adult moth is nocturnal and is most active at nightfall. Its flight is rather slow and clumsy; flight is not powerful but can be sustained. The moths rest away from draughty places, on shaded store structures or surfaces of bag stacks; they are thus most commonly found in dark, sheltered corners of a store.

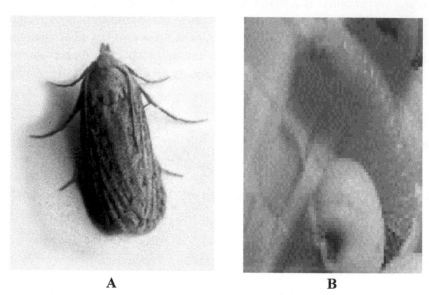

A   **B**

FIGURE 4.9   Adult (A) and larvae (B) of *Corcyra cephalonica.*

4.1.2.3   INDIAN MEAL MOTH, PLODIA INTERPUNCTELLA (HbBNER) (LEPIDOPTERA: PYRALIDAE)

Distribution

Plodia interpunctella is a cosmopolitan pest of stored products; they infest most of the stored commodities. Besides India it is reported from Sri Lanka, Pakistan, Canada, USA, Bangladesh, and Africa (Spitler & Clark, 1970).

Host range

P. interpunctella infest the several stored grain and processed products, but they also attack on milled cereal products, nuts, spices, peas, beans, lentils, chocolate, and other commodities.

Nature of damage

The larvae of *P. interpunctella* feed on germ of kernels near the surface. They completely web over the surface by making the surface grains together with silken threads and at times they can be seen webbing over the bag surface in store. It is a surface feeder, restricted to the top 20 cm, in grain bins and elevators (Bell, 1975).

Life cycle

The adult of *P. interpunctella* (Fig. 4.10) lays eggs just after matting, single female lays on an average 200 eggs either single or in group on stored commodities. The egg hatch within 2–4 days at 20 °C and 65% relative humidity (Bell, 1975). The full-grown larvae are yellowish white in color, depend upon food materials. Larvae pass through several molt and last larval skin converted into cocoon and took 55 days at 20 °C and 65% relative humidity, but it take 16–20 days at 30 °C (Bell, 1975). The adult moth is small, 5–10 mm long with wing held close together at resting, the anterior one third of fore wings is grayish while the outer two third has a coppery luster so that at resting stage of moth the wings appears to be marked with a prominent brown band. The hind wings are uniformly silvery grey and fringed with fine hairs.

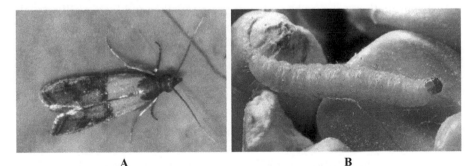

A   B

FIGURE 4.10   Adult (A) and larvae (B) of *Plodia interpunctella.*

4.1.2.4   ALMOND MOTH, CADRA CAUTELLA WALKER (LEPIDOPTERA: PYRALIDAE)

Distribution

Cadra cautella is a serious insect of stored wheat and other cereals products (Tuli & Mookherjee, 1962). Besides India it is also occurs in Egypt, Purtgal, and Britain. In temperate countries, it can develop during summer but needs heated storage to survive the winter.

Host range

C. cautella is a serious insect of several stored commodities, especially maize, rice, wheat, sorghum, millet, oats, oilseeds, cereal products but also they infest dried cassava, groundnuts, cocoa beans, dried mango, dates, nutmeg, mace, cowpeas, and so forth.

Nature of damage

The damage is caused by larvae just after hatching it cut small circular holes in part of grain and consume at a later stage a large part of food. It start feeding on germ point, sometimes they eat whole content of grain. In the infested grain they spun webs and matt grain together and form a silken web over the top. As a result of feeding by the larvae an appreciable amount of frass and nibbled grains appear besides the webbed materials.

Life cycle

The adult of *C. cautella* (Fig. 4.11) starts egg laying after matting. Single female lays up to 250 eggs at 25 °C and 75% relative humidity in entire lifetimes (Mullen & Arbogast, 1977). The eggs are translucent yellow with a distinctly sculptured surface pattern. The eggs hatch in about 3–4 days. The larvae range from 1.5 to 15 mm in length and are light brown in color with dark brown spots on the cuticle. They have a sparse covering of hair and in the males the testes can be seen through the cuticle as a dark patch in the posterior region. The pupae are dark-brown and found within a relatively light pupal case. The forewings of the adult are greyish-brown with an indistinct pattern. The wing span is 11–20 mm and both fore- and hind-wings have broadly rounded tips and only short fringes of hairs. Under optimum conditions, development from egg to adult takes 29–31 days, and the estimated rate of increase is about 50 times per lunar month.The limiting levels of moisture at 30 °C are 20 and 90% RH and development is only possible within a range of 15–36 °C. Adult emergence from the cocoon usually occurs during the late afternoon, and the adults show a marked periodicity in their flight activity and egg laying, both of which show a major peak at around dusk and a minor peak just before dawn.

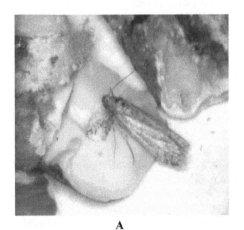

A   B

FIGURE 4.11   Adult (A) and larvae (B) of *Cadra cautella.*

KEYWORDS

- **stored product Coleoptera**
- **stored product Lepidoptera**
- **cereals**
- **pulses**
- **seed insects**

REFERENCES

Alfeiri, A. Les Insect de la Tombe de Tautan Khamon. *Bull. Soc. Ent.* **1931**, *115*, 188–189.

Atwal, A. S.; Bains, S. S. *Ecological Studies on Trogoderma granarium Everts and Methods of its Control;* Final Technical Report, PL- 480 Project (A7-MQ-24), Punjab Agricultural University: Ludhiana, Punjab, India, 1974; 137.

Ayyar, P. N. K. A Very Destructive Pest of Stored Products on South India *Corcyra cephalonica* Staint (Lep). *Bull. Ent. Res.* **1954**, *25*, 155–169.

Bains, S. S. Effect of Temperature and Moisture on the Biology of *Rhyzopertha dominica* Fabricius Bostrichidae Coleoptera. *Bull. Grain Tech.* **1971**, *9* (4), 257–264.

Bare, C. O.; Jenhet, J. N.; Brubaker, R. W. Improved Techniques for Mass Rearing of the Cigarette Beetle and Tobacco Moth. *U. S. D. A. Bur. Ent. Plant Quar.* **1947**, *247*, 1–11.

Bell, C. H. Effect of Temperature and Humidity on Development of Pyralid Moth Pests of Stored Products. *J. Stored Prod. Res.* **1975**, *11*, 167–175.

Birch, L. C. The Influence of Temperature on the Development of the Different Stages of *Calandra oryzae* (L) and *Rhyzopertha dominica* (F). *Aust. J. Exp. Biol. Med. Sci.* **1945**, *23*, 29–35.

Burges, H. D. Studies on the Dermesid Beetle *Trogoderma granarium* Everts I Identification and Duration of Development Stage. *Ent. Month mag.* **1957**, *93*, 105–110.

Burges, H. D. Studies on the Dermestid Beetle *Trogoderma granarium* Everts Π reactions of Diapauses Larvae to Temperature. *Bull. Ent. Res.* **1962**, *53* (1), 193–213.

Caldoron, M. The Bruchidae (Col) of Israel. Ph. D. Thesis, Submitted to the Senate of the Hebrew University, Jerusalam, 1958; p 217.

Commodity Online; http://www.comodityonline.com/commodities/future-trading/market/wheat-prices-to-improve-on-crop-loss/1406.html. 2007.

Cotton, R. T. *Insect Pests of Stored Grain and Grain Products, Identification, Habits and Method of Control;* Burges Pub. Co.: Minneapolis, MN, 1941; p 242.

Cotton, R. T. *Pests of Stored Grain and Grain Products;* Burges Pub. Co.: Minneapolis, MM, 1960; p 289.

Fletcher, T. B. Reports of the Imperial Entomologist. *Agric. J. India.* **1911**, 333–344.

Freeman, J. A. Prediction of New Storage Pest Problem. In *Origins of Pest, Parasite, Disease and Weed Problems;* Cherret., Sagar, G. R., Eds.; Blackwell Sci. Pub.: London, 1977; p 203.

Goncalves, L. I. Ogorglho Japones Callosobruches Chinensis Pic. Observacoes Sove. A Sua Biologia Habitos, Danus Causados e Meios de Combati. *Publ. Div. Def. Sanit veg Rio de J.* **1939**, *14*, 44.

Harein, P. K.; Soderstron, E. L. *Coleoptera Infesting Storing Products: Insects Colonisation and Mass Production;* Academic Press Inc.: New York, NY, 1966; pp 241–257.

Hodges, R. J. A Review of the Biology and Control of Rice Moth *Corcyra cephalonica* Stant. (Lepidoptera : Pyralidae). *Trop. Stored Prod. Inst. Pub. G.* **1979**, *20*, 125.

Howe, R. W.; Curri, J. E. Some Laboratory Observations on the Rate of Development, Mortality and Oviposition of Several Species of Bruchidae Breeding in Stored Pulses. *Bull. Ent. Res.* **1964**, *55*, 437–477.

Howe, R. W. The Development of *Rhyzopertha dominica* F. Under Constant Conditions. *Ent. Monthly Mag.* **1950**, *85*, 1–5.

Howe, R. W. The Effect of Temperature and Humidity on the Rate of Increase of an Insect Population. *Ann. Appl. Biol.* **1956**, *40*, 134–151.

Kamel, A. H.; Hassanein, M. H. Biological Studies on *Corcyra cephalonica* Stainton. *Bull. Soci. Entomol. Egypt.* **1967,** *51,* 175–196.

Kamel, A. H.; Hassanein, M. H. Instas and Ecdysis in Two Larvae Associated with Stored Milled Products. *Bull. Soci. Entomol. Egypt.* **1968,** *52,* 1–8.

Khare, B. P.; Agrawal, N. S. Effect of Emperature, Relative Humidity, Food Material and Density of Insect Population of *Sitophilus oryzae* L. and *Rhyzopertha dominica* F. *Bull. Grain Tech.*1963, *1,* 48–60.

Khare, B. P.; Agrawal, N. S. Seasonal Variation and the Peak Period of Occurrence of *Sitophilus oryzae* and *Rhyzopertha dominica*. *Indian J. Ent.* **1962,** *24* (2), 137–139.

Khare, B. P.; Mills, R. B. Development of Agoumois Grain Moth in Kernals of Wheat, Sorghum and Corn Affected by Site of Feeding. *J. Eco. Ent.* **1968,** *61* (2), 450–452.

Kurup, A.; Annamma.; Parkhe, D. P. Some Observations on the Biology of the Cigarette Beetle. *Indian J. Ent.* **1961,** *23* (4), 274–278.

Laudani, H.; Biology and Habits of Dermestids. *Pest. Control Mag.* **1961,** *29* (10), 58–61.

Le Cato, G. L. Infestation and Development by the Cigarette Beetle in Spices. *J. Georgia Entomol. Soc.* **1978,** *13* (2), 100–106.

Mahihu, S. K.; Depth of Infestation by *Sitotroga cerealella* into Grain Layers of Wheat, Maize and Sorghum. *Trop. Stored. Prod. Infor.* **1984,** *47,* 34–38.

Mullen, M. A.; Arbogast, R. T. Influence of Substrate on Oviposition by Two Species of Stored Product Moths. *Environ. Entomol.* **1977,** *6* (5), 641–642.

Pheloung, P.; Macbeth, F. *Export Inspection Adding Value to Australia Grain*, Proceeding of the Australian Postharvest Technical Conference Adelaide, Aug 1–4, 2000; CSIRO Stored Grain Research Laboratory: Canberra, Australia. 2000; pp15–17.

Pruthi, H. S.; Singh, M. Pests of Stored Grain and their Control. *Indian J. Agric. Sci.* **1950,** *18* (4), 1–87.

Rahman, K. A.; Sohi, G. S.; Sapra, A. N. Studies on Stored Grain Pests in the Punjab. Biology of *Trogoderma granarium* E. *Indian J. Agric. Sci.* **1945,** *15,* 85–92.

Raina, A. K. *Callosobruchus* spp. Infesting Stored Pulses in India and Comparative Study of their Biology. *Indian J. Ent.* **1970,** *32* (4), 303–310.

Rajendran, S. Postharvest Pest Losses. In *Encyclopedia of Pest Management;* Pimentel, D., Ed.; Marcel Dekker Inc.: New York, NY, 2002; pp 654–656.

Richards, O. W. Observations on Grain Weevils Calandra General Biology and Oviposition. *Proc. Zool. Soc.* **1947,** *117,* 1–43.

Soderston, E. L.; Wilbur, D. A. Biological Variations in Three Geographical Populations of Rice Weevil Complex. *J. Kansas Ent. Soc.* **1966,** *39* (1), 32–41.

Solomon, M. E.; Adamson, B. E. The Power of Survival of Storage and Domestic Pests Under Winter Condition in Britain. *Bull. Ent. Res.* **1955,** *46,* 311–355.

Spitler, G. H.; Clark, J. D. Laboratory Evaluation of Malathion as a Protectant for Prunes during Storage. *J. Econ. Entomol.* **1970,** *63,* 1668–1669.

Srivastava, B. K.; Bhatia, S. K. Development of *Callosobruchus chinensis* L. in Certain Vegetable Seeds. *Madras Agric. J.* **1958,** *45* (10), 392–395.

Surtees, G. Laboratory Studies on Dispersion Behavior of Adult Beetles in Grain. *Bull. Ent. Res.* **1964,** *54* (4), 715–722.

Tuli, S.; Mookherjee, P. B. Ecological Studies on *Cadra cautella* W. *Indian J. Ent.* **1962,** *25,* 379–380.

Uberoi, N. K. Nutritional Requirements of the Larvae of the Rice Moth *Corcyra cephalonica* S. Studies on Feeding Response to Various Natural Food. *Proc. Indian Agri. Sci.(B).* **1991,** *5* (6), 284–347.

Van Emden, F. I. Egg Bursters in Some Families of Polyphagus Beetles and Some General Remarks on Egg Bursters. *Proc. Royl. Ent. Soc*. **1946,** *21,* 89–97.

White, N. D. G. Insects, Mites and Insecticides in Stored Grain Ecosystem. In *Stored Grain Ecosystems;* Jayas, P., White, N. D. G., Muir, W. E., Eds.; Marcel Dekker: New York, NY, 1995; pp 123–167.

Zacher, F. Vorrisschadlinge Und Vorra Isnhutzchre Boleutongfur. *Volksernohrung Und Weltwirts Charf- Z hyg zool quoted by Aderson.***1937,** 1–11.

CHAPTER 5

DETECTION OF INFESTATION IN STORED PRODUCT

CONTENTS

ABSTRACT

Stored product is damaged by several species of insects. It may cause qualitative and quantitative losses to grain and seeds, besides causing direct losses they are responsible for contamination in stored commodities by toxin, feces, webbing, and body part of insects. This leads to economic losses of stored products of nation. In order to minimize such types of losses and improve quality and storability of stored commodities, detection of insects' infestation is very important. The detection of infestation is most crucial part of stored product management. The commodities stored for either domestic utilization needs or commercial utilization needs to take preventive and curative treatments for maintaining standard quality, so there are needs of correct detection of insect infestation. Generally visual inspection of storage structure and stored commodities is very popular, and therefore any results can only be recorded in a descriptive way. So long as the methodology remains uniform, direct comparison and therefore useful appraisals of different situations can be used. For more accurate and quantitative measurements it becomes necessary to sample the grain whereby the results can be interpreted by physical or chemical methods. There are several techniques for detection of insect infestation available for commercial or small level of storage. These techniques are manual sampling, probe and insect traps, sieving, floatation, X-ray, utilization of specific chemicals for hidden infestation, and Berlese funnels. Molecular techniques are very useful for the detection of infestation or contamination in stored commodities.

5.1   INTRODUCTION

Stored product is damaged by several species of insects. It may cause qualitative and quantitative losses to grain and seeds, besides causing direct losses they are responsible for contamination in stored commodities by toxin, feces, webbing and body part of insects. This leads to economic losses of stored products of nation. In order to minimize such types of losses and improve quality and storability of stored commodities, detection of insects' infestation are very important. The detection of infestation is most crucial part of stored product management. The commodities stored either for domestic or commercial utilization needs to take preventive and curative treatments for maintaining standard quality, so there are needs of correct detection of insect infestation. Stored commodities is vulnerable to both types of infestation either external or internal, the detection of internal infestation is very difficult (Pedersen, 1992). The inspection for insect damaged is expensive and many of stored grain insects developed inside the grain.

5.2   INSPECTION AND DETECTION TECHNIQUES

There are several techniques for detection of insect infestation available for commercial or small level of storage. These techniques are manual sampling, probe and insect traps, sieving, floatation, X-ray, utilization of specific chemicals for hidden infestation, and Berlese funnels (Neethirajan et al., 2007). The application of molecular techniques for detection of insect infestation is reported by Chambers et al. (1998). The measurement of uric acid content of cereals (Pixton, 1964) is also use in some countries. The detection of immature stage of insects by X-ray imaging and nuclear infrared reflectance spectroscopy have been extensively studies since this techniques are rapid and nondestructive and provide accurate results (Milner et al., 1950; Karunakaran et al., 2003; Karunakaran et al., 2004; Rajendran & Steve, 2005; Fornal et al., 2007; Neethirajan et al., 2007; Haff & Toyofuku, 2008). In USA some researcher investigates odor detection techniques for insect infestation detection, which are very useful tools in terms of costs, sensitivity, and accuracy (Ram et al., 1999).

Generally visual inspection of storage structure and stored commodities is very popular, and therefore any results can only be recorded in a descriptive way. So long as the methodology remains uniform, direct comparison and therefore useful appraisals of different situations can be used. For more accurate and quantitative measurements it becomes necessary to sample the grain whereby the results can be interpreted by physical or chemical methods. The following insect infestation techniques are well adopted.

5.2.1   MANUAL SAMPLING

The sampling of stored commodity is very important to detect infestation of insects. The size and frequency of sampling depends upon nature of commodities. Sampling is carried out with the objective of assessing the degree of insect infestation, although all forms of deterioration should be subsequently identified. Freeman (1948) first developed the need for standardization of estimates of infestation into defined categories for all insect species and all stored commodities, and in some cases his estimate have been modified for the tropics where the degrees of infestation are somewhat higher than in Britain (Hall, 1953). The manual sampling also provide living or dead insects for their correct identification.

5.2.2 INSECT TRAPS AND PROBES

This is very common method used for detection of adult insects in any storage condition. The insect traps and probes are simple in structure and easy in used. There are so many types of traps developed from Tamil Nadu Agricultural University, India, for detection of stored grain insects and very commonly used across the country at farmers as well as commercial levels. The commercial available traps for adults of stored grain insects are pitfall and probe trap and used worldwide (White et al., 1990). The regular inspection and removal of insects from traps are necessary to maintain storage hygiene. White and Loschiavo (1988) developed a stacked version of probe trap for monitoring insect density at different depths of grains. The results provided by different types of traps are only based on estimation so they are not accurate.

5.2.3 INSECT PHEROMONES

Insect infestation detection by pheromone is potentially useful because they are involved in chemical communication of stored grain insects. Generally sex and aggregation pheromones utilized in monitoring and matting disruption. The pheromone is impregnated in rubber septa so easily used in pheromone trap to detect the infestation. Stored grain insects may be detected with a several types of traps either by attractant or aggregation (Vick et al., 1990). The pheromone of *Rhyzopertha dominica, Tribolium castaneum*, and *Trogoderma granarium* are commercially available and used worldwide (Suzuki & Mori, 1983; Cross et al., 1976). The population density indicated by pheromone trap needs several factors to draw the actual population of insects. The environmental factor like temperature, rainfall, wind direction, and seed influence the trap catch.

5.2.4 LIGHT TRAP

The adult stored grain insects may trap by light trap. Insects attract to incandescent, fluorescent, and ultra violet light so we can use this method of trapping for detection of infestation. The wavelength and strength of light are crucial factor in trapping.

5.2.5   ACOUSTICAL METHODS

The application of insect feeding sounds for detection of external and internal stored grain insect infestation without sampling suggested by Neethirajan et al. (2007). In this method insect can be detected acoustically by amplification and filtering of their movement and feeding sounds. Hagstrum et al. (1988) reported that sounds of *R. dominica* grub can be detected population of grub. This method needs quantitative understanding of physical and biological factors that affect sound production and insect distribution (Neethirajan et al., 2007). The physical factors may be intensity, duration, and spectral feathers of sound at the source and distance to receiver (Beranek, 1988). The biological factors include insect response, unfavorable environment, and insect inactivity. This method cannot uses in detection of immature stage and dead insects.

5.2.6   BERLESE FUNNEL METHODS

The detection of live insects in commercial grain storage structures, this method is commonly used. This method is based on repellency against heat, so it catches free-living adults (Smith, 1977). This method is time taking and not provide accurate results.

5.2.7   ELECTRICAL CONDUCTANCE

The detection of hidden internal infestation in wheat this method used by Pearson et al. (2003) and their finding showed that accuracy varied from 87 to 88%. In this method only single kernel is used for weight, moisture content, diameter, and hardness. This is based on electrical conductance and compression force. This method is cost expensive and their results are not much more accurate.

5.2.8   NEAR INFRARED REFLECTANCE SPECTROSCOPY

This technique is economical fast and authenticated based on composition of grains (Kim et al., 2003). This method used for both qualitative and quantitative analysis of grains. It is based on the absorption of electromagnetic wavelengths in the range 780–2500 nm. The composition of grain is determined by traditional spectroscopy. This technique is best for detecting any stage of insects by a single kernel (Elizabeth et al., 2002). This method may not detect all types of infestation and do not differentiate the live and dead insects.

5.2.9   URIC ACID DETECTION IN CEREALS

This method is most commonly used for counting the number of insect fragments in sample of flour or ground wheat. This is apparently a promising technique for measuring of uric acid produced by insect metabolism has been described by Subrahmanyan et al. (1955) and Venkat et al. (1957).

5.2.10   X-RAY DETECTION TECHNIQUE

This technique is very fast and authenticated in detection of insect infestation in very short time. Karunakaran et al. (2003) reported that identified infested sample provide image of immature stage of insect with accuracy of 97%. For this purpose X-ray machine are used and operated at 15 kV capacity, grain observe this light and produce image of immature stage on soft X-ray. The cost of X-ray machine is high as compare to other chemical detection methods.

5.2.11   NUCLEAR MAGNETIC RESONANCE

The application of Nuclear Magnetic Resonance is similar to the X-ray technique but suffers from the same limitations of time constraints, sampling efficiency, and relative cost.

5.3   RAPID DETECTION OF HIDDEN INFESTATION

The infestation of *Sitophilus oryzae*, *R. dominica*, *Sitotroga cerealella*, *Callosobruchus* spp., and *Tribolium* spp. detected by following staining methods. Campbell (2004) reported that contamination of stored grain insects is matter of concern in milling industries.

5.3.1   STAINING METHOD

This is rapid method used for detection of egg plugs of insect with the help of various dye.

5.3.1.1   ACID FUCHSIN

The egg plugs of rice weevil are stained differently as compared to grain when treated with acid fuchsin. The stain solution prepare by mixing 10 ml glacial acetic acid, 190 ml distilled water, and 0.1 g acid fuchsin. Soak ~100 grain in warm water in a beaker. Drain off the water and cover grain with stain solution for 2–5 min. Drain off stain solution and wash the grain in water thoroughly to remove the excess stain. Weevil egg plugs are stained deep cherry red while feeding punctures and mechanical damage stain a light color under microscope.

5.3.1.2   GENTIAN VIOLET

In this method prepare 1% gentian violet aqueous stock solution and mix 10 drops of stock solution to 50 ml of 95% ethanol. Soak ~100 grain in warm water containing a pinch of detergent. Dip the soaked grain in stain solution for 2 min. Wash the stained grain in water for 5 s. The egg plugs of wet kernel will appear purple under microscope.

5.3.1.3   BERBERINE SULFATE

The 20 ppm berberine sulfate solution prepares in water. Soak ~100 grains in berberine sulfate solution for 1 min. Examine the grain under UV light, the egg plug will appear as fluorescence intense yellow color in microscope.

5.3.2   FLOATATION METHOD

This method consists from two solutions of chemicals sodium silicate and methyl chloroform at different specific gravity. The sodium silicate remains on the top and healthy grain separated at bottom and infested grain on the top.

5.3.3   GELATINIZATION METHOD

For this purpose prepare 100 ml of 10% sodium hydroxide solution. Stirring frequently with glass rod, boil ~100 grains in sodium hydroxide solution for 10 min. Drain off sodium hydroxide solution and wash the grain thoroughly in water. Examine the semi-transparent grain under microscope the insects developing inside grain may be seen in semi-transparent grain.

5.3.4 NINHYDRIN METHOD

The amino acids of insects react with ninhydrin and give purple spots when the infested grain with any stage of insect is crushed between coated papers. For this purpose prepare 0.7% ninhydrin solution in acetone. Treat Whatsman filter paper No. 1 or 2 in ninhydrine solution and air dry the filter papers. Place some grain between two treated filter papers and pass through crushing machine or crush manually. Dry the filter papers in oven at 100 °C. Purple spots on the filter papers indicate the presence of insect inside the grains.

5.3.5 INSECT PHENOL

The detection of insect phenols test based on spectrophotometry analysis for the concentration of a hydroxyphenol occurring in insect cuticle, which produces phenolidophenol dyes when chemically treated with 2,6- dichloroquinone chlorimide was proposed. Bailey (1975) stated that the method showed particular merit based on Food and Drug Administration evaluations, but at that stage required more work to perfect the method.

5.3.6 CARBON DIOXIDE DETERMINATION METHOD

The quantity of carbon dioxide produced in 24 hours in a representative sample of wheat can be measured. Generally at 25 °C, 0.3% carbon dioxide is released from grain with 14% moisture content, it is sign of insect free grain, if the amount of carbon dioxide increased than insect infestation is increased.

5.4 VOLATILE DETECTION TECHNIQUES BASED ON HEADSPACE ANALYSIS

5.4.1 DYNAMIC HEADSPACE EXTRACTION

This is a conventional extraction method for analysis of grain odor. This technique is operated by flushing a stream of air or inert gas to purge volatiles in the headspace, and then an adsorption tube is used to collect organic compounds carried by gas. The tenax resin is widely used to observe organic compound as well as aromatic compound from headspace. Besides it is made of porous polymers which are similar to the packing material of gas chromatography (GC) (Wampler, 1997). Subsequently the volatile compounds from dynamic headspace can be directly introduced into the GC or gas chromotography and mass spectra (GCMS) (Wampler, 1997; Rouseff & Cadwallader, 2001).

5.4.2 SOLID PHASE MICRO EXTRACTION

The solid phase micro extraction technique is inexpensive, rapid, and heat sensitive for detection of insect infestation (Richter & Schellenberg, 2007). In this method a coated silica fiber fixed on the sample and after some time the volatile compound absorbed by it. Subsequently, the solid phase micro extraction needle is removed and inserted into the GC and finally volatile compound separated by GC or GCMS analysis (Reineccius, 2002; Turner, 2006).

5.4.3 ELECTRONIC NOSE

This is the newer and excellent approach for grain odor detection. This technique is widely used in food industries (Rajendran & Steve, 2005). Electronic nose first used by Persaud and Dodd (1982) for the mimicking and discrimination of human olfactory system. It is consisted by chemical sensor to detect odor from infested sample (Persaud & Dodd, 1982; Marti et al., 2005; Rock et al., 2008). Zhang and Wang (2007) reported that 15% of insect damage may be detected by electronic nose.

5.5 MOLECULAR TECHNIQUE FOR DETECTION OF INFESTATION

The infestation of stored grain insects can be detected with the help of molecular tools. The damage caused by *Sitophilus granarius*, *Oryzaephilus surinamensis*, and *Cryptolestos ferrugineus* has been detected by molecular techniques (Chambers et al., 1998).

Sample preparation
Antigen release from insect

Homogenization of specific stored grain insects performed by plastic homogenizer, insects were crushed in 50 µl of phosphate buffered saline for 30 s. The protein concentration of the extract was determined and diluted into phosphate buffered saline then used in ELISA.

Antigen release from grain

For this purpose 5 g of infested sample of wheat mixed with 1 ml of phosphate buffered saline. The sample placed in plastic bag (12 cm × 12.5 cm) to retain the grain from antigen supernatant and removal of antigen when pipetting.

Antibodies

The monoclonal antibodies are used for this purpose (Kohler & Milstein, 1975).

5.5.1　INDIRECT ELISA

The microtiter plates coated overnight at 4 °C with 50 µl insect homogenates. After incubation, excess antigen was removed from the wells and the plates were rinsed in three changes of 0.1 phosphate buffered saline and Tween 20 for 3 min. The plates were incubated for 30 min on a plate shaker at room temperature with 100 µl per well of phosphate buffered saline, Tween and 5% Marvel milk powder. The plates rinsed as before and incubated with either tissue culture supernatant. After three more rinses, 50 µl of second antibody conjugate at 1:1000 dilution in phosphate buffered saline and Tween added to the wells and the plates were shaken for 45 min at room temperature, finally after three rinses with phosphate buffered saline and Tween, 100 µl of alkaline phosphatase substrate was added to each well and incubated at room temperature in the dark.

5.5.2　DOUBLE ANTIBODY SANDWICH

The antibody which recognizes antigen bound directly to ELISA plate and used to trap antigen from solution. The amount of antigen determined by second antibody with enzyme label. In this method microtiter plates coated with 100 µl purified mAb diluted in phosphate buffered saline and keep to incubate at 4 °C overnight. After washing wells incubated with 200 µl of blocking buffer for 30 min at room temperature on plate shaker, than after washing the wells incubated with 100 μl of insect homogenate or grain extract for one hour at room temperature on plate shaker.

5.5.3 WESTERN BLOT

Western blot chemiluminescent detection system regent and method used in conjugation with a polyvinylidene difluoride membrane for the detection of insect infestation.

KEYWORDS

- **infestation**
- **contamination**
- **detection technique**
- **stored product standardization**
- **pheromones**

REFERENCES

Bailey, S. W. Detection and Management of Contamination. ADAB Int. Training Course in the Preservation of Stored Cereals. *Selected Ref. Papers.* **1975,** *1,* 307–322.

Beranek, L. L. *Acoustial Measurements;* American Institute of Physics: Woodbury, NY, 1988; pp 1–7.

Campbell, J. F. Stored Product Insect Management in Flour Mills. *Outlooks Pest Manag.* **2004,** *15,* 276–278.

Chambers, J.; Dunn, J. A.; Thind, B. B. *A Rapid, Sensitive, User Friendly Method for Detecting Storage Mites and Insects;* HGCA Report No. 208: Sand Hutton,York, April 1998, pp 1–14.

Cross, J. H.; Byler, B. C.; Cassidy, F.; Silverstein, R. M.; Greenblatt, R. E.; Burkholder, W. E. Porpak-Q Collection of Pheromone Components and Isolation of (Z) and (E)- 14- Methyl- 8-hex-adecenal Potent Sex Attractant Components, from Females of 4 Species of *Trogoderma* (Coleoptera: Dermestidae). *J. Chem. Ecol.* **1976,** *2,* 457–468.

Elizabeth, B. M.; Dowell, F. E.; Baker, J. E.; Throne, J. E.; *Detecting Single Wheat Kernels Containing Live or Dead Insects Using Near-Infrured Reflectance Spectroscopy,* ASAE Conference, Chicago, Jul 28–31, 2002.

Fornal, J.; Jelinski, T.; Sadowska, J.; Grundas S.; Nawrot, J.; Niewiada, A.; Warchalewski, J. R.; Blaszczak, W. Detection of Granary Weevil Sitophilus Granaries Eggs and Internal Stage in Wheat Grain Using Soft X-Ray and Image Analysis. *J. Stod. Prod. Res.* **2007,** *43,* 142–148.

Freeman, J. A. World Foci of Infestation and Principal Channels of Dissemination to Other Points, with Suggestions for Detection and Standards of Inspection. *U.N., F.A.O., Agric. Studies Bull*. **1948,** *2,* 15–34.

Haff, R.; Toyofuku, N. X-Ray Detection of Defects and Contaminants in the Food Industry. *Sens. Instrum. Food Qual. Saf*. **2008,** *2,* 262–273.

Hagstrum, D. W.; Webb, J. C.; Vick, K. W.; Acoustical Detection and Estimation of *Rhyzopertha dominica* F. Larval Populations in Stored Wheat. *Flo. Entomol*. **1988,** *71,* 441–447.

Hall, D. W. Definitions for Reporting the Degrees of Infestation in Stored Products in Colonial Territories. *D.S.I.R, R.I.L*. **1953,** 3–5.

Karunakaran, C.; Jayas, D. S.; White, N. D. G. Detection of Internal Wheat Seed Infestation by Rhyzopertha Dominica Using X-Ray Imaging. *J. Stod. Prod. Res*. **2004,** *40,* 507–516.

Karunakaran, C.; Jayas, D. S.; White, N. D. G. Soft X-Ray Inspection of Wheat Kernels Infested by *Sitophilus oryzae. Trans. ASAE*. **2003,** *46* (3), 739–745.

Kim, S. S.; Phyu, M. R.; Kim, J. M.; Lee, S. H.; Authentication of Rice Using Near Infrared Reflectance Spectroscopy. *Cereal Chem*. **2003,** *80* (3), 346–349.

Kohler, G.; Milstein, C. Continuous Cultures of Fused Cells Secreting Antibody of Defined Specificity. *Nature*. **1975,** *256,* 495–497.

Marti, M. P.; Busto, O.; Guasch, J., Bouque, R. Electronic Nose in the Quality Control of Alcoholic Beverages. *TAC Trends Anal. Chem*. **2005,** *24,* 57–66.

Milner, M.A.X.; Lee, M. R.; Katz, R. Application of X-Ray Technique to the Detection of Internal Insect Infestation of Grain. *J. Econ. Entomol*. **1950,** *43,* 933–935.

Neethirajan, S.; Karunakaran, C.; Jayas, D. S.; White, N. D. G. Detection Techniques for Stored Product Insects in Grain. *Food Control*. **2007,** *18,* 157–162.

Pearson, T. C.; Brabec, D. L.; Schwartz, C. R. Automated Detection of Internal Insect Infestations in Wheat Kernel Using a PERTEN SKCS 4100. *Appl. Engg. Agric*. **2003,** *19* (6), 727–733.

Pedersen, J. R. Insects: Identification, Damage and Detection. In *Storage of Cereal Grains and their Products;* Sauer, D. B., Ed.; American Association of Cereal Chemists, Inc.: St. Paul, MN, 1992; pp 435–489.

Persaud, K. C.; Dodd, G. Analysis of Discrimination Mechanisms in the Mammalian Olfactory System Using a Model Nose. *Nature*. **1982,** *299,* 352–355.

Pixton, S. W. Detection of Insect Infestations in Cereals by Measurement of Uric Acid. *Pest. Infestation*. **1964,** *42,* 315–322.

Rajendran, S.; Steve, L. T. Detection of Insect Infestation in Stored Food. *Adv. Food Nutr. Res*. **2005,** *49,* 163–232.

Ram, M. S.; Seitz, L. M.; Rengarajan, R. Use of an Autosampler for Dynamic Headspace Extraction of Volatile Compounds from Grains and Effect of Added Water on Extraction. *J. Agric. Food Chem*. **1999,** *47,* 4202–4208.

Reineccius, G. Instrumental Methods of Analysis. In *Food Flavour Technology;* Taylor, A. J., Ed.; Sheffield Academic Press: U K, 2002; pp 211–227.

Richter, J.; Schellenberg, I. Comparison of Different Extraction Methods for the Determination of Essential Oils and Related Compounds from Aromatic Plants and Optimization of Solid Phase Microextraction/ Gas Chromatography. *Anal. Bioanal. Chem*. **2007,** *387,* 2207–2217.

Rock, F.; Barsan, N.; Weimar, U. Electronic Nose Current Status and Future Trends. *Chemical Rev*. **2008,** *108,* 705–725.

Rouseff, R.; Cadwallader, K. Headspace Techniques in Foods, Fragrances and Flavors. In *Headspace Analysis of Food and Flavors: Theory and Practice;* Kluwer Academic/Plenum Publisher: New York, NY, 2001; pp 1–6.

Smith, L. B. Efficiency of Berlese-Tullgren Funnels for Removal of the Rusty Grain Beetle *Cryptolestes ferrugineus* from Wheat Samples. *Can. Entomol.* **1977,** *109* (4), 503–506.

Suzuki, T.; Mori, K. (4R, 8R)- 4,8 Dimethyldecanal: The Natural Aggregation Pheromone of the Red Flour Beetle *Tribolium castaneum* (Coleoptera: Tenebrionidae). *Appld. Ento. Zool.* **1983,** *18,* 134–136.

Subrahmanyan, V.; Pingale, S. V.; Kadkol, S. B.; Swaminathan, M. Assessment of Insect Infestation and Damage to Stored Grain and their Products. *Bull. Cent. Food Technol. Res. Inst. Maysore India.* **1955,** *5,* 113–116.

Venkat, R. S.; Nuggehalli, R. N.; Swaminathan, M.; Pingale, S. V.; Subrahmanyan, V.; A Simple Method for Assessing the Extent of Insect Damage in Commercial Samples of Stored Grains. *Food Sci.* **1957,** *6,* 102–103.

Turner, C.; Overview of Modern Extraction Techniques for Food and Agricultural Samples. In *Modern Extraction Techniques Food and Agricultural Samples;* American Chemical Society: Washington, DC, 2006; pp 3–19.

Vick, K. W.; Mankin, R. W.; Cogburn, R. R.; Mullen, M.; Throne, J. E.; Wright, V. F. Review of Pheromone Baited Sticky Traps for Detection of Stored Product Insects. *J. Kansas Entomol. Soc.* **1990,** *63,* 526–532.

Wampler, T. P. Analysis of Food Volatiles Using Headspace Gas Chromatographic Techniques. In *Techniquea for Analyzing Food Aroma;* Marsali, R., Ed.; Marcel Dekker: New York, NY, 1997; pp 27–54.

White, N. D. G.; Loschiavo, S. R. Effects of Localized Regions of High Moisture Grain on Efficiency of Insect Traps Capturing Adult *Tribolium castaneum* and *Cryptolestes ferrugineus* in Stored Wheat. *Tribolium Inf. Bull.* **1988,** *28,* 97–100.

White, N. D. G.; Arbogast, R. T.; Field, P. G.; Hillmann, R. C.; Loschiavo, S. R.; Subramanyam, B. The Development and Use of Pitfall and Probe Traps for Capturing Insects in Stored Grain. *J. Kansas Entomol. Soc.* **1990,** *63,* 505–525.

Zhang, H.; Wang, J. Detection of Age and Insect Damage Incurred by Wheat, with an Electronic Nose. *J. Stod. Prod. Res.* **2007,** *43,* 489–495.

CHAPTER 6

ESTIMATION OF LOSSES DUE TO INSECTS IN STORED PRODUCTS

CONTENTS

ABSTRACT

The losses in stores are seldom a result of a single variable. They are of multiple factors including losses in weight, quality, and quantity. Each of these types of losses may have different significances which vary with people, time, place, and in the face of existing methods and technologies. The socio-economic realities play an important role in the estimation of storage losses particularly in developing countries. The estimation of losses at post harvest levels could vary widely depending upon the technologies adopted. The losses are assessed with respect to its biological and economic values. The economic loss refers to changes in values that may take place as a result of physical alternations of a produce while it is in threshing, handling, and transportation thereby, resulting into a loss normally termed as economic loss. From the farmers' point of view, the crop when harvested is threshed, handled, stored, and transported, ultimately what one gets is the difference in the quantity harvested and sold, there might have some loss. The weight loss is evolving degradation of quantity harvested and ultimately sold by the farmers. Apart from suffering a quantitative loss, it is possible that grain having been damaged at a micro level during threshing, handling, and transportation. The losses caused by immature stages if stored grain insects are most spectacular, and not easily seen. The estimation of losses is more complex if shrinkage due to loss in moisture and chemical changes. In seeds man minds the quality of grain kernel by keeping its germination at national and international standards, whereas in food man will like to have sound food. Losses in stores may be by weight and changes in physical state of grain chemical due to biological activities. The biological activities may be on dry basis and wet basis; wet basis is due to moisture. Biochemical changes result into increase in fat acidity, reduction of non-reducing sugar, heating, and loss of nutritional constituents. In dry form, the moisture increases due to insect, microbial, or fungal activity, such types of changes occur at latter stage, these all are termed as quality loss. The deterioration in quality of food grain due to insect damage is very important from availability of nutritional constituents as well as human health, while for seed man it is germinability of a kernel. But both need a grain kernel which is alive as it is easy to keep alive grain for long compared to dead grain. Dead grain is to be turned to flour and consumed soon. Food grains also undergo metabolic activities as other living things and are governed by the presence of moisture, temperature, insects, and fungi.

6.1 INTRODUCTION

The production technology as obvious has long received recognition and funds but stored grain protection technologies still lack sufficient appreciation and funds, if not all over the world, at least it is true in developing countries. Presumably the storage of grain and seeds has not been appreciated, because the damage done is an insidious nature and is often gone unnoticed until the commodities are about to be consumed or sold. In stored product entomology, the information on commodity losses caused during threshing, handling, transportation, and storage is scanty and scattered. Therefore, this attempt is to summarize the information on the losses and estimation of losses from the literature.

The estimation of losses at post harvest levels could vary widely depending upon the technologies adopted. The losses are assessed with respect to its biological and economic values. The economic loss refers to changes in values that may take place as a result of physical alternations of a produce while it is in threshing, handling, and transportation thereby, resulting into a loss normally termed as economic loss. From the farmers' point of view, the crop when harvested, is threshed, handled, stored, and transported, ultimately what one gets is the difference in the quantity harvested and sold, there might have some loss. The weight loss is evolving degradation of quantity harvested and ultimately sold by the farmers. Apart from suffering a quantitative loss, it is possible that grain having been damaged at a micro level during threshing, handling, and transportation. The losses caused by immature stages if stored grain insects are most spectacular, and not easily seen.

The losses in stores are seldom a result of a single variable. They are of multiple factors including losses in weight, quality, and quantity. Each of these types of losses may have different significances which vary with people, time, place and in the face of existing methods and technologies. The socio-economic realities play an important role in the estimation of storage losses particularly in developing countries. Khare et al. (1973) made several efforts to estimate losses of storage at farmers' level, where social factors plays an important role. The storage losses are of a multiple nature and all the factors cannot be easily combined. A single method can be developed by taking important basic components and a formula is developed for obtaining the damage index of different storage commodities. This would indicate the actual position of a given lot of grain and seeds with regard to various concepts of losses and prospects of its storage ability.

Storage losses if grouped efforts are made by experienced storage entomologists, the principal component analysis would be most useful in generating hypothesis from exploratory data, it may provide valuable clues to the pattern of relationship among basic components and other variables likes moisture, insects, fungi, and others.

Though the nutritional loss is an important aspect, unfortunately evaluation at these levels is difficult. The information of nutritional matters to provide subjective assessment of difference required is not adequate yet. This cannot be overcome at the market level. Nevertheless, where nutritional losses occur an attempt is made to evaluate them.

The losses in storage is almost final and cannot be compensated whereas growth losses of crops partially or more is compensated by increased yield from the surviving plants. There are several attempts are made to minimizing the possibilities of contaminating stored products with insects and this is accomplished by official inspection of the processing plants and storage warehouses as well as of the raw of finished products (Randal, 1957).

6.2   ESTIMATION OF LOSSES

Losses of stored products are calculated by several methods. A trader represents the loss as difference in the quantities received and quantities disposed off. It may also be at different stages, for example during transportation, loading and unloading, heaping, cleaning, weighment, packaging, and storage. All these stages give rise to greater losses on account of dampness, weevils, and other organisms. The estimation of losses is more complex if shrinkage due to loss in moisture and chemical changes. In seeds man minds the quality of grain kernel by keeping its germination at national and international standards, whereas in food man will like to have sound food. Losses in stores may be by weight and changes in physical state of grain chemical due to biological activities. The biological activities may be on dry basis and wet basis; wet basis is due to moisture. Biochemical changes result into increase in fat acidity, reduction of non-reducing sugar, heating and loss of nutritional constituents. In dry form, the moisture increases due to insect, microbial, or fungal activity, such types of changes occur at latter stage, these all are termed as quality loss.

6.2.1   INTELLECTUAL ESTIMATES

First time in 1916 Fletcher reported losses in store up to 33%. Food and Agricultural Organization (FAO) experts estimated as 10% loss of harvested crop by biological attack in storage, their earlier publication indicate 5% loss annually through insect infestation of all harvested cereals, pulses, and oilseeds. Cotton (1950) reported up to 10% losses in the USA. More than 5% losses of rice, corn, wheat, barley, and sorghum in storage were reported by Haeussler (1952).

6.2.2 EXPERIMENTAL ESTIMATE

The actual estimate is also termed as macro estimate. Find out of experiments is conducted and that gives true data of losses under storage. Though information at present, is scanty and fragmentary, but still very few is on records. The worth mentioning papers are of Cotton (1954); Rahman (1944); Parkin (1956); Hall (1963); Parpia (1967); Pingale (1970); Freeman (1968). Several scientists reported storage losses from different parts of the world ranging from 1 to 28% (Harries, 1950; Kockum, 1953; Cockbill, 1953). Sorghum showed 8% kernel infestation in three months after harvest and about 94% after nine months in Belgium (Lefevre,1953). Howe (1952) reported that net loss of 4.5% in 12 months due to insect infestation in Nigeria. Pingale (1954) reported that insect infestation increased from 2.7 to 48.5%, when wheat was stored in untreated bags for five months. Argentina has put its wheat losses in storage as equaling 400 million kilograms, in central Africa, a scientifically controlled test showed that 50% of sorghum was being lost due to insect during 12 months of storage. Ramasivan et al. (1986) reported 2.03 to 9.5% losses under different storage levels, while Willson et al. (1970) accounted losses for indigenous method of storage from Punjab as related to time-to-time survey. Kumar and Tiwari (2015) reported 2.7 and 3.4% damage due to insect infestation after 12 months of storage in natural conditions.

The micro estimate of losses appeared in literature in the year 1941 when Gay and Ratclife found that adult of *Sitophilus oryzae* L. consumed in a week an amount of wheat approximately equal to their own body weight and *Rhyzopertha dominica* F. consumed five to six times of their weight. Crombie(1943) found flour consumed by grub and adult male and female of saw toothed grain beetle *Oryzaephilus surnamensis* L. as 6.75, 2, and 2.25 mg by *Tribolium confusum* . Similar estimates were given by Richards (1947), Hurlocks (1965), and Davies (1960). Hall (1963) accounted weight loss as 8% in jowar as against 28% kernel damage while 4% in maize against 16.7% kernel damage. Apparent direct loss in weight due to insect attack is usually hidden because broken grain, dust, and leaving dead insects may have been included in weight sold. If the grain is screened as aspirated to render it comparable with its original condition the losses would be revealed. Maize stored in central Africa for 18 months showed an apparent loss in weight of 3% which increased to 17.6% when dust and insects were removed (Hurlock, 1967; Rao & Wilbur, 1972).

6.2.3   QUALITATIVE LOSSES

The deterioration in quality of food grain due to insect damage is very important from availability of nutritional constituents as well as human health, while for seed man it is germinability of a kernel. But both need a grain kernel which is alive as it is easy to keep alive grain for long compared to dead grain. Dead grain is to be turned to flour and consumed soon. Food grains also undergo metabolic activities as other living things and are governed by the presence of moisture, temperature, insects, and fungi. Increase in fat acidity and change in non-reducing sugars are two main deteriorating indices (Anderson & Alcock, 1954). According to a report of National Nutrition Advisory Committee 1969 in rural areas the food grains are stored in underground structures and above ground structures which are far from ideal, they are invariably infested by insects and micro organisms causing caloric losses and production of toxic and foul materials. This not only results in heavy quantitative and qualitative losses but also presents some form of health hazards. Fat acidity during storage was considered to be deteriorating the quality (Fenton & Swanson, 1930; Pomeranz et al., 1956) while finding of Greer et al. (1954) indicated that the free fat acidity increased up to 70% as compared to 5 to 10% in freshly milled flour, yet flour retained good packing quality even long after hydrolysis of fats.

Generally grain and seeds carry spores of storage fungi and a developing insect infestation will provide both the temperature and moisture to accelerate growth of storage fungi. Wheat stored at 14.6–14.8% moisture resulted in the increase of moisture to 17.6–23.0% due to damage caused by *S. oryzae* (Agrawal et al., 1957). Similar increase in moisture content accompanied by rapid increase in storage fungi of grain infested with weevils was reported by Christensen and Hodson (1960). On the other hand grain infesting mites *Acarus siro* and *Tyrophagus castellanii* were found in some abundance in samples of commercially stored wheat, the moisture contents of which ranged from 13.5 to 15.0%, at this range of moisture so many fungi are predominated (Griffiths et al., 1959).

The grain and seeds harvested in the field pass through many steps before reaching to final consumer. These can be headed under post harvest operations and broadly clubbed for threshing, processing, transportation, and storage. It is well known that losses and damage to grain take place in all these steps. The losses can be accessed from the point of view of the individual and the society, sum of losses encountered by the individual need not equal the loss to the society as a whole. Further the quantum of loss can vary widely according to their usages, the grain can be accepted or rejected as food or feed. While grains which are slightly damaged will be generally considered totally lost as food to the people who are well off materially. Such types of grain used for livestock.

The movement of food grain from the grower to his home storage or to market involves various modes of transport in this process losses of weight are generally observed. Garg and Agrawal (1966) reported that harvesting losses were 1–2% and during processing 10%. The qualitative damage caused by insects is quantified on the basis that each affected grain kernel has on an average lost half of its value in terms of food quality. However, it is extremely difficult to assess qualitative losses of food grains amounting from physical, chemical, biological, and engineering factors which can be subdivided into losses due to unfavorable temperatures, lack of moisture, or of oxygen content and damage due to rodents, insects, mites, micro organisms, handling equipments, and storage structures. Qualitative changes in food grains frequently remain invisible. It is not possible to assess the extent of losses emanating from changes in the chemical composition of stored food.

6.2.4 INDEX OF DETERIORATION

It is well known fact that deterioration in grain and its milled products in storage is accompanied by an increase in acidity. Hydrogen ion concentration tends to increase with age but because of the buffer action of proteins and other constituents of grain marked changes in hydrogen ion concentration ordinarily do not occur until deterioration is fairly well advanced. Titratable acidity tends to increase significantly even in very early stage of deterioration. The titratable acidity of grain and its milled products may be determined by following methods.

6.2.4.1 BESLEY AND BASTON METHOD

The corn meal obtained is digested with 80% alcohol, filtered and diluted the aliquot of filtrate with water and titrated the acid with standard alkali using phenolphthalein as indicator. The results so obtained are expressed as the number of milliliters of normal potassium hydroxide required to neutralize the 1000 g of corn.

6.2.4.2   GREEK OR BALL METHOD

Acidity of flour is determined by extracting the flour with 85% alcohol. Extract is filtered and filtrate is titrated with alcoholic potash using curcuma as an indicator. Results are expressed as percent sulfuric acid.

6.2.4.3   SCHULERUD METHOD

In this method flour is digested with 75% alcohol and filtrate is titrated with standard alkali using phenolphthalein. Results are expressed as milliliters of normal alkali required to neutralize the acid with 100 g of flour.

6.2.4.4   DETERMINATION OF FREE FATTY ACIDS

Fat and free fatty acids are extracted with suitable solvent and determined the free fatty acid content either of definite weight of extracted material or of material extracted from definite weight of original grain or flour.

The fat acidity, phosphate acidity, and titratable acidity increased with the deterioration of wheat in store and that the viability of wheat has direct relationship.

6.2.5   DISAPPEARANCE OF NON-REDUCING SUGAR

The disappearance of non-reducing sugar on corn stored at high moisture content, non-reducing invertase reducing sugars due to which several molds develop rapidly at 75% relative humidity (Christensen & Hodson, 1960). There appears to be good evidence that the activity of enzymes that split non-reducing sugars is less variable than the activity of fat splitting enzymes among different kinds of molds. The non-reducing sugar content of corn and other grains may therefore be a better index of the degree of moldiness and perhaps of overall deterioration than the fat acidity of grain. Christensen and Hodson (1960) reported non-reducing sugar of corn dropped from initial values as high as 174 mg of sugar per 10 g to as zero for badly deteriorated corn.

6.2.6 DETERMINATION OF MOISTURE CONTENT

The following methods are generally used to determine moisture content of grain or seed.

1. Oven method
2. Air oven method
3. Vacuum oven method
4. Drying with desiccants
5. Toluene distillation method
6. Brown Quvel distillation method
7. Direct heating method
8. Calcium carbide method
9. Dichromate method
10. Method for relative humidity measurement
11. Universal digital moisture meter method

Among all the methods for determination of moisture the oven method has been well adopted but recently universal digital moisture meter is being used all over the world.

In oven method the removal of all types of moisture from biological mass takes place even when finely ground seems to be due to the fact that water is present in various forms. If the inorganic object such as wet sand is heated at 100 ° C or higher at atmospheric pressure the water loss is quick and uniform until the object is completely dried. The drying of ground grain, however, requires a much longer time to remove moisture. The rate of moisture loss decreases with the presence of drying. The air oven method is a standard method of Association of Official Analytical Chemists and this method provides for heating a weight mass of finely ground grain for two hours at 135 °C.

Vacuum oven method for moisture content analysis of grain is based on drying the ground grain at temperature of 98–100 °C in an oven chamber at 25 mm moisture pressure. The heating is progressed up to about five hours until appreciable further loss of weight does not take place, then final moisture content works out by following formula.

6.2.6.1 ONE-STAGE

For samples containing less than 16% moisture, except soybean and rough rice the moisture content of which is less than 10 and 13%, respectively, flour and semolina.

1. Take two properly cleaned moisture dishes, dry for 1 h at 130 °C, cool in desiccator and obtain tare weight.
2. Grind 30–40 g sample in mill leaving minimum possible amount in the mill. Mix rapidly with spoon or spatula and transfer immediately a 2–3 g portion to each of two tared moisture dishes.
3. Cover and weigh dishes immediately and after subtracting tare weight record weight of the sample.
4. Uncover dishes and place them with cover beneath on shelf of oven.
5. Heat exactly for 60 min at 130 °C counting the time after oven recovers its temperature.
6. After one hour heating, remove shelf and dishes from oven, cover rapidly, and transfer to desiccator as quickly as possible.
7. Weigh dishes after they reach to room temperature.
8. Determine loss of weight as moisture.
9. Calculate the moisture content of grain putting the values in following equations:

$$\text{Percent moisture} = \frac{A}{B} \times 100$$

   A = Moisture loss in g

   B = Original weight of sample

6.2.6.2 TWO-STAGE

For samples containing 16% or more moisture (10 and 13% for soybean and rough rice), two-stage procedure should be followed as loss of moisture during grinding is likely to be excessive.

1. Fill two tared moisture dishes nearly full with representative portion of unground sample. Cover and weigh dishes. Subtract tare weights and record weight of sample.
2. Uncover dishes and place them with cover beneath in warm, well-ventilated place, preferably on top of heated oven protected from

dust, for 14–16 h so that the sample will dry reasonably fast. In all cases except soybean and rough rice the moisture content must be reduced to 16% or less (10 and 13% for soybean and rough rice).

3. Cover dishes containing air-dried samples and weigh them soon after they cool to room temperature. Determine the loss in weight and record it as moisture loss due to air drying.

4. Using air-dried sample, follow one-stage procedure described above, starting with the grinding step.

5. Calculate total moisture loss by using following equation :

$$\text{Percent moisture} = \frac{\left(\dfrac{EB}{D} + C\right) \times 100}{A}$$

A = Weight of original sample used for air-drying

B = Weight of sample after drying

C = Moisture loss due to air drying

D = Weight of sub-portion of air-dried sample used in 130 °C oven

E = Moisture loss due to oven-drying

The brown Duvel Distillation method was developed in 1907 and utilized in North America. In this method six flasks are heated with sample containing 100 g and 150 ml non-volatile oil. The flasks are heated to a temperature of 180 °C for wheat and cooled at 160 °C. The moisture thus released is collected in graduated cylinders and reading is taken which is later converted into moisture percent.

Direct heating is the method observed by the use of special heating equipment which shortens the time needed as compared to standard air oven method. The grain mass is heated to a temperature considerably higher than those employed in usual oven method so far more than 100 °C. it is accomplished by ordinary electric heating coils, may be by high frequency high voltage field or radiation from infra red radiators. The time of heating and temperature required are to be predetermined and tested while using this method.

Calcium carbide is used for determination of moisture which reacts completely with water to form Ca (OH)$_2$ and acetylene. This method was first described by Park 1941 to utilize in determination of moisture content of plant materials. The moisture content of mass is calculated from weight loss due to release of gas. A typical weighing balance is used to find out the moisture content in terms of percentage, but these methods do not provide accurate data. The moisture content of grain be kept appropriately low enough so that relative humidity is considerably below 75% and possibly below 65%.

The determination of moisture contents in grains helps to have an advance information on the durability of stored product, high moisture levels are bound to bring about certain biochemical and nutritive changes in grain during storage. The biochemical changes in grain take place and carbohydrate at 14% moisture is acted upon by the amylase to turn up to dextrins and maltose obviously, starch of grain is acted upon amylase. The moisture increase in grain is responsible for the quantity of soluble carbohydrates in wheat, this is stored at a particular moisture level and temperatures produced germ damaged kernel. Wheat stored for eight days at moist levels from 9 to 25% and temperatures from 29 to 50 °C produced characteristic increase of reducing sugars at expense of non-reducing sugar (Linko et al., 1960). The grain has proteolytic enzyme and organisms associated with grain too have this enzyme. This enzyme hydrolyzes the proteins into polypeptides and ultimately to amino acids, though the reaction becomes slow and may take place only when grain reaches to advance stage of deterioration. The kjeldahl protein has been demonstrated to be higher in insect infested and moldy grain (Khare et al., 1972; Daftary et al., 1970).

6.2.7   FAT CHANGES

The fat and oil are changed either by hydrolysis which may result in free fatty acids, if only flour is kept in storage for short duration, it will give some soury odor even at minimum moisture content. Grain fats are affected by lipases which turn to free fatty acids or glycerol with the length of storage. Fat hydrolysis is faster as compared to carbohydrates and proteins. Khare (1972) reported that free fatty acid increased in jute bags in which insect infestation was more while in stores it was less.

6.2.8   VITAMIN CHANGES

The most of cereal grain contains thiamin, niacin, pyridoxin, inositol, biotin vitamin E, panthothenic acid, and para amino benzoic acid. The reduction in thiamin content in storage is quite rapid, even place at 12% moisture level.

Further at time, reduction may take place at later stage of storage. If rice is stored for four years, by first two years there will be no change in thiamin while in rest period sharp reduction is observed (Fraps & Kemmerer, 1937).

6.2.9   FREE FAT ACIDITY

At the time of storage value of free fat acidity is generally 40.33 mg of KOH/100 g, but this increases as storage duration is increased. The free fat acidity values varied from 45.22 to 56.66, 59.88 to 72.88, 62.88 to 98.33, and 68.77 to 110.44 mg of KOH/ 100 g of grains in 3, 6, 9, and 12 months of storage, respectively (Khare, 1972).

6.2.10   NON-REDUCING SUGARS

In initial stage of flour contain non-reducing sugar is 82.66 sucrose in mg/10 g of flour, but value of non-reducing sugar changes as per duration of storage varies from 77.11 to 89.11, 69.88 to 86.55, 65.56 to 87.55, and 59.22 to 83.77 sucrose in mg/10 g of flour in 3, 6, 9, and 12 months of storage, respectively (Khare,1972).

6.2.11   LOSSES IN GERMINATION PERCENTAGE

Generally in seeds more than 90% germination is well accepted but it can be reduced due to infestation of insects and other biotic factors, the duration of storage and condition of store house are directly responsible for losses in germination percentage (Ramasivan et al., 1968; Khare, 1973; Grish et al., 1974).

6.3   QUANTITATIVE LOSSES

The quantitative losses of stored product can be easily observed at farmers' levels, by calculating the percent infestation and percent weight loss with the help of following procedures (Adams, 1976).

1. Collect the sample in polyethylene bag.
2. Clean the sample with the help of sieve.

3. Remove the mechanically damaged grain or broken grain from the sample.
4. Spread the sample on white paper and separate the healthy and bored or eaten grain.
5. Cut the healthy grain transversely to find out the insect developing inside it.
6. With the help of forceps take out the insect developing inside grain.
7. With the help of hair brush clean the excreta accumulated in the feeding gallery.
8. Count the healthy and infested grain separately.
9. Take the weight of healthy and infested grain.

$$\text{Per cent infestation} = \frac{Nd}{Nu + Nd} \times 100$$

where

Nd = Number of damaged grain

Nu = Number of undamaged grain

$$\text{Per cent weight loss} = \frac{(Wu \times Nd) - (Wd \times Nu)}{Wu(Nd + Nu)} \times 100$$

where Wu = Weight of undamaged (healthy) grain

Wd = Weight of damaged grain

Nu = Number of undamaged (healthy) grain

Nd = Number of damaged (healthy) grain

KEYWORDS

- nutritional loss
- quantitative loss
- estimation
- stored grain
- non-reducing sugar

REFERENCES

Adams, J. M. A Guide to the Objective and Reliable Estimation of Food Losses in Small Scale Farmer Storage. *Trop. Stod. Proct.* **1976,** *32,* 15–32.

Agrawal, N. S.; Christensen, C. M.; Hodson, A. C. Grain Storage Fungi Associated with the Granary Weevil. *J. Econ. Entomol.* **1957,** *50,* 650–663.

Anderson, J. A.; Alcock, A. W. *Storage of Cereal Grain and their Products;* American Association of Cereal Chemists: Minnesota, USA, 1954; p 105.

Christensen, C. M.; Hodson, A. C. Development of Granary Weevils and Storage Fungi in Columns of Wheat. *J. Econ. Entomol.* **1960,** *53,* 375–380.

Cockbill, G. F. Investigation on the Control of Insect Pest of Stored Grains and Pulses. *Rhod. Agric. J.* **1953,** *50,* 294.

Cotton, R. T. *Insect Pests of Stored Grain and Grain Products;* Burgs Publishing Co.: Minneapolis, MN, 1950; p 267.

Cotton, R. T. *Storage of Cereal Grains and their Products;* American Association of Cereal Chemists: Minnesota, USA, 1954.

Crombie, A. C. The Effect of Crowding Upon the Natality of Grain Infesting Insects. *Proc. Zool. Soc. London.* **1943,** *133,* 77–98.

Daftary, R. D.; Pomeranz, Y.; Sawer, D. B. Changes in Wheat Flour Damaged by Mold during Storage. *Agric. Food Chem.* **1970,** *18,* 613–616.

Davies, J. C. Experiments on the Crip Storage of Maize in Uganda. *East Afr. Agric. J.* **1960,** *26,* 71–77.

Fenton, E. G.; Swanson, C. O. Qualities of Combined Wheat as Affected by Types of Bin, Moisture, and Temperature Conditions. *Cereal Chem.* **1930,** *7,* 428–448.

Fraps, G. S.; Kemmerer, A. R. *Losses on Vitamin A and Carotene from Feeds during Storage;* Bulletin No. 557: Taxas Agricultural Experiment Station, 1937.

Freeman, J. A. *F A O Agric. Stud.* **1968,** *2,* 15–34.

Garg, O. P.; Agrawal, N. S. Quantitative and Qualitative Losses in the Production of Rice. *Bull. Grain. Tech.* **1966,** *4,* 24–27.

Greer, E. N.; Jones, C. R.; Moran, T. The Quality of Flour Stored for Periods up to 27 Years. *Cereal Chem.* **1954,** *31,* 439–450.

Griffiths, D. A.; Hodson, A. C.; Christensen, C. M. Grain Storage Fungi Associated with Mites. *J. Econ. Entomol.* **1959,** *52,* 514–518.

Grish, G. K.; Birewar, B. R.; Goyal, R. K.; Tomar, R. P. S.; Krishnamurthy, K. *Evaluation of Some Modern Rural Storage Bins for Storage of Wheat;* Indian Grain Storage Institute: Hapur, India, 1974; pp 201–212.

Haeussler, G. J. *Losses Caused By Insects. Insects the Year Book of Agriculture;* U. S. Department of Agriculture: Washington, D. C., 1952; pp 141–146.

Hall, D. W. *The Storage of Food Grains in Tropical Africa;* Informal Working Bulletin No. 24: Farm Products Processing, FAO: Rome, 1963.

Harries, D. V. *Report Deptt. Agric. Uganda.* **1950,** *48,* 1–5.

Howe, R. W. Miscellaneous Experiments with Grain Weevils. *Entomol. Non. Mag.* **1952,** *88,* 252–255.

Hurlock, E. T. Some Observations on the Amount of Damage Caused by *Oryzaephilus suranmensis* on Wheat. *J. Stod. Prod. Res.* **1967,** *3,* 75–78.

Hurlocks, E. T. Some Observations on the Loss in Weight Caused by *Sitophilus oryzae* to Wheat under Constant Experimental Conditions. *J. Stod. Prod. Res.* **1965,** *1,* 193–195.

Khare, B. P.; Choudhary, R. N.; Singh, K. N.; Senger, C. S. Loss in Protein Due to Insect Feeding on Maize. *Indian J. Ent.* **1973,** *36* (4), 312–315.

Khare, B. P.; Sengar, C. S.; Singh, K. N.; Agrawal, R. K.; Singh, H. N. Losses in Grain Due to Insect Feeding Wheat. *Indian J. Agric. Res.* **1972,** *6* (2), 125–133.

Khare, B. P. In *Technology of Pest Control and Pollution Problem in Storage Post Harvest Tech of Cereals and Pulses,* Proceeding of the National Academy of Science, 1972; pp 146–151.

Kockum, S. E. X Control of Insects Attacking Maize on the Cob in Crib Stores. *East. Afr. Agric. J.* **1953,** *23* (4), 275–279.

Kumar, R.; Tiwari, S. N. Fumigant Toxicity of Some Essential Oils Against Insect Pests of Stored Wheat. *Indian J. Ento.* **2015,** Accepted.

Lefevre, P. C. *Bull. Inform. Natt. Extend. Agron. Belg.* **1953,** *2,* 263– 368.

Linko, P.; Cheng, Y.Y.; Milner, M. Changes in the Soluble Carbohydrates during Browning of Wheat Embryos. *Cereal Chem.* **1960,** *37,* 548–556.

Parkin, E. A. Stored Product Entomology (The Assessment and Reduction of Losses Caused by Insects to Stored Foodstuffs). *Annu. Rev. Entomol.* **1956,** *1,* 223–239.

Parpia, H. B. A. Possible Losses of Food Grains in India. *Vijnan Karmee.* **1967,** *18* (4), 82.

Pingale, S. V. Effect of Damage by Some Insects on the Viability and Weight of Stored Grain. *Bull. Cent. Food Technol. Res. Inst.* **1954,** *2* (6), 153–154.

Pingale, S. V. Prevention of Losses in Storage. *Bull Grain Tech.* **1970,** *8* (1–2), 3–13.

Pomeranz, Y.; Halton, P.; Peers, F. G. The Effects on Flour Dough and Bread Quality of Mold Grown in Wheat and Those Invaded to Flour in the Form of Specific Culture. *Cereal. Chem.* **1956,** *33,* 157–169.

Rahman, K. A. Studies on the Stored Grain Pests in the Punjab Biology of Pulses Beetles. *Indian J. Agric. Sci.* **1944,** *12* (6), 851–854.

Ramasivan, T.; Krishnamurthy, K.; Pingale, S. V. Studies on Preservation of Food Grains in Rural Storage. *Bull. grain Tech.* **1986,** *6* (2), 69–75.

Randal, L. Insect Problem in the Marketing of Agricultural Products. *J. Eco. Entomol.* **1957,** *50* (4), 519–521.

Rao, G.; Wilbur, D. A. Loss of Wheat Weight from Feeding of Lesser Grain Borer. *J. Kansal Ento. Soc.***1972,** *42* (2), 238–241.

Richards, O. W. Observation on Grain Weevil General Biology and Observation. *Proc. Zool. Soc.* **1947,** *117,* 1–43.

Willson, H. R.; Singh, A.; Bindra, O. S.; Everett, T. R. *Rural Wheat Storage in Ludhiana District Punjab;* Staff Document, Ford Foundation: New Delhi, India, 1970; p 39.

CHAPTER 7

SOURCES AND KIND OF INFESTATION

CONTENTS

ABSTRACT

The food of insect is same as that of humans and other animals; insects came into being earlier than man. Hence, insects started feeding first on grains. The natural habitats of insects feeding grain must have been field. Stored product insects congregate in small or large numbers and maintain grain temperature high enough for winter survival while other survives in cracks and crevices and dusts or nearby farm buildings. The occurrence and spread of several insects are directly related to environmental conditions, but the methods of storage and the risks of cross infestation from imported commodities are responsible for infestation. Thus, infestation of stored grain and grain products may be supplemented in many ways. Most infestations of stored-grain insects originate from immigration of the insects into the bin from outside. All species of stored-grain insects have numerous food sources on which they survive when stored products are not available. Other common sources of stored-grain insects are old grain, grain spills, feeds, seed, and grain debris. Insects often move to new grain from carryover grain, from grain not cleaned from "empty" bins, from existing storage buildings, and from grain debris beneath perforated bin floors. Some stored-grain insects infest maturing grain in the field.

7.1 SOURCE AND KIND OF INFESTATION

The insect population is, maintained under adverse conditions until conditions are favorable for spread. Some insects congregate in small or large numbers and maintain grain temperature high enough for winter survival while other survive in cracks and crevices and dusts or nearby farm buildings. The occurrence and spread of several insects are directly related to environmental conditions, but the methods of storage and the risks of cross infestation from imported commodities are responsible for infestation.

Thus infestation of stored grain and grain products may be supplemented in many ways. Most infestations of stored-grain insects originate from immigration of the insects into the bin from outside. All species of stored-grain insects have numerous food sources on which they survive when stored products are not available. Other common sources of stored-grain insects are old grain, grain spills, feeds, seed, and grain debris. Insects often move to new grain from carryover grain, from grain not cleaned from "empty" bins, from existing storage buildings, and from grain debris beneath perforated bin floors. Some stored-grain insects infest maturing grain in the field. The sources of infestation are:

A. Field infestation
B. Infestation from already used storage bags
C. Infestation from machineries
D. Infestation from migration
E. Infestation through transportation
F. Infestation from threshing yards
G. Infestation through storage building
H. Infestation through birds' nests and rodent burrows

7.1.1   FIELD INFESTATION

The food of insect is same as that of humans and other animals; insects came into being earlier than man. Hence, insects started feeding first on grains. The natural habitats of insects feeding grain must have been field. The field infestation of storage pests like rice weevil, lesser grain borer, angoumois grain moth, pulse beetle, and potato tuber moth have been reported and all are found to be primary pests with good mobility of adults. Field infestation of secondary pests like red flour beetle, cigarette beetle, drug store beetle, saw toothed beetle, and flat grain beetle are not noticed as they are poor fliers. The field infestation of grain depends upon various factors, like the nearby area having godowns, people working in farm area dwell nearby godowns with grain and grain products, and climatic conditions favorable for spread and infestation of ripening grain.

The adults of *Sitophilus oryzae*, *Sitophilus granary*, and *Sitophilus zeamais* infest the grains either in the field or in the store before harvest (CABI, 2010). Extent and intensity of field infestation of maize by grain weevil in the field depend upon the husk covering (Girma et al., 2008). Maize genotypes with good husk characteristics like extended tip, tight husk, and flint grains resulted in low number of weevils and damaged ears. Field infestation of corn by rice weevil and angoumois grain moth was reported by Cotton and Winburn (1941) at Texas, Oklahoma, Kansas, and Missouri of USA during as early as 1938–1939.

The adults of *Rhyzopertha dominica* are infesting the grain in field by eggs laying on the cereal grains. Both grub and adult bore through the grain usually causing characteristic round tunnels. In later stages of infestation, these beetles may also hollow out the grains. Pupation usually takes place within the eaten grain. The lesser grain borer is primarily a pest of cereal grains, other seeds, cereal products, and dried cassava.

Callosobruchus spp. are the most common and widespread insect pests of pulse. They infest both pods in the field and seeds in the storage. They attack nearly mature and dried pods. Infested stored seeds can be recognized by the round exit holes and the white cemented eggs on the seed surface (Minja et al., 1999; Olubayo & Port, 1997).

Prostephanus truncatus is a serious pest of stored maize and dried cassava roots, and will attack maize on the cob, both before and after harvest (Saxena, 1995).

Phthorimaea operculella is the most serious pest of potatoes and it also attacks tobacco, eggplants, and tomatoes. Caterpillars of the potato tuber moth are feed as leaf miners, causing silver blotches on leaves, and bore into the petiole or a young shoot or main leaf vein, and later into the tuber. This causes wilting of plants. When eggs are laid on tubers, caterpillars begin feeding on the tubers immediately upon hatching making long irregular black tunnels, which are filled with excreta (feces), where disease-causing microorganisms grow. Major damage is caused by caterpillars burrowing in the tubers. Infestations start in the field. The pest is transferred with the harvested tubers to the potato store, where it can reproduce and infest other tubers. This may lead to total destruction of the stored produced.

Sitotroga cerealella may also infest the crop in the field before harvesting, and damage can reach serious levels, before the grains are stored. Female moths lay ovoid and pinkish eggs at night in clumps on the outside of cereal grains, in cracks, grooves, or holes made by other insects. Eggs are initially white turning red near hatching. The eggs which are white when first laid, soon change to a reddish color and are laid on wheat heads, exposed tips of corn ears in the field, or in stored grain. The larvae are caterpillars of dirty white color and about 8 mm long when fully grown. Caterpillars penetrate into and feed inside whole grains. They prepare a round exit hole for the month, leaving the outer seed wall only partially cut as a flap over the hole, resembling a trap door. The adult pushes its way out through this "window" leaving the trap door hinged to the grain. Infested grains can be recognized by the presence of these small windows. They are pests of whole cereal grains like paddy, sorghum, maize, and wheat (Ragumoorthy & Gunathilagaraj, 1988).

7.1.2   INFESTATION FROM ALREADY USED STORAGE BAGS

The most of stored grain insects survive in storage bags and having potential to re infest the stored commodities in the same bags. In many cases, the
empty bags are thrown into a pile to await the accumulation of enough stores
next time. The infested bags have potential to cause infestation (Cotton et
al., 1960). Most of the flour mills, farmers and bakery industries refilling the
commodities in used bags.

7.1.3   INFESTATION FROM MACHINERIES

The farm machineries also serve as source of infestation of stored grain
insects. The survival stage of insects may remain inside the post-harvest
machineries. The infestation of Mediterranean flour moth is known to have
carried from one mill to another mill through machineries.

7.1.4   INFESTATION FROM MIGRATION

The movement of adult insects from their population territory to other places.
Cotton et al. (1960) reported that in the warmer periods of summer, insects
were in flight during daytime. The flight migration mainly by Indian meal
moth, angoumois grain moth, rice weevil, and red flour beetles are generally
seen under field condition. Rice weevils were found migrating from store to
field through seeds.

7.1.5   INFESTATION THROUGH TRANSPORTATION

The production of grain and seed is aimed to distribute to places required
to feed people and grow the crop where production is insufficient for the
population. This made the geographical distribution of food grain in a nation
having wide climatic variations. The effectiveness of transportation and
distribution of food grain plays an important role in the national economy.
The transportation and distribution systems are different in different countries; developing countries have their traditional system combined with
some modern technologies. The bullock carts, camel carts and horse, or
other animal driven carts are used to transport grain from threshing yards to
store, and then moved to markets. The absence of such facilities delays the

transportation and insects already present in cartload get an opportunity to multiply and take the process of movement to other places. Similarly, trucks and rail wagons unattended also carry infestation from one place to another. Pruthi and Singh (1950) reported that railway wagons, lorries, or trucks, which are used to carry grain, constitute an important source of grain infestation. The most spectacular introduction among these was the khapra beetle *Trogoderma granarium* into most of the countries (Howe, 1962). Freeman (1974) reported the occurrence of stored grain insects in cargoes and plant or animal products imported from different countries. Storage mites and insects from the structures of the ship move with grain products, oilseed, and pulses (Hurlock, 1963).

7.1.5 INFESTATION FROM THRESHING YARDS

Threshing yards play a vital role in post-harvest operations and are the major source of infestation to grain and seeds. The insects reach the threshing yards either from empty godowns of the nearby area or from insect inoculums already present on the threshing yards (Khare & Agrawal, 1964).

7.1.6 INFESTATION THROUGH STORAGE BUILDING

The insect lives in cracks and crevices along with residual material lying loosely in existing storage building. At times, the insects may penetrate into the structure of building where hidden from sight and inaccessible during normal cleaning. This residues, thus act as a breeding site for insects. Coombs and Freeman (1955) and Khare (1963) reported that insects were found in cracks and crevices, webbing, dusts, and gunny bags used for storage of grain.

7.1.7 INFESTATION THROUGH BIRDS' NESTS AND RODENT BURROWS

Stored grain insect may colonies in the nests of several birds and rodent burrows, which may be built on or storage structures (Howe, 1962; Khare & Agrawal, 1964). Rodent burrows had on an average 2.50 kg grain on which several stored grain insects may survive during adverse period. The importance of bird nests as sources of infestation of carpet beetles. Stored product insects from bird nests also observed by the Woodroffe and Southgate (1951).

7.2 KINDS OF INFESTATION

There are following kinds of infestation of stored grain insects are:

7.2.1 HORIZONTAL INFESTATION

Infestation of stored grain insect on mature crop, which comes to the storage through post-harvest operations.

7.2.2 VERTICAL INFESTATION

In bulk storage, spreading of infestation from top to bottom and vice versa through gradual migration.

7.2.3 LATENT INFESTATION

Infestation by stored grain pests which are already present in the storage structures.

7.2.4 CROSS INFESTATION

The infestation of polyphagous insects from one source to another source.

7.3 MANAGEMENT OF SOURCES AND KIND OF INFESTATION

In order to minimize the infestation of stored grain insects the following measures should be taken.

7.3.1 STORE DISINFECTION

After the store has been cleaned completely and all residues of grain and dusts have been removed, it is good practice to dust the whole store with diatomite earth, lime, or ashes as a further prevention of problems. Where larger grain borer has attacked the wood in the construction, the wood should be treated with any of the approved wood preservatives or thoroughly sprayed with kerosene, oil mixture to get rid of any surviving grain borers.

7.3.2   STORAGE HYGIENE

Always keep the store and its surroundings clean. Before newly harvested crops are stored, the store should be carefully prepared well ahead of time. Old stored products should be removed and the room completely cleaned up. The whole building should be well aired and if possible fumigated or disinfected. The walls roof and floor should be both watertight and rat proof, and small holes and cracks, which are potential breeding places for storage insects, should be sealed.

7.3.3   STORE LOCATION

Site of stores always away from any potential source of infestation. The grain and tuber moths are good flyers and adults from infested stores often infest growing crops in the field. Separations of stores from fields may help to reduce attack. Prevent pest entry by sealing the store with insect-proof gauze (Golob & Muwalo, 1984).

7.3.4   SORTING AND CLEANING THE PRODUCE

Removal of infested grains or seeds and pests can also be done by hand, sieving, winnowing, or moving the grain. When using methods that merely separate the pests from the stored product ensure that the pests removed from the produce are killed to avoid reinfestation.

7.3.5   CHOICE OF VARIETY

There is a widespread perception that modern, high-yielding varieties of maize may be more susceptible to storage pests. These varieties often have open cob husks, allowing insects to easily attack maize in the field, whereas some of the traditional varieties have closed husks, thus effectively protecting the crop from insect attack. The same varieties have been observed with some sorghum varieties. Therefore, the increased yield offered by some varieties should be weighed against the susceptibility to storage pests, the expected period of storage, and the price to be expected for grain of a particular damage level. Efforts are going on to develop high-yielding varieties with resistance to storage pests.

7.3.6 CHOOSING HARVEST TIME

Care should be taken when cultivating new high yielding and early ripening varieties, since the harvest may fall in the wetter parts of the year, and this may create problems of storage. Some storage infest beans and grains in the field only when the crop is almost dry. Therefore, timely harvest can ensure that these pests are not carried into the store along with the beans or grain. Thus, timely harvesting significantly reduced infestation by the bean bruchid and cowpea bruchids (Olubayo & Port, 1997). As a rule, do not leave crops in the field when they are ready for harvest, this increases the chances of infestation by some storage pests.

7.3.7 INSPECTING THE STORE

The regular inspection of store is always preventing the infestation of store grain insects. Brush stacks of bags with a stick or broom to disturb the resting moths. Lift bags in order to detect moth cocoons along the line where bags touch each other. Insects can also be sieved out using a box sieve with a mesh of 1–2 mm. Identify the insect found in order to perform the correct treatment. These measures should prevent the breeding of carry-over insects from former crops.

KEYWORDS

- infestation
- stored grain
- pest
- threshing yards
- post harvest
- source and kind

REFERENCES

C A B I; *Sitophilus zeamais (Maize Weevil) Datasheet. Crop Protection Compendium;* 2010 Edition, CAB International Publishing: Wallingford, UK, 2010.

Coombs, C. W.; Freeman, J. A. The Insect Fauna of Empty Granary. *Bull. Ento. Res.* **1955,** *46* (2), 397–417.

Cotton, R. K.; Winburn, T. F. Field Infestation of Wheat by Insects Attacking it on Farm Storage. *J. Kans. Ent. Soc.* **1941,** *14* (1), 12–16.

Cotton, R. T.; Walken, H. H.; White, G. D.; Wilbur, D. A. *Causes of Outbreak of Stored Grain Insects;* Bulletin No. 416; Agricultural Experiment Station, Kansas State University and U. S Department of Agriculture Cooperating: Manhattan, USA, 1960; pp 1–34.

Freeman, J. A. Infestation of Stored Food in Temperate Countries with Special Reference to Great Britain. *Outlook Agr.* **1974,** *8* (1), 34–41.

Girma, D.; Tadele, T.; Abraham, T. Importance of Husk Covering on Field Infestation of Maize by *Sitophilus zeamais* Motsch (Coleoptera:Curculionidae) at Bako, West Ethiopia. *Afr. J. Biotechnol.* **2008,** *7* (20), 3777–3782.

Golob, P.; Muwalo, E. Pirimiphos-Methyl as a Protectant of Stored Maize Cobs in Malawi. *Int. Pest Control.* **1984,** *26* (4), 94–96.

Howe, R. W. Notes on the Biology of *Trogoderma versicolor* C. *Ento. Month Maga.* **1962,** *98,* 181–184.

Hurlock, E. T. The Infestation of Canadian Produce Inspected in United Kingdom Ports between 1953–1959. *Can. Ent.* **1963,** *95* (12), 1263–1284.

Khare, B. P.; Agrawal, N. S. Rodent and Ant Burrows as a Source of Insect Inoculums in the Threshing Floors. *Indian J. Ent.* **1964,** *26,* 97–102.

Khare, B. P. Insect Fauna of Jobner Godowns and Seasonal Variation in Occurrence. *Bull. Grain Tech.* **1963,** *1* (4), 83–91.

Minja, E. M.; Shanower, T. G.; Songa, J. M.; Ongaro, J. M.; Kawonga, W. T.; Mviha, P.; Myaka, F. A.; Slumpa, S.; Okurut-Akol, H.; Opiyo, C.; Surveys of Insect Pest Damage and Pest Management Practices on Pigeon Pea in Farmers' Fields in Kenya, Malawi, Tanzania and Uganda. *Afr. Crop Sci. J.* **1999,** *7* (1), 59–69.

Olubayo, F. M.; Port, G. R. The Efficacy of Harvest Time Modification and Intercropping as Methods of Reducing the Field Infestation of Cowpeas by Storage Bruchids in Kenya. *J. Stored Prod. Res.* **1997,** *33* (4), 271–276.

Pruthi, H. S.; Singh, M. Pests of Stored Grain and their Control. *Indian J. Agric. Sci.* **1950,** *18* (4), 1–87.

Ragumoorthy, K. N.; Gunathilagaraj, K. Field Incidence of and Host Resistance to Angoumois Grain Moth (AGM). *Int. Rice Res. Newslett.* **1988,** *13* (4), 12.

Saxena, R. C. Pests of Stored Products. In *The Neem Tree, Azadirachta indica A. Juss., and Other Meliaceous Plants: Source of Unique Natural Products for Integrated Pest Management, Medicine, Industry and Other Purposes;* Schmutterer, H., Ed.; VCH: Weinheim, Germany, 1995; pp 418–432.

Woodroffe, G. E.; Southgate, B. J. Bird Nests as Pests a Source of Domestic. *Proc. Zool. Soc. London.* **1951,** *121*(1), 55–62.

CHAPTER 8

FACTORS AFFECTING INFESTATION OF STORAGE INSECTS

CONTENTS

ABSTRACT

The infestation of stored grain insects depends upon both biotic and abiotic factors, but abiotic factors are directly related to storage stability. The optimum temperature and moisture content required for growth and development of stored grain insects. The moisture content considerably affects pest status, but it is not a factor which can be cost-effectively manipulated, in most cases, to achieve sufficient control of insect pests. The development of stored grain insect occurs within a fairly narrow range of 5–10 degrees around the optimal temperature which, for most storage insects, is in the region of 30 °C. At temperatures nearer to 20 °C, development and activity of stored grain insects are slower and population growth may be considerably reduced. Insect populations will certainly not be eliminated at these temperatures and, while grain may often be held safely in cold storage, any eventual transfer to warmer conditions will bring about a resurgence of the suppressed infestation. There is new evidence that insects may adapt to low oxygen tensions and evolve strains with considerable resistance to suboptimal levels even down to about 1%. However, it seems very improbable that any storage insect would multiply rapidly in such conditions.

8.1   INTRODUCTION

The infestation of stored grain insects depends upon both biotic and abiotic factors, but abiotic factors are directly related to storage stability. The optimum temperature and moisture content required for growth and development of stored grain insects. The moisture content considerably affects pest status, but it is not a factor which can be cost-effectively manipulated, in most cases, to achieve sufficient control of insect pests. Cost-effective drying is greatly reducing the spectrum of pest species. However, it is not responsible to prevent significant damage by one or more of the major insect pests.

Insect development and population growth rates are directly affected by temperature, moisture, and relative humidity. Upper limits for development and survival vary from insect to insect; the grain borers are more resistant than the grain weevils, but temperatures above 45 °C are eventually fatal to all storage insects. At 50 °C, most species will die quite quickly, within a matter of hours, and complete disinfestation of wheat grain can be achieved rapidly, economically, and without damage to the grain by very short exposures to air heated to 60 °C (Evans, 1948).

The development of stored grain insect occurs within a fairly narrow range of 5–10 degrees around the optimal temperature which, for most storage insects, is in the region of 30 °C. At temperatures nearer to 20 °C, development and activity of stored grain insects are slower and population growth may be considerably reduced. Insect populations will certainly not be eliminated at these temperatures and, while grain may often be held safely in cold storage, any eventual transfer to warmer conditions will bring about a resurgence of the suppressed infestation. Even in cold storage (at 6–9 °C), at least some of the important insect pests of stored grain can survive longer than one year (Wohlgemuth et al., 1976). The traditional concept of airtight storage as a means of controlling insect infestation depends upon conditions of storage structures. Most storage insects will die when the oxygen in the storage atmosphere is reduced, by the insects' respiration, to 2% (Hyde, 1969). There is new evidence that insects may adapt to low oxygen tensions and evolve strains with considerable resistance to sub-optimal levels even down to about 1%. However, it seems very improbable that any storage insect would multiply rapidly in such conditions. The durability of stored commodities is mostly dependent on the following factors.

8.2   EFFECT OF ABIOTIC FACTORS

8.2.1   PHYSICAL QUALITY OF GRAIN

If stored grain contains some inert materials or broken grain, it would provide favorable condition for insect infestation, particularly, development of hot spot. The broken grain acts as harbor of greater number of mould spore and storage fungi. This grain tends to saturate inter granular air with humidity and availability of moisture content increases for insects and storage fungi (Anonymous, 1957). Before storage, such types of inert materials and broken grain must be avoided for storage durability (Balasubramani et al., 2003).

The turning of grains either from one bin to another or within bins, can reduce infestation of grain weevil or check its further development (Joffe & Nolte, 1957). Bruchid pests of stored pulses may be particularly susceptible to control by physical turning. In the bean bruchid *Acanthoscelides obtectus*, this may be because the hatched first instar larvae require approximately 24-h period to penetrate the bean testa. However, the practicability of these proposed techniques, for routine use by farmers or traders in developing countries is yet to be established .

8.2.2   TEMPERATURE

Temperature may affect the stored commodities, if the initial temperature of freshly harvested grain in storage is equal to or considerably higher than atmospheric air temperature, or if the temperature reduced rapidly, it may encourage rapid deterioration of grain. When the difference in temperature inside and outside bin due to very high temperature during day and low temperature during night, or any other reason due to difference in temperature during day and night, it may lead to condensation and moisture migration. Solar radiation falling due to false location may cause the temperature of the bin wall to rise considerably above atmospheric temperature. During winter, center of the bin below upper surface of the grain remains warmer than the remainder of the bin as convective air currents flow upward through the center of the bin, whereas during summer, center of the bin near bottom is coolest as air currents move upward along the warm walls and downward through the center of the bin.

The saturation of moisture content of air depends on the temperature, for example, if the temperature is low in atmosphere, then more moisture content in grains may be absorbed. Increase in temperature leads to increase in respiration rate resulting in quality deterioration of grains and seeds as well as germination of seeds.

Temperature during storage may be managed by storage at low moisture content or selection of suitable method of storage.

- Grain drying
- By selecting proper bin size
- Insulation
- Underground bin
- Selecting proper material for wall and roofing
- Shading
- Grain transfer
- Ventilation
- Cold storage

8.2.3   MOISTURE

Increase in moisture reduces bulk density due to which wet grain occupies more storage space. Moisture decides storability as deterioration is fast at high moisture content—the quantity of water held by grain or seeds.

Moisture content is expressed as mass of water per unit of wet grain (wet weight basis) or mass of water per unit of dry grain (dry weight basis).

Water may be associated with grain as free water or bound water. The moisture content held in intergranular space or water molecule observed on surface of grain is called as absorbed water. The water molecules held by molecular attraction is known as adsorbed water.

All produce has its own characteristic balance (or equilibrium) between the moisture it contains and water vapor in the air with which it is in contact.

The moisture content of the air may vary, as well as that of the stored commodities. The moisture absorbed by the air in the form of water vapor is referred to as absolute humidity and expressed in g/m^3 air. The air is, however, not able to absorb an unlimited amount of moisture. There is a maximum amount the atmosphere can absorb at any specific temperature. If the atmosphere does actually contain this maximum amount, we speak of saturation and the saturation moisture content of the air. If the absolute humidity is only half the saturation moisture content, the relative humidity is 50%; if it is only a quarter of it, the relative humidity is 250% (Cotton, 2013).

Relative humidity thus expresses the degree of saturation of the air with vapor in it. This means that saturation is reached with different amounts of water vapor at different temperatures. The absolute moisture content of the air will change, for example, when it rains. There will be more moisture available for the air to absorb, thus causing a rise in the relative humidity. On sunny days, the absolute humidity will remain more or less constant. What will then occur if the temperature fluctuates? If the air gets warmer, its ability to absorb moisture increases, that is, the saturation moisture content will be higher. If the amount of moisture in the air remains constant, the degree of saturation will then drop. If the air gets cooler, its ability to absorb moisture decreases, that is, the saturation moisture content will be lower. If the amount of moisture in the air remains constant, the degree of saturation will go up. This means that on days without rain, the relative humidity is at its highest in the early morning, and at its lowest shortly after midday when the temperatures are highest, increasing again toward the evening as the air cools down.

The moisture content in given relative humidity may have two meanings, depending on whether the material is adsorbing or desorbing the moisture. For materials such as cereal grain and their products, the moisture content in equilibrium with a given relative humidity is higher when the product is desorbing than when it is adsorbing the moisture. This phenomenon is known as hysteresis. It may make a difference of as much as 1.5–2.0% in the moisture content represented by a particular meaning of equilibrium relative humidity.

Moisture will move in a grain bulk of uniform moisture content if temperature gradient exists in the bulk. Thus, a relative humidity gradient exists parallel to temperature gradient in a grain bulk of uniform moisture content. Due to increasing temperature and RH, partial pressure and concentration of water vapor in air increases. Therefore, when a temperature gradient exists in a grain bulk of uniform moisture content, pressure and concentration gradient of water vapor exist parallel to temperature gradient causing water vapor to diffuse from a high-temperature region to a low-temperature region. This process is known as moisture migration.

Water vapor is also transferred within the bin by convective air currents, which develop in the grain bulk due to temperature differences throughout the bin. When air, which is in equilibrium with warm grain passes into cooler grain, the air gives up both moisture and heat to cooler grain to come in equilibrium with it. During winter, convective air currents pick up moisture from warm grain at the center of the bin and deposit it in cooler grain near the bin surface. During summer, moisture migrates from surface down into the grain bin. Moisture migration appears to be dependent on initial moisture content. Moisture migration is slower in small bin than the larger bins because at most times of the year, difference in temperature between walls and the center of the bin will be less in small bins because heat has a shorter flow path from center to the wall in small bin than large bin.

In a mixture of dry and wet kernel, where convective air currents through the grain were not present, moisture exchange is mainly by vapor diffusion across the intergranular air space. The rate of moisture migration through grain is limited by rate of transfer of moisture through inter granular air because this transfer occurs more slowly than the exchange of moisture between air and grain. In unaerated grain bins, moisture is transferred more by convective air currents than by vapor diffusion.

8.3   EFFECTS OF BIOTIC FACTORS

Biotic factors determine the diversity and abundance of stored grain insects. Biotic factors are numerous, diverse, and in some cases, highly complex. Only a few of the better-known factors are discussed in this section.

8.3.1 INSECT FEEDING AND FOOD WEBS

Stored grain insects have different types of damaging natures. The feeding habit of a species determines its position in food chains that are mainly based on the food resource of the stored commodity. These chains extend from the major commodity-feeding pests up to hyperparasites or to pathogens of predators. The interlinking of such chains forms complex food webs. The main categories of feeding habits are noted below.

Those species that feed directly on the commodity comprise most of the pest species, though some of the other feeding types sometimes achieve pest status. Most commodity-feeders are beetles, but several are moths whose larvae feed on the commodity. Not all these pests feed in the same way or are able to attack the same type of food, and these feeding (and related) differences in behavior affect a pest's importance in a particular situation. Many of these differences are at the generic or specific level. In relation to grains or seeds, however, there is an arbitrary but useful distinction between two general categories of pests: the so-called primary pests and secondary pests.

Primary pests are capable of successfully attacking and breeding in previously undamaged solid grains, for example, whole cereals and pulse grains. That is, they are capable of penetrating an undamaged seed coat (and sometimes also a pod or sheath) in order to feed on the embryo, endosperm, or cotyledons of the seed. Such pests are also sometimes capable of feeding on other solid but non-granular commodities, for example, dried cassava, but they are rarely successful on milled or ground commodities.

Secondary pests are not capable of successfully attacking previously undamaged solid commodities. They can only attack and breed in commodities that have been previously damaged by some other agency: that is, by (a) other pests (especially primary pests); (b) bad threshing, drying, handling, etc.; or (c) intentional processing of the commodity.

Some pests do not fall easily into either category: particularly, those pests that are just capable of attacking undamaged commodities but develop much more rapidly if some previous damage is present. In such cases, it is best to classify these species as secondary pests, partly because they do not develop successfully on undamaged grains and also because they usually exhibit other secondary pest characteristics, for example, a wide range of food preference.

The inclusion of intentional processing in the definition of the damage required by secondary pests is based on the fact that the pests that breed on a processed product are usually the same as those that are secondary pests on the original (unprocessed) product. However, the definition should not be carried to illogical extremes. It is useful to distinguish between primary and secondary pests of cereal grains, but it is inappropriate to refer to a species as a "secondary pest of flour" because flour, by definition, cannot have primary pests.

On whole undamaged grains, primary pests usually have to attack the commodity first, before there can be any significant secondary pest attack. Therefore, on any batch of stored whole grain, there is usually a succession of pest attack. The first attack is by the primary pests, some of which may start to damage the grain before harvest. Invasion of secondary pests follows this, and primary and secondary attack then continues in the absence of pest control, until the damage is so high that the primary pests start to die out.

The word "primary" does not mean "more important": it refers to the order of succession in the infestation of whole grains. In different situations, primary and secondary pests differ in their relative importance.

8.3.2   PREDATION

Predators kill and eat other animal species. There are two distinct divisions of predators: those that are facultative, for example, *Tenebroides mauritanicus* and *Tribolium castaneum*, which eat both the stored commodity and other insects, and those that are obligate, which feed only on other insects and mites. Obligate predators in stores can be sub-divided into: general predators, for example, spiders, Carabidae and certain Staphylinidae, that consume a wide range of prey both within and outside stores; and specific predators that are adapted to the storage environment and prey on storage insects and mites, for example, the mite *Blattisocius tarsalis*, which mainly eats moth eggs, *Thaneroclerus buqueti*, which consumes anobiid larvae and certain Histeridae, which live on the larvae of Lyctidae and Bostrichidae.

Obligate specific predators can cause significant reductions in the rate of population growth of stored-product pests. Each female *B. tarsalis* can kill up to four or five moth eggs a day, and the hemipteran *Xylocoris flavipes*, in simulated warehouse conditions, has been found to cause a significant reduction in numbers of *T. castaneum* and *Oryzaephilus surinamensis*. *X. flavipes* will also eat other species of beetle and moth larvae, but its efficiency as a predator depends largely on its ability to move through the food medium. In dense media such as flour, it is ineffective, and in kibbled grains, it is only partially effective, whereas whole grain with large inter granular spaces allows efficient predation.

8.3.3   CANNIBALISM

Several species of stored grain insects are cannibalistic. The larvae of primary pests are often cannibalistic when developing in overcrowded grains and the adults and large larvae of some Tenebrionidae, especially *T. castaneum*, are known to eat the small larvae and pupae of their own kind. Cannibalism can sometimes have a significant effect on population growth.

8.3.4   PARASITISM

A parasite lives in or on the body of another organism, its host. Only the parasite gains an advantage from this relationship. Stored-product insects are hosts to several different types of parasites, many of which eventually cause the death of the host. The most widely known parasites are species of parasitic wasp (order Hymenoptera). The adult female seeks out the developing stages of stored-product Coleoptera and Lepidoptera and lays its eggs in or on the host.

Stored-product insects may also be parasitized by mites. The best-known parasitic mite species are *Pyemotes* spp. and *Acarophenax tribolii*, which belong to the order Prostigmata. They live on the surface of the insects, attaching themselves to soft cuticle which they pierce; the body fluids are then sucked from the insect. Although these parasites may exert considerable control on insect populations, a few can also burrow into human skin and cause great irritation. Consequently, they would not be of practical use in biological control. These mites can destroy experimental insect cultures.

8.3.5   PATHOGEN

Certain microorganisms such as the bacterium *Bacillus thuringiensis* and the protozoans *Triboliocystis, Mattesia, Nosema, Adelina*, etc., are parasites (or, more strictly, pathogens) of stored-product insects. They are most frequently encountered in dense populations of insects and may cause high levels of mortality. None of these organisms is known to infect humans; they offer potential for use in biological control.

8.4 TOXIC SUBSTANCES

The presence of toxins in the habitat may have serious consequences for storage arthropods. The toxins may be unnatural, that is, those introduced by humans, such as contact insecticides and fumigants, or they may occur naturally. The latter include the quinones produced by adult tenebrionid beetles; in dense populations, these chemicals reach such high concentrations that developmental abnormalities occur in the larvae and pupae. Lectins and trypsin inhibitors that are found in pulses confer partial or total protection against infestation by insect pests. Carbon dioxide can also act as a toxin if it accumulates as a result of poor ventilation or intentional hermetic storage.

Insecticides and fumigants may completely destroy pest populations or simply upset the ecological balance. For instance, the use of insecticidal fogging as a space treatment will kill the flying insects in a store and leave those in the stack unharmed, while the use of fumigation under gas-tight sheets will kill all the arthropods in and on the sacks but will leave those on store structures free to return to the produce once the fumigation sheets are removed. The survivors may then multiply rapidly, for a period, in the absence of competition and natural enemies. The presence of strains of insects resistant to an insecticide or fumigant may result in a predominance of those strains after pest control treatment. However, the proliferation of the resistant strains may be slower than expected as their reproductive potential is often below average.

KEYWORDS

- **stored grain insects**
- **insect infestation**
- **storage stability**
- **biotic and abiotic factors**
- **insect development**

REFERENCES

Agriculture and Agri-Food Canada, Cereal Research Centre, res2.agr.ca/winnipeg/cgs_e.htm Canola Council of Canada, Growing Canola, canola-council.org/podnew/index.aspxagriculture, res2.agr.ca/winnipeg/storage/pubs/aeration.pdf

Anonymous. *Cereal Laboratory Methods.* American Association of Cereal Chemists: St. Paul, Minnesota, 1957; pp 102.

Balasubramani, V.; Mohan, S.; Ragumoorthi, K. N. *Storage Entomology - An Introduction.* Department of Agricultural Entomology, Tamil Nadu Agricultural University: Coimbatore, India, 2003; pp 100.

Cotton, R. T. Insect Pests of Stored Grain and Grain Products, Revised Edition. Biotech Books: Delhi, India, 2013; pp 241.

Evans, J.W. Recent Developments in the Control of Insect Infestation of Stored Wheat in Australia. In *FAO, Preservation of Grains in Storage. FAO Agricultural Studies No. 2.* FAO: Rome, 1948; pp 84–87.

Hyde, M. B. Hazards of Storing High Moisture Grain in Airtight Silos in Tropical Countries. *Trop. Stored Prod. Inf.* **1969,** *18,* 9–12.

Joffe, A.; Nolte, M. C. A. Methyl Bromide Fumigation of Horizontally Stored Bulk Maize. *J. Entomol. Soc. S. Afr.* **1957, 20,** 144–153.

Wohlgemuth, R.; Drosihn, J.; Ellakwah, F. Tests for Fumigation of Bulk-Loaded Expeller in Barges Against Khapra Beetle *Tronoderma granarium* E. *Mitt. Biol. Bundesanst. Land Forstwirtsch.* **1976,** *173,* 1–24.

CHAPTER 9

METHODS OF STORAGE AND DIFFERENT STORAGE STRUCTURES

CONTENTS

ABSTRACT

Grain and seed storage is an essential component in the marketing and distribution chain, especially in the event of fluctuation of production in one season. The storage of grain and seeds at farmer, trader, and commercial or at government levels is very common all over the country. At the farmer level, storage is normally inter-seasonal and the purpose for food supply throughout the year to their family and for better price during off season. Traders generally store the commodities for very short period to meet out the profit quickly. Commercial storage of grains is provided by government agencies and cooperative societies to meet out the food supply requirements throughout the year in the country; in this process they hold the grain and seeds for longer period. The problem of storage in the developing countries is more severe when compared to that in the developed countries. All populations need a food supply that should be free from insect, birds, rodents, fungi and mycotoxin, and harmful chemical residues. There are two types of storage needs: one at the time of harvesting and the other during dearth periods. The need for storage has always arisen for distributing the food in deficit production area. In this way, grower, processor, transporters, warehouseman, government agencies, and private agencies must cooperate along each other to comply with these requirements. Food grains undergo a series of operations such as harvesting, threshing, winnowing, bagging, transportation, storage, and processing before they reach the consumer, and there are appreciable losses in crop output at all these stages. The post-harvest losses in India are estimated to be 12–16 million metric tons of food grains every year.

9.1   INTRODUCTION

The storage of grain and seeds has been in practice as a basic need of human beings for the continuous food supply throughout the year. The problem of storage of food grain has been increasing fast day by day population increase. The feeding of world population neither begins nor ends at farmers' level, because of the involvement of the whole social and economic structure. The mission of food security cannot be achieved only by growing more food grains; the essential part is to save the produce from any type of deterioration and waste and distribute it effectively among the population. The benefit of applying storage technology is a necessity to select a proper method of storage after measuring the condition of grain and seeds, micro environment, socio-economic status, and availability of the resources.

During the human civilization, the storing of commodities for further uses was started at some point. Some of the storage structures are still in use in several parts of the world like clay pots underground pits mud bins, bamboo bins, woven basket, etc, to store stocks ranging from a few kilograms to several quintals.

The problem of storage in the developing countries is more severe when compared to that in the developed countries. All populations need a food supply that should be free from insect, birds, rodents, fungi and mycotoxins, and harmful chemical residues. There are two types of storage needs: one at the time of harvesting and the other during dearth periods. The need for storage has always arisen for distributing the food in deficit production area. In this way, grower, processor, transporters, warehouseman, government agencies, and private agencies must cooperate along each other to comply with these requirements. Food grains undergo a series of operations such as harvesting, threshing, winnowing, bagging, transportation, storage, and processing before they reach the consumer, and there are appreciable losses in crop output at all these stages. The post-harvest losses in India are estimated to be 12–16 million metric tons of food grains every year. The monetary value of these losses is estimated to be more than Rs. 50,000 crores per year (Singh, 2010). Ramesh (1999) reported that high wastage and value loss are due to lack of storage infrastructure at the farm level. The losses during storage are quantity losses and quality losses. Quantity losses occur when insects, rodents, mites, and birds consume the grain. Infestation causes reduced seed germination, increase in moisture and free fatty acid levels, and decrease in pH and protein contents, etc., resulting in total quality loss. Quality losses affect the economic value of the food grains, fetching low prices to farmers (Ipsita et al., 2013). The insect problem in tropics has been reported to be critical and related to the methods and conditions of storage structures, the weather, poor sanitation, and economic condition (Khare et al., 1972). Government of India estimated on an average 10% of storage losses. Hall (1969) reported that there were enormous losses of food grain in Africa, Asia, and Latin America.

9.2 NEEDS OF STORAGE

Grain and seed storage is an essential component in the marketing and distribution chain, especially in the event of fluctuation of production in one season. The storage of grain and seeds at farmer, trader, and commercial or at government levels is very common all over the country (Mushira, 2000).

At the farmer level, storage is normally inter-seasonal and the purpose for food supply throughout the year to their family and for better price during off season. Traders generally store the commodities for very short period to meet out the profit quickly. Commercial storage of grains is provided by government agencies and cooperative societies to meet out the food supply requirements throughout the year in the country; in this process they hold the grain and seeds for longer period (Hall, 1980). Government's active involvement in grain storage, through its own special departments, agencies, or government grain marketing boards, focuses on the intervention in the staple grain market to balance national supply and demand over a time. The purposes are to create a national food reserve especially for the urban population, national food security reserves, stimulation of productivity, and price stabilization (Proctor, 1994).

9.3   STORAGE STRUCTURES

The food grains in India are stored at farmer, trader, and commercial levels. Adequate scientific storage facilities are very few in our country for food grains and seeds. In India only 30% of the produce is handled by commercial level of storage and rest 70% is handled by the farmers and traders. Grain storage structures for food grains are used after harvesting of crops to store grains safely till their consumption or transportation (Tiwari et al., 2012). The indigenous storage structures do not permit cent percent protection against deterioration from any types of sources, particularly due to post-harvest insect pests and grain fungi (Tefera et al., 2011). The storage structures are classified as follows.

9.3.1   INDIGENOUS STORAGE STRUCTURE

The different storage structures used by farmers depend upon their needs and locally available materials like bamboos, straw, wood, clay, brick, and jute bags.

9.3.1.1   TEMPORARY STORAGE

This method is used for very short time storage of food grain, for example during drying of grain, it is a very common method used at farmer levels just after harvesting of produce.

9.3.1.1.1 Arial Storage

In some part of country cob of maize or panicle of sorghum or millets are tied in bundles which are suspended on a tree or on a rope for the purpose of drying of grains. This method do not provide any types of protection to grains and only used during drying of produce.

9.3.1.1.2 Storage on Ground

This is only a provisional method since grain is exposed to all pests and animals as well as weather changes; In fact, this used when the producer is compelled to attend other works or lack transportation facilities.

9.3.1.1.3 Open Timber Platform

The grain is stored on platforms in heaps, or in woven baskets or in bags, for the purpose to dry the produce and deter insects or other pests. The open timber platform is horizontal and flat or may be conical and up to three meter in diameter, it provides drying due to their shape. The locally available timber is used for this purpose. Platforms with roofs (but no walls), of whatever shape or form, may be regarded as transitional types between temporary and long-term stores. In Southern Benin, Togo, and Ghana, for example, maize cobs in their sheaths are laid in layers on circular platforms with their tips pointing inward.

9.3.1.2 LONG-TERM STORAGE

9.3.1.2.1 Storage Basket Made up from Plant Materials

In humid countries, where grain cannot be dried adequately before storage and needs to be kept well ventilated during the storage period, traditional granaries are usually constructed from locally available plant materials, for example timber and bamboo. This method of storage is not adequate for protection of insets and rodents.

9.3.1.2.2   Earthen Pods

These are small capacity containers commonly used for storing seed and pulse grains. They have a small opening, can be made hermetic, by sealing the walls inside and out with liquid clay and closing the mouth with stiff clay, cow dung, or a wooden. If the moisture content of grain is less than 12% this structure may be used for short-term storage of grain.

9.3.1.2.3   Jars

These are large clay structures whose shape and size vary from place to place. The upper part is narrow and is closed with a flat stone or a clay lid, which is sealed in position with clay or other suitable material. Generally it is kept in house, and serves for storing food grain.

9.3.1.2.4   Solid Wall Bins

This type of storage structure is used in African counties, under which it is possible to reduce the moisture content of the harvested grain to a satisfactory level. The base of a solid wall bin may be made of timber, earth, or stone. Generally earth is not recommended because it permits termites and rodents to enter. The better base is made of stone. Mud or clay silos are usually round or cylindrical in shape, depending on the materials used. To give it added strength, certain straw materials such as rice straw may be mixed with it. The storage capacity of solid wall bins is varying from 150 kg to 10 tons.

9.3.1.2.5   Underground Storage

Underground storage is common in India, Turkey, and Southern Africa, this method of storage is used in dry regions where the water table does not affect the commodities. For long-term storage of food grain the size of pits varies from 100 kg to 200 tons. They are usually cylindrical, spherical, or amphoric in shape (Gilman & Boxall, 1974).

9.3.1.3 ALTERNATIVE STORAGE STRUCTURE AT FARMER LEVEL

9.3.1.3.1 Stack of Bags

After the harvesting and threshing so many buying agencies issue empty sacks of bags to producers so that they may be filled grain on the farm. The buying agency may collect the bagged grain from the farmers, or the farmers have to deliver it to the nearest collection point. In either case, the farmers have to store the sacks of grain for some time before they are sold. During this period precautions have to be taken to ensure the safety of the grain and maintain its quality.

9.3.1.3.2 Metal or Plastic Bins

Metal or plastic bins are most convenient storage structures at small farmer level. Bins are often used as storage containers in houses and serve notably for the storage of stored grain or seeds and pulses. Plastic bins are used after cutting of upper half portion in order to facilitate loading and unloading. Metal or plastic bins may be adopted for domestic storage of grains. These structures provide safety against insects and rodents. However, they should be protected from direct sunlight and other sources of heat to avoid condensation.

9.3.1.3.3 Pusa Bin

The Pusa bin was invented at Indian Agricultural Research Institute, New Delhi. This bin is made of earth or sun-dried bricks. They are rectangular in shape and have a capacity of 1–4 tons. A typical "Pusa" bin has a foundation of unburnt bricks, two walls were made, and a layer of polythene line was sandwiched thereby protecting abrasive damage (Pradhan & Mookherjee, 1969). The "Pusa" bin has been widely adopted in India, and it has been demonstrated in some African countries. It gives good results when loaded with well-dried grain.

9.3.1.3.4 Burkina Silo

Burkina silo is a small sized storage structure constructed from bricks. The foundation of burkina silo is made of bricks resting on the ground or on concrete pillars. The dome shaped roof is also made of bricks, using special wooden frames.

9.3.1.3.5   Usaid Silo

This silo is same as burkina silo used in Nigeria and other countries, holding 1 ton of maize grain, the silo rests on stone or concrete pillars supporting a reinforced concrete slab 1.5 m in diameter. The walls are made of bricks and are plastered inside and out with cement.

9.3.1.3.6   Concrete or Cement Silos

This is a permanent structure made up from cement and other materials, very useful for storing the food grain at domestic level. It provides protection against rodent and insect infestation.

9.3.1.3.7   Ferrocement Bin

This bin was developed in Cameroon in 1977, and tested in a number of African countries, this bin is similar to the Burkina bin in shape but consists mainly of chicken wire plastered inside and out with cement mortar. The capacity of this bin is 1.5–5.0 tons.

9.3.1.3.8   Dichter Silo

This is a cylindrical silo developed in Benin, and is constructed with trapezoidal section concrete blocks supported externally by tightened steel wire. Both internal and external surfaces are plastered with cement, and the outside is painted with coaltar to provide water-profess.

9.3.1.4   MODERN STRUCTURES OF GRAIN STORAGE

9.3.1.4.1   Temporary Storage Structure

Temporary storage structures are used for storing of huge amount of grain in short period due to lack of transportation and permanent storage structure.

This type of structure is not used every year, because it is difficult to maintain quality of the stored commodities. Such type of structure is used during bumper production of produce. There are following types of temporary storage structures.

9.3.1.4.1.1 Existing Buildings

Existing buildings may be used for temporary storage of grain and seed. Before storing the food grain make sure the location of building is well drained. If the building does not have a concrete floor, place the grain on plastic or wood to prevent moisture moving from the ground to the grain. Even with a concrete floor, it is advisable to cover the concrete with plastic, especially if the concrete is cracked. Moisture vapor will move through concrete and into the grain if the soil below the concrete is wet.

9.3.1.4.1.2 Steel Bin Rings

Round steel bin rings can be conveniently and economically erected to provide temporary grain storage. Two or three steel rings from round grain bins can be set on the existing floor.

9.3.1.4.1.3 Round Plywood Bins

Plywood bins, 4 or 8 feet deep and 10–37 feet in diameter, are made up of attached plywood sheets and 2 × 4 m into a long strip and bending to form a circle. Fasteners and nails are critical to keep the bin from bursting.

9.3.1.4.2 Movable Wood Grain Walls

These are built of plywood and lumber, are self-supporting, and can convert all or part of a building to grain storage.

9.3.1.4.3 Commercially Temporary Grain Storage Structure

Commercially temporary storage structures are used for storing the produce in a particular region. These units vary in construction, some are using polyethylene sheeting, reinforced-fiber sheeting, wire mesh, wood, or metal panels for holding the grains. The costs of repairs, freight, and erection are high as compared to other structures. Several types of plastic sheeting are available to be used with grain piled on the ground to prevent rain and wind loss. Usually the covering will need repair or replacement after one season of use. Rodents, birds, and chewing insects may damage to such types of structures.

9.3.1.4.4   Bales Bins

Bales bins are formed by large round bales in a circular bin wall. With a five-foot grain wall depth, there is a force of about 115 kg of force on each foot at the bottom pushing the bale outward, so they will likely need to be restrained by wrapping with ware. Plastic along the inside of the bales is recommended to keep grain from leaking out and to prevent water from entering.

9.3.1.4.5   Outside Piles

The grain can be stored outside on the ground, outside pile should be on high ground and the earth crowned under the pile. Plastic of 6 mm in thickness should be placed on the ground to avoid ground moisture. Generally plastic or tarp covering is used to reduce wetting by rain and to minimize damage by wind and birds. To facilitate drainage from the top surface should be smooth. The direction of pile should be in North and South to allow the sun ray to dry off the sloping sides.

9.3.1.4.6   Cover on Plinth Storage (Cap)

Cover on plinth storage is an improved arrangement for storing food grains in the open space. The plinth which is free from rat and damp is used for this purpose. The grain bags are stacked in a standard size on wooden dunnage. The stacks are covered with 250 micron low-density polyethylene (LDPE) sheets from the top and all four sides. Food grains such as wheat, maize, gram, paddy, and sorghum are generally stored in cover on plinth for 6–12 month periods. It is the most economical storage structure and is being widely used by the Food Corporation of India (FCI) and other government agencies for bagged stored commodities.

9.3.1.4.7 Bunker Storage

The word bunker storage is used to describe large bulks of produce stored on the ground in the open space surrounded by low retaining walls, and polythene sheets are used to protect from the weather. Bunker storage is sound, convenient, and low cost method for storing commodities for medium to long-term storage of grain. Capacities of bunkers vary as per need, but generally its capacity is a few hundred tons to 20–30 thousand tons. However, bunker storage is still only recommended in the area where the harvest period coincides with the dry season of the year, since loading the grain may take many days during which time the grain is exposed to several damaging agencies. This method is only recommended for the storage of well dry grain.

9.3.1.4.8 Mobile Silo

Mobile silo are weld mesh walled foldable plastic silos. It is suitable as medium sized grain storage up to 1000 tons capacity. However, it was designed for bulk storage but also used for bag storage. Bag or bulk storage is contained within a vertical wall consisting of units of galvanized weld mesh bolted together. After loading to a centered peak, the plastic roof-section is then brought over the grain and the roof is zipped to the wall to obtain airtight seal. This silo is well equipped with aeration systems, when not under aeration; the hermetic seal is used to control stored grain insect inside the silo.

9.3.1.4.9 Volcani Cubes

Vocani cubes are frameless flexible storage structures consist of poly vinyl chloride (PVC). It is suitable for bag storage in small quantities of up to 50 tons of cereal grains, but some companies also been manufactured in sizes capable of storing grains up to 150 tons. Since the stack itself provides the mechanical strength of the structure, so no rigid frame is required in this structure. The liner is made of an upper and a lower section which can be zipped together to form the airtight seal. These structures are very simple to reuse.

9.3.1.5   LONG-TERM STORAGE STRUCTURE

9.3.1.5.1   Warehouses

The warehouses are very important long-term storage structures for storage of commodities, protection of produce, in order to continue supply food materials, and food security in the country. In this system storage commodities are kept in bags, it may also include materials and equipment required for the packaging and handling of bagged grain, and pest control operation in warehouses.

9.3.1.5.1.1   Location and Orientation

Warehouse is a raised commercial storage site with well drainage to ensure that there is dampness of moisture in the vicinity of stores. Establishment of warehouse in longitudinal side in the East–West axis provides less sun radiation on the building or exposed to the main wind direction creates balanced temperature conditions, thus reducing infestation of stored grain. Compact soil and well connected to road and railways are prerequisite for warehouse.

9.3.1.5.1.2   Roofing

The roof of warehouse should be properly sealed and well connected from walls to prevent any types of infestations from insects, rodents, or birds. Insulation under the roof in case of corrugated iron sheets reduces the effect of the sun radiation and maintains optimum storage temperature.

9.3.1.5.1.3   Flooring

The floor of warehouse must be at a height of 1 m above the ground level with an isolation to prevent underground moisture. Floor also prevents damage by vehicles during loading and unloading of trucks. To provide moisture proof insert polyethylene foil of at least 0.2 mm thickness or a 5 cm layer of bitumen in the floor and in the first 25 cm of the walls. A smooth surface without any cracks, crevices, and hole is easy to clean and does not afford insects any place to hide.

9.3.1.5.1.4 Walls

The walls of warehouse should be smooth, without any cracks either inside or outside, to avoid any invasion of insects. All walls may be white, water-resistant, and plastered from both sides.

9.3.1.5.1.5 Doors

In warehouse tight-sealing hinged doors protect from the entry of rodents from outsides. Sliding doors always leave a gap between door and wall so they are generally not preferred. Roll-up doors rust and often become defective when older. Metal doors are most resistant against any damage by rodents. Wooden doors should be fitted at the bottom with a panel of steel sheet of half a meter in height.

9.3.1.5.1.6 Ventilation Openings

Control ventilation openings are always set in warehouse for the evacuation of heat from the store and maintain optimum temperature and moisture level. Ventilation openings should have a size of 0.5 m^2/100 m^2 storage area for incoming air in lower ventilation openings and 1.5 m^2 /100 m^2 storage area for outgoing air in upper ventilation openings. The lower ventilation openings should be situated approximately 1.5 m above the floor, the upper ones approximately 1.5 m below the roof on both sides of the store. Tightly-sealing ventilation openings permit fumigation activity. Fine wire and iron grilles must be fitted in the ventilation openings to prevent the entry of insects, rodents, and birds (FAO, 1994).

9.3.1.5.2 Stacking of Bags in Warehouse

Stacking of bags on the floor of warehouse for safe storage of food commodities depends upon the nature of produce. Ears of bags should be pointing inward the stack in order to prevent grain spilling. Stack the bottom layers with larger intervals than the top ones in order to enable the stack to slightly taper. On higher stacks, occasionally leave out some bags in the upper layers to retain the conical form.

9.3.1.5.2.1 Pallets

Always use pallets for stacking the bags, place the pallets in such positions which do not prevent air flow under the stack. The pallets should be 10 cm high in order to facilitate aeration from below. As an additional advantage rodent infestation can easily be observed. The overall surface area of the supporting bars should not be less than approximate height of pallets.

9.3.1.5.2.2 Size of Stacks

The size of stack directly depends upon the stored commodities, for the more stability, jute bags should not be stacked any higher than 4 m and plastic bags not higher than 3 m. Plastic bags are more slippery and the stacks are thus less stable. When determining the size of stacks, take into account the store's capacity, the ratio of its length to its breadth and its height, the position of the doors and the size of the fumigation sheets available, under any circumstances height of stack must be 1 m below the roof in order to provide space for insect management program.

9.3.1.5.2.3 Positioning of Stacks

All stacks must in the position which is easily accessible from each side in order to apply insect management operation, surface treatment, and fumigation. Keep minimum space of 1 m from sidewall and between two stacks.

9.3.1.5.2.4 Marking the Stacks

The identification of sack of an individual stack is marked with numbers or sign. These markings may be made on the bags, walls, the floor, or the roof pillars, as long as they are always clearly visible. They must also be entered on the stack card.

9.3.1.5.2.5 Stack Cards

The most important information should be entered in stack cards and attach it in a clearly visible position to every stack of bags. The date insect management measures also entered in stack card.

9.3.1.5.3 Bulk Storage in Silos

The silos are made up from metal or concrete but plastic silos are also being used for small scale storage of grains and seeds. The silo system is well equipped with cleaning and drying equipments.

In Indian conditions the silos are expensive as compared to warehouse, but their initial cost may be realized after two to three years. In terms of structural cost per ton of storage, round silos are generally more economical than rectangular silos. They have fewer joints due to which they can be sealed more easily to make the structure air-tight and suitable for fumigation. In the case of flat bottom stores, structural efficiency is also increased by minimizing the height of the structure. Silos are generally used on commercial scale storage of food grain as well as seeds.

- Advantages of bulk storage
- Low running costs
- Less labor requirements
- Rapid handling
- Rodents proof
- Efficient and effective fumigation operation
- More grain storing capacity on less land
- Complete control of internal environment
- Store the grain for longer periods
- Possibility to store grain after harvesting for short periods.

9.3.1.5.3.1 Metal Silo

Metal silos are often quicker to erect and cost effective as compared to concrete silos. Steel silos provide excellent storage where grain is dry. Where grain moisture content is high, there may be a greater likelihood of moisture migration developing due to the heat conductivity of the steel causing temperature gradients to develop between the inside and outside of the grain mass. Steel bins are less robust than concrete bins and require careful design. Welded steel silos will remain gas-tight throughout their lives since they are not subjected to cracking or differential movements. There are two types of steel silos are generally in use.

9.3.1.5.3.1.1   Flat Bottom Steel Silo

In India flat bottom silos are established by FCI. The size of flat bottom steel silo is about 22 feet in diameter and 76 feet in high and at the head of these silos stands the head house which is 30 feet in diameter and 156 feet high. In the head house the grain elevators, auxiliary shipping, tanks, grain-cleaners, automatic weighing machines, and man lift are located.

9.3.1.5.3.1.2   Hopper Bottom Steel Silo

Hopper bottoms silos are more useful where high throughputs are required. Such silos are equipped with automatic dry, handling, and storage devices.

9.3.1.5.3.2   Concrete Silo

Concrete silos are most suitable in that area where moisture content of environment is very high and risk of corrosion to metal is high. Concrete silos are usually preferred where bins have to be above 30 m. In concrete silo we never use CO_2 fumigation because they react to concrete.

9.3.1.5.3.3   Plastic Silo

Now a days plastic is most popular, cheaper, and easily available anywhere. So it is also used for storing food grain and seeds. It is suitable for small-scale storage of grain. Plastic silos do not use for long-term storage and chance of rodent infestation is very high.

9.3.1.5.4   Cold Rooms for Grain Storage

Many developed countries have the facility for storing food grain and seeds in fully temperature-controlled room. Such type of storage is very expensive and not suitable for all types of storage commodities.

9.3.1.5.5   Hermetic Storage

The meaning of hermetic storage is making air tight by fusion or sealing. It is simple and cheap method of storage as compared to other methods of storage. The application of hermetic storage became popular to protect dry and wet grain without use of hazardous chemicals. The basic concept of air tight storage is the depletion of the oxygen in the air inside the structure to a level which inactivates the organisms that destroy the grain. The respiratory process in dry grain and organisms present in structure cannot be ordinarily distinguished. The moisture and temperature influence the rate of respiration in grain as well as in damaging organisms. In hermetic storage development of insects is delayed or completely suppressed.

KEYWORDS

- storage structures
- indigenous
- modern
- bins
- insect problem
- timber
- bamboo

REFERENCES

Food and Agricultural Organization Technical Bulletin. **1994**, *109*, 17–20.

Gilman, G. A.; Boxall, R. A. Storage of Food Grains in Traditional Underground Pits. *Trop. Stored Prod. Inf.* **1974**, *28*, 19–38.

Hall, D. W. Food Storage in the Developing Countries. *J. R. Soc. Arts.* **1969**, *117* (5156), 562–573.

Hall, D. W. *Handling and Storage of Food Grains in Tropical and Subtropical Areas;* FAO: Rome, Italy 1980; p 350.

Ipsita, D.; Grish, K.; Narendra, G. S. Microwave Heating as an Alternative Quarantine Method for Disinfestations of Stored Food Grains. *Int. J. Food. Sci.* **2013**, *2013*, 13.

Khare, B. P.; Sengar, C. S.; Singh, K. N.; Agrawal, R. K.; Singh, H. N. Loss in Grain due Toinsect Feeding on Wheat. *Indian J. Agric. Res.* **1972**, *6* (2), 125–133.

Mushira, M. A. *Manual on Grain Management and Equipment Maintenance in Silos;* FAO: Nigeria, 2000; p 42.

Pradhan, S.; Mookherjee, P. B. Pusa Bin for Storage of Grain. In *Council Agric. Res. Tech. Bull.* **1969**, *21*, 11.

Proctor, D. L. *Grain Storage Techniques Evolution and Trends in Developing Countries;* FAO Agricultural Services Bulletin No. 109; FAO: Rome, Italy, 1994; p 154.

Ramesh, A. *Priorities and Constraints of Post Harvest Technology in Asia;* Japan International Research Center for Agricultural Science: Tokyo, 1999; p 37.

Singh, P. K.; A Decentralized and Holistic Approach for Grain Management in India. *Current Sci.* **2010,** *9* (9), 1179–1180.

Tefera, T.; Kanampiu, F.; DeGroote, H.; Hellin, J.; Mugo, S.; Kimenju, S.; Beyene, Y.; Boddupalli, P. M.; et al. **2011**.

Tiwari, B. K.; Gowen, A.; Mckenna, B. *Pulse Foods: Processing, Quality and Nutraceutical Applications;* Academic Press: US, 2012; pp 172–192.

CHAPTER 10

BEHAVIORAL MANAGEMENT OF STORAGE INSECTS

CONTENTS

ABSTRACT

In the early days of development we completely depended upon chemical insecticides, but they cause resistance, resurgence, and secondary outbreak of storage insects; besides these insecticides also contaminate the environment with high residual effect. Scientist from several countries reported the dangerous effects of methyl bromide and aluminum phosphide on the ozone, but now days there are many safe, cost-effective methods, which have the efficiency to manage the storage insects several methods, without the need to use chemical fumigants. In developed countries, the presence of insects in food grain or the contamination in any form is a factor for rejection, and consignments were rejected on this basis. Stored grain insects cause infestation to the several stored commodities resulting in serious food and monetary losses. The behavior of stored grain insects is associated with food, mating, egg laying, and defensive mechanism in order to protect themselves from natural enemies and adverse environmental effects. The behavior of insect is due to their own needs and host plants upon which they feed. To minimize the use of insecticides and to maximize the application of eco-friendly and sustainable approach would be the aims for managing stored grain insects below economic threshold level. In recent years, this concept has been gaining good acceptance, with special reference to the application of semiochemicals.

10.1   INTRODUCTION

The infestation of insects to stored commodities has been occurring since human civilization. As per archaeological reports, there are so many insects associated with stored produce (Solomon, 1965; Buckland, 1981; Levinson & Levinson, 1994). There are several methods applied for the management of stored grain insects, but nowadays we are in need to save the lives through safe food and safe environment. In the early days of development we completely depended upon chemical insecticides, but they cause resistance, resurgence, and secondary outbreak of storage insects; besides these insecticides also contaminate the environment with high residual effect. Scientist from several countries reported the dangerous effects of methyl bromide and aluminum phosphide on the ozone layer (Price, 1994). There are many safe, cost-effective methods, which have the efficiency to manage the storage insects several methods, without the need to use chemical fumigants. In developed countries, the presence of insects in food grain or the contamination in any form is a factor for rejection, and consignments were rejected on this basis (Pinniger et al., 1984). In the UK, the Food Safety Act was amended that all stores containing grain which might be destined for human consumption are treated as food premises and subject to inspection to ensure that food is not contaminated.

Stored grain insects cause infestation to the several stored commodities resulting in serious food and monetary losses (Ali et al., 2004). The quantitative and qualitative damage to stored grains and grain products from the insect pests may amount to 20–30% in the tropical zone and 5– 10% in the temperate zone (Talukder, 2006; Rajendran & Sriranjini, 2008). The production of food grain in India has reached 250 million tons in 2010–2011, out of which nearly 20–25% food grains are infested by stored grain insect (Rajashekar et al., 2010). Scientists from all over the world are searching the alternate ways of chemical fumigants for the effective management of stored grain insects.

The behavior of stored grain insects is associated with food, mating, egg laying, and defensive mechanism in order to protect themselves from natural enemies and adverse environmental effects (Cox & Collins, 2002). The behavior of insect is due to their own needs and host plants upon which they feed. To minimize the use of insecticides and to maximize the application of eco-friendly and sustainable approach would be the aims for managing stored grain insects below economic threshold level. In recent years, this concept has been gaining good acceptance, with special reference to the application of semiochemicals. Law & Regnier (1971) first time reported the term semiochemicals that modified interactions between organisms. The importance of semiochemicals in pest management program under field conditions (Phillips, 1997; Jones, 1998; Agelopoulos et al., 1999; Phillips et al., 2000; Throne et al., 2000; Weaver & Subramanyam, 2000) is greatly understood, but not so in storage conditions., Many types of traps are used for monitoring populations of stored product insect populations such as moth and beetles in warehouses and mills (Burkholder, 1990; Cogan, 1990; Mullen, 1992; Phillips, 1997; Plarre, 1998).

The technique for management of stored grain through behavioral manipulation using semiochemicals may be classified into two types, according to their activities: long-range and short-range techniques. In long-range management technique, these volatile chemicals that act as insect attractants or repellents are applied at some distance from the source of host that is to be protected. In short-range management technique, these chemicals which act as stimulants or deterrents are applied in order to affect feeding of insects.

The semiochemicals have additional properties to enhance activity of predators, parasites, and parasitoids of stored grain insects.

10.2 LONG RANGE MANAGEMENT TECHNIQUES

10.2.1 ATTRACTANTS AND MASS-TRAPPING

Generally, traps are used for mass-trapping; whether insects are captured in traps or not depends upon the position of traps and population of insects. Food volatiles are source of lures, some of which have the advantage of being attractive to the adults and larvae of both sexes of a wide range of species, although none appear to have been used for mass-trapping programs for stored grain insects. Wheat germ composed from 15% lipid, in which 60% are triglyceride; unsaturated triglycerides are particularly attractive and elicit aggregation responses in storage insects such as *Sitophilus granarius* and *Oryziphilus surinamensis* (Quartey & Coaker, 1992). If the food lures used in combination with pheromones may enhance the effectiveness of mass-trapping for stored grain insects. A combination of three grain volatiles—valeraldehyde, maltol, and vanillin—and sitophilure was more attractive to the rice weevil *S. oryzae* than either the pheromone or the grain volatile alone. Addition of carob volatiles to the aggregation pheromone 4S, 5R-sitophilure has been shown to be more effective than either carob or pheromone alone in attracting all three species of *Sitophilus*. In Kenyan field trials, traps baited with 4S, 5R-sitophilure and cracked wheat gave higher catches of *S. zeamais* and *S. oryzae* than those baited with pheromone or cracked wheat alone. Fleurat-Lessard et al. (1986) reported that the mass-trapping of *Plodia interpunctella* H. in several years from seed store reduces the infestation without any application of insecticides. The application of sex pheromone for the mass-trapping of *Lasioderma serricorne* F. is not much effective as compared to sticky trapping (Bauchelos & Levinson, 1993).

A cup of readymade tea or coffee was effective for attracting and killing the almond moths of *Cadra cautella*, before egg laying (Temerak, 2010). The number of moths entangled inside spider webs was greater than those that captured in the tea or coffee solutions. Naturally pheromones are species specific. So different types of these may be required for stored grain insect management; however, some pheromone aggregated the other species (Cox & Collins, 2002), for example, *Sitophilus oryzae*, *Sitophilus zeamais*, and *Sitophilus granaries* (Walgenbach et al., 1983); *Oryzaephilus surinamensis*, *Oryzaephilus Mercator*, and *Ahasverus advena* (Pierce et al.,

1991); *Rhyzopertha dominica* and *Prostephanus truncates* (Williams et al., 1981); and *Tribolium castaneum*, *Tribolium confusum*, and *Tribolium free-mani* (Suzuki et al., 1987). They produce selective advantage of interspecific attraction to the emitting species. The stored produce may be attacked by a multitude of insect species simultaneously and so to be cost-effective any lure must be attractive to a range of target species (Collins et al., 2008). The behavioral responses of *C. pusillus, O. surinamensis*, and *P. truncatus* to different doses of carob extract evoked a quick directional response and induced high attraction to treated pitfall area as compared to untreated control.

10.2.2 ATTRACT AND KILL TECHNIQUES

Attracting and killing the stored grain insects are very useful in storage conditions. Recently, slurry containing pheromone and permethrin has been applied for managing *Cydia pomonella* in Switzerland (Charmillot et al., 2000). The attract and kill technique is used in Indian flour mills for the management of *Efestia kuehniella* in combination of pheromone and cypermethrin (Trematerra & Capizzi, 1991). In this way, the utilization of semiochemicals is likely to reduce the pressure of insecticides and adverse effect on natural enemies. This technique reduces cost of management of stored grain insects and may apply in small-scale protection. This technique used in combination of protozoa *Mattesia* sp for the management of *Trogoderma glabrum* (Burkholder & Boush, 1974). Some vegetable fat pellets with combination of aggregation pheromone and entomopathogenic fungi *Beauveria bassiana* have been reported to attract and kill the *Plodia truncates* (Smith et al., 1999).

10.2.3 MATING DISRUPTION TECHNIQUES

This technique disrupts mating and prevents laying of fertile eggs. The mode of action reported by Cox (2004) involves continuous exposure of an insect to high concentration of pheromone, antennal receptors, and habituation of central nervous system. The insect was unable to respond to a potential mate, as camouflaging of natural pheromone plume from calling mate disrupted the trail or made a false trail. This technique has been evaluated against *Pectinophora gossypiella* by Campion (1989) and *C. pomonella* by Howell et al. (1992), which showed a disruption in mating, reducing the infestation. Very little work has been reported about managing the stored grain insect by such techniques. Hodges et al. (1984) reported the population reduction with mating disruption of *E. cautella* by synthetic pheromone in the form of encapsulated formulation. This technique involves the application of inhibitors; *P. interpunctella* can inhibit the male response to female of *S. cerealella*. Mating inhibitors have been identified for *Stegobium paniceum* and *L. serricorne* (Kodama et al., 1987; Levinson & Levinson, 1987). Practically, the cost of semiochemicals is more as compared to chemical.

10.2.4 INSECT REPELLENTS

Several insects produce the chemicals from their odor gland or from body that repel the other insects (Norris, 1990). The Tribolium species secrete quinines from special types of glands under overcrowding and scarcity of food that elicit a reaction in their larvae to disperse (Mondal, 1985; Faustini & Burkholder, 1987; Mondal & Port, 1994). Quinines may be utilized as repellent in pest management program under storage conditions during packing of food or seed to minimize insecticide pressure but they are not successful at commercial storage level. The repellents of plant origins are recently used in pest management due to their safety and are free from pesticide residues in order to ensure the safety of people, food, environment, and wildlife. Repellent methyl salicylate has received regulatory approval in the USA recently for the use in packing (Mullen & Pedersen, 2000). The essential oil of *Mentha arvensis* L. showed repellency and toxicity against *T. castaneum* and *S. oryzae* even at low concentration, but its repellency was more marked toward *S. oryzae* (Mishra et al., 2012). The highest repellency (93.30%) of *T. castaneum* occurred at the highest concentration (5.0% suspension) of acetone extracts from *Toona sureni* (Blume Merr); while the lowest (0.0%) repellency occurred at 0.5% suspension after 1 day of treatment (Parvin et al., 2012). Methanolic plant extracts of *Ambrosia tenuifolia* Spreng., *Baccharis trimera* (Less.) DC, *Brassica campestris* L., *Jacaranda mimosifolia* D. Don, *Matricaria chamomilla* L., *Schinus molle* (L.) var. *areira* (L.) DC., *Solanum sisymbriifolium* Lam., *Tagetes minuta* L., and *Viola arvensis* Murray exhibited repellent effectiveness against the red flour beetle, *Tribolium castaneum* at 30 minutes which reduces the possibilities of contamination of grains by the insect pests (Padin et al., 2013). Scientists from all over the world reported several medicinal and aromatic plants and spices in the form of essential oils, extract, powder, and ash, having great potential to act as repellents against stored grain insects, their details are discussed in the next chapter.

10.3   SHORT RANGE MANAGEMENT TECHNIQUES

10.3.1   DETERRENTS

Some chemicals are responsible for the inhibition of insect behavior: for example, feeding, mating, and egg laying. The mandibular gland of *E. kuehniella* and *P. interpunctella* caterpillar secrete semiochemicals which are deposited on their food and silk (Corbet, 1971). About 25 compounds have been isolated from this secretion which act as territorial markers to reduce cannibalism (Mudd, 1981; Kuwahara et al., 1983; Nemoto et al., 1987). The feeding inhibitor has advantages to break life cycle of any insects and mating inhibitor suppresses mating and vital eggs production, while egg-laying inhibitor deters the female from the site of egg deposition. Anderson (1988) reported insects able to produce their own egg-laying deterrent from the 50 species of hymenoptera, most of them feed upon crops. Very few works have been reported as production of deterrent in stored grain insects, female of *Acanthoscelides obtectus* and *Callosobruchus maculatus*, (Credland & Wright, 1990), *L. serricorne* (Imai et al., 1990), larvae of *Oryziphilus surinamensis* (Pierce et al., 1990), *Ephestia kuehniella* at very high dose levels (Corbet, 1973), and egg-laying deterrent produce in *Sitophilus* granaries (Cox et al., 2000). There are so many plants and their constituents reported to act as deterrents, which are discussed in next chapter.

10.3.2   STIMULANTS

For the effectiveness of insecticides or bioagents, some semiochemicals are used to modified the behavior of feeding and movement. Dawson et al. (1990) and Hockland et al. (1986) reported that in aphids alarm pheromone induce the mobility which is helpful in efficacy of insecticides. The application of microbial insecticides along with stimulants for management of stored grain insects is registered in the USA in the form of feeding stimulants to enhance the activity of microbial insecticides (Cox & Wilkin, 1998).

10.4   TRAPS USED IN BEHAVIORAL MANAGEMENT OF STORED GRAIN INSECTS

The presence of stored grain insects in any storage structures is detected only when they are hovering and flying around, by which time enormous loss and population buildup of insects might have been occurred in the grain. The timely detection of stored grain insects will help to prevent severe damage. There are several ready-to-use traps are available for detection of these insects.

10.4.1   PROBE TRAP

The detection of infestation in bulk and bag storage is very crucial; for this purpose probe trap is used by keeping them inside the grain surface. Probe trap is simple to use and provides a mechanism for continuous monitoring of stored grain insects. Probe trap consists of a main tube, insect trapping tube, and detachable cone at the bottom of the main tube with equi-spaced perforations of 2 mm diameter. It is used for the detection of all types of beetle and weevils infesting the stored produce.

10.4.2   PULSES BEETLE TRAP

This trap is a modified form of probe trap for collection of pulses beetles. In this trap, perforations are 3.5 mm in diameter, as the pulse beetles are active and climb narrow surface, in the bottom of the trap they are captured.

10.4.3   PITFALL TRAP

Pitfall traps are generally used to capture the soil inhibiting insects, but it is practically used in bulk storage condition for the detection of stored grain insects. This trap captures the crawling insects inside the store.

10.4.4   LIGHT TRAP

The majority of stored grain insects respond to light at wavelengths ranging from 300 to 700 nm, stored grain insects gave maximum response at 350 nm (ultraviolet rays) and 500–550 nm (green light), according to the behavioral studies reported by Balasubramani et al. (2003). These traps are used for mass-trapping and population monitoring under bulk and bag storage conditions.

10.4.5 STICKY TRAP

For monitoring and mass-trapping of stored product moths, sticky traps are used in the form of sticky boards and strips. Such types of traps are widely used in flour mills and processing plants (Cotton, 1960).

10.4.6 BAIT TRAP

Several bait traps are used for the monitoring of stored grain insects inside the warehouse. The main purpose of this trap is to detect the presence of insects. Bait traps are also used in the form of perforated cylindrical rod of 20 cm long, with one end opened and other side closed.

KEYWORDS

- storage insects
- attractants
- mass-trappings
- repellents
- deterrents
- stimulants
- traps

REFERENCES

Agolopoulos, N.; Birkett, M. A.; Hick, A. J.; Hooper, A. M.; Pickett, J. A.; Pow, E. M.; Smart, L. F.; Smiley, D. W. M.; Wadhams, L. J.; Woodcock, C. M. Exploiting Semiochemicals in Insect Control. *Pest. Sci.* **1999,** *55,* 225–235.

Ali, S. M.; Mahgoub, S. M.; Hamed, M. S.; Gharib, M. S. A. Infestation Potential of *Calloso-bruchus chinensis* and *C. maculatus* on Certain Broad Bean Seed Varieties. *Egypt J. Agric Res.* **2004,** *82* (3), 1127–1135.

Anderson, P. Oviposition-Deterring Pheromones in Insects. *Entomol. Tidskr.* **1988,** *109,* 14–18.

Balasubramani, V.; Mohan, S.; Ragumoorthi, K. N. *Storage Entomology an Introduction;* TNAU: Coibatore, India, 2003; p 100.

Bauchelos, C. T.; Levinson, A. R. Efficacy of Multi Surface Traps and Lasiotraps with and without Pheromone Addition for Monitoring and Mass Trapping of *Lasioderma serricorne* in Insecticide Free Tobacco Stores. *J. Appld. Ent.* **1993,** *116* (1–5), 440–448.

Buckland, P. C. The Early Dispersal of Insect Pests of Stored Products as Indicated by Archaeological Records. *J. Stored Prod. Res.* **1981,** *17,* 1–12.

Burkholder, W. E. Practical Use of Pheromones and other Attractant for Stored Product Insects. In *Behavior Modifying Chemicals for Insect Management*; Ridgway, R. L., Silver-stein, R. M., Inscoe, M. N., Eds.; Marcel Dekker Inc.: New York, NY, 1990; pp 497–515.

Burkholder, W. E.; Boush, G. M. Pheromones in Stored Product Insect Trapping and Pathogen Dissemination. *Bull. OEPP.* **1974,** *4,* 455–461.

Campion, D. G. Semiochemicals for the Control of Insect Pests. In *British Crop Protection Council Monograph Progress and Prospect in Insect Control;* British Crop Protection Council: Farnham, UK, 1989; Vol. 43, pp 119–127.

Charmillot, P. J.; Hofer, D.; Pasquier, D. Attract and Kill a New Method for Control of the Codling Moth *Cydia pomonella. Entomol. Exp. Appl.* **2000,** *94,* 211–216.

Cogan, P. M.; Wakefield, M. E.; Pinniger, D. P. In *A Noval and Inexpensive Trap for the Detection of Beetle Pests at Low Densities in Bulk Grain,* Proceeding of the Fifth International Working Conference on Stored Product Protection, Bordeaux, France, Sep 1–9, 1990; Fleurat Lessard, F., Ducom, P., Eds.; 1990; pp 1322–1330.

Collins, L. E.; Bryning, G. P.; Wakefield, M. E.; Chambers, J.; Fennah, K.; Cox, P. D. Effec-tiveness of a Multi-Species Attractant in Two Different Trap Types Under Practical Grain Storage Conditions. *J. Stored Prod. Res.* **2008,** *44,* 247–257.

Corbet, S. A. Mandibular Gland Secretion of Larvae of the Flour Moth *Anagasta kuehniella,* Contains an Epideictic Pheromone and Elicits Oviposition Movements in a Hymenopteran Parasite. *Nature.* **1971,** *232,* 481–484.

Corbet, S. A. Oviposition Pheromone in Larval Mandibular Glands of *Ephestia kuehniella. Nature.* **1973,** *243,* 537–538.

Cotton, R. T.; Walken, H. H.; White, G. D.; Wilbur, D. A. *Causes of Outbreak of Stored Grain Insects;* Bulletin No. 416; Agricultural Experiment Station, Kansas State University and US Department of Agriculture Cooperation: Manhattan, USA, 1960; pp 1–34.

Cox, P. D. Review Potential for Using Semiochemicals to Protect Stored Products from Insect Infestation. *J. Stored Prod. Res.* **2004,** *40,* 1–25.

Cox, P. D.; Collins, L. E. Factors Affecting the behavior of Beetle Pests in Stored Grain, with Particular Reference to the Development of Lures. *J. Stored Prod. Res.* **2002,** *38,* 95–115.

Cox, P. D.; Wilkin, D. R. *A Review of the Options for Biological Control against Invertebrate Pests of Stored Grain in the UK;* IOBC/WPRS Bulletin No. 21(3); International Organiza-tion for Biological and Integrated Control of Noxious Animals and Plants, Dijon, France, 1998; pp 27–32.

Cox, P. D.; Collins, L. E.; Bryning, G.; *Studies towards the Identification of an Oviposition Deterrent for the Grain Weevil Sitophilus granaries;* Central Science Laboratory Report, Sand Hutton, York, 2000, *108,* 1–24.

Credland, P. F.; Wright, A. W. Oviposition Deterrent of *Callosobruchus maculatus*. *Physiol. Entomol.* **1990**, *15*, 285–298.

Dawson, G. W.; Griffiths, D. C.; Merritt, L. A.; Mudd, A.; Pickett, J. A.; Wadhams, L. J.; Woodcock, C. M. Aphid Semiochemicals a Review and Recent Advances on the Sex Pheromone. *J. Chemical. Ecol.* **1990**, *16*, 3019–3030.

Faustini, D. L.; Burkholder, W. E. Quinone Aggregation Pheromone Interaction in the Red Flour Beetle. *Anim. Behav.* **1987**, *35*, 601–603.

Fleurat-Lessard, F.; Siegfried, M.; Le Torch, J. Utilization dun Attractif de Synthese Pour la Surveillance et le Piegeage des Pyrales Phycitinae Dans les Locaux de Stockage et de Conditionnement de Denrees Alimentaires Vegetales. *Agronomie.* **1986**, *6*, 567–573.

Hockland, S. H.; Dawson, G. W.; Griffiths D. C.; Marples, B.; Pickett, J. A.; Woodcock, C. M. The Use of Aphid Alarm Pheromone to Increase Effectiveness of the Entomophilic Fungus *Verticillium lecanii* in Controlling Aphids on Chrysanthemums under Glass. In *Fundamental and Applied Aspects of Invertebrate Pathology;* Samson, R. A., Vlak, J. M., Peters, R., Eds.; The Netherlands Society of Invertebrate Pathology: Netherlands, 1986; p 252.

Hodges, R. J.; Benton, F. P.; Hall, D. R. Control of Ephestia Cautella by Synthetic Sex Pheromones in the Laboratory and Store. *J. Stored Prod. Res.***1984**, *20*, 191–197.

Howell, J. F.; Knight, A. L.; Unruh, T. R.; Brown, D. F.; Krysan, J. L.; Sell, C. R.; Kirsch, P. A. Control of Codling Moth in Apple and Pear with Sex Pheromone Mediated Mating Disruption. *J. Econ. Ent.* **1992**, *85*, 918–925.

Imai, T.; Kodama, H.; Chuman, T.; Kohno, M. Female Produced Oviposition Deterrent of the Cigarette Beetle *Lasioderma serricorne*. *J. Chemical Ecol.* **1990**, *16*, 1237–1247.

Jones, O. T. The Commercial Exploitation of Pheromonones and Other Semiochemicals. *Pest. Sci.* **1998**, *54*, 293–296.

Kodama, H.; Mochizuki, K.; Kohno, M.; Ohnishi, A.; Kuwahara, Y. Inhibition of Male Response of Drugstore Beetles to Stegoninone by its Isomers. *J. Chemical. Ecol.* **1987**, *13*, 1871–1879.

Kuwahara, Y.; Nemoto, T.; Shibuya, M.; Matsuura, H.; Shiraiwa, Y. 2-Palmitoyl-and 2-Oleoyl-Cyclohexane-1,3-Dione from Feces of the Indian Meal Moth *Plodia interpunctella*, Kairomone Components Against a Parasitic Wasp *Venturia canescens*. *Agril. Biol. And Chem.* **1983**, *47*, 1929–1931.

Law, J. H.; Regnier, F. E. Pheromones. *Ann. Rev. Biochem.* **1971**, *40*, 533–548.

Levinson, H. Z.; Levinson, A. R. Pheromone Biology of the Tobacco Beetle with Notes on the Pheromone Antagonism between 4S, 6S, 7S, and 4S, 6S, 7S Serricornin. *J. Applied Ento.* **1987**, *103*, 217–240.

Levinson, H. Z.; Levinson, A. R. Origin of Grain Storage and Insect Species Consuming Desiccated Food. *Anzeiger fur Schadlingskunde Pflanzenschutz Umweltschutz.* **1994**, *67*, 47–59.

Mishra, B. B.; Tripathi, S. P.; Tripathi, C. P. M. Response of *Tribolium castaneum* (Coleoptera: Tenebrionidae) and *Sitophilus oryzae* (Coleoptera: Curculionidae) to Potential Insecticide Derived from Essential Oil of *Mentha arvensis* Leaves. *Biol. Agri. Horti.* **2012**, *28* (1), 34–40.

Mondal, K. A. M. S. H. Response of *Tribolium castaneum* Larvae to Aggregation Pheromone and Quinines Produced by Adult Conspecifics. *Int. J. Pest Control.* **1985**, *27*, 64–66.

Mondal, K. A. M. S. H.; Port, G. R. Pheromones of *Tribolium* spp and their Potential in Pest Management. *Agric. Zool. Rev.* **1994**, *6*, 121–148.

Mudd, A. Novel 2-Acylcyclohexane-1,3-Diones in the Mandibular Glands Lepidopteran Larvae Kairomones of *Ephesia kuehniella*. *Chem. Soc. Perkin Trans.* **1981**, *1*, 2357–2362.

Mullen, M. A. Development of a Pheromone Trap for Monitoring *Tribolium castaneum*. *J. Stored Prod. Res.* **1992**, *28*, 245–249.

Mullen, M. A.; Pedersen, J. R. Sanitation and Exclusion. In *Alternative to Pesticides in Stored Product IPM;* Subramanyam, B., Hagstrum, D. W., Eds.; Kluwer Academic Publishers: London, 2000; pp 45–92.

Nemoto, T.; Kuwahara, Y.; Suzuki, T. Interspecific Differences in Venturia Kairomones in Larval Feces of Four Stored Phycitid Moths. *Appl. Ent. Zool.* **1987**, *22*, 553–559.

Norris, D. M. Repellents. In *VI Insect Attractants and Repellents: CRC Handbook of Natural Pesticides;* Morgan, E. D., Mandava, N. B. Eds.; CRC Press Inc.: Boca Raton, Florida, **1990**, 135–149.

Padin, S. B.; Fuse, C.; Urrutia, M. I.; Bello, G. M. Toxicity and Repellency of Nine Medicinal Plants Against *Tribolium castaneum* in Stored Wheat. *Bull. Insectol.* **2013**, *66* (1), 45–49.

Parvin, S.; Islam, M. T. Bioactivity of Indonesian Mahogany, *Toona sureni* (Blume) (Meliaceae), Against the Red Flour Beetle, *Tribolium castaneum* (Coleoptera, Tenebrionidae). *Rev. Brasileira de Entomol.* **2012**, *56* (3), 354–358.

Phillips, T. W. Semiochemicals of Stored Product Insects Research and Applications. *J. Stored Prod. Res.* **1997**, *33*, 17–30.

Phillips, T. W.; Cogan, P. M.; Fadamiro, H. Y. Pheromones. In *Alternative to Pesticides in Stored Product IPM;* Subramanyam, B., Hagstrum, D. W., Eds.; Kluwer Academic Publishers: London, 2000; pp 273–302.

Pierce, A. M.; Borden, J. H.; Oehschlager, A. C. Suppression of Oviposition in *Oryzaephilus surinamensis* Following Prolonged Retention in High Density Cultures or Short Term Exposure to Larval Volatiles. *J. Chem. Ecol.* **1990**, *16* (2), 595–601.

Pierce, A. M.; Pierce, H. D.; Oehschlager, A. C.; Borden, J. H. 1-Octen-3-ol, Attractive Semiochemicals for Foreign Grain Beetle *Ahasverus advena*. *J. Chem. Ecol.* **1991**, *17* (3), 567–580.

Pinniger, D. B.; Stubbs, M. R.; Chambers, J. In *The Evaluation of Some Food Attractants for the Detection of Oryzaephilus surinamensis and otherstorage Pests,* Proceeding of the Third International Working Conference on Stored Product Entomology, Oct 1984; KS, USA, pp 640–650.

Plarre, R. Pheromones and Other Semiochemicals of Stored Product Insects a Historical Review, Current Application and Perspective Needs. *Mitt. Boil. Bundesanst. Land Forstw.* **1998**, *342*, 13–84.

Price, L. H. In *Using Pheromones for Location and Suppression of Phycitid Moth and Cigarette Beetles in Hawaii a Five Year Summary,* Stored Product Protection Proceeding of the Sixth International Working Conference on Stored Product Protection, Apr 17–23, 1994; Canberra, Australia, Highley, E., Wright, E. J., Banks, H. J., Champ, B. R., Eds.; CAB International: Wallingford, UK, 1994; Vol. 1, pp 439–443.

Quartey, G. K.; Coaker, T. C. The Development of an Improved Model Trap for Monitoring *Ephestia cautella* . *Entomol. Exp. Appl.* **1992**, *64*, 293–301.

Rajashekar, Y.; Bakthavatsalam, N.;Shivanandappa, T. Botanicals as Grain Protectants. *Psyche.* **2012**, *2012*, 13.

Rajendran, S.; Sriranjini, V. Plant Products as Fumigants for Stored-Product Insect Control. *J. Stored Prod. Res.* **2008**, *44*, 126–135.

Smith, S. M.; Moore, D.; Karanja, L. W.; Chandi, E. A. Formulation of Vegetable Fat Pellets with Pheromone and *Beauvaria bassiana* to Control the Larger Grain Borer *Prostephanus truncates*. *Pest. Sci.* **1999**, *55*, 711–718.

Solomon, M. E. Archaeological Records of Storage Pests *Sitophilus granaries* from an Egyptian Pyramid Tomb. *J. Stored Prod. Res.* **1965,** *1,* 105–107.

Suzuki, T.; Nakakita, H.; Kuwara, Y. Aggregation Pheromone of *Tribolium freeman* Identification of the Aggregation Pheromone. *Appl. Ento. Zool.* **1987, 22,** 340–347.

Talukder, F. A. Plant Products as Potential Stored Product Insect Management Agents–a Mini Review. *Emir. J. Agric. Sci.* **2006,** *18,* 17–32.

Temerak, S. A. Attraction of the Almond Moth, *Cadra cautella* (Walker), to Readymade Tea or Coffee: An Alternation Method of Methyl Bromide in Storehouses of Date Fruit in Egypt. *Acta Hort.* **2010,** *882,* 563–568.

Throne, J. E.; Baker, J. E.; Messina, F. J.; Kramer, K. J.; Howard, J. A. Vertical Resistance. In *Alternative to Pesticides in Stored Product IPM;* Subramanyam, B., Hagstrum, D. W., Eds.; Kluwer Academic Publishers: London, 2000; pp 165–172.

Trematerra, P.; Capizzi, A. Attracticide Method in the Control of *Ephestia kuehniella* Studies on Effectiveness. *J. Appl. Entomol.* **1991,** *111,* 451–456.

Walgenbach, C. A.; Phillips, J. K.; Faustini, D. L.; Burkholder, W. E. Male Produced Aggregation Pheromone of the Maize Weevil *Sitophilus zeamais* and Interspecific Attraction between Three *Sitophilus* Species. *J. Chem. Ecol.* **1983,** *9,* 831–841.

Weaver, D. K.; Subramanyam, B.; Botanicals. In *Alternative to Pesticides in Stored Product IPM;* Subramanyam, B., Hagstrum, D. W., Eds.; Kluwer Academic Publishers: London, 2000; pp 303–320.

Williams, H. J.; Silverstein, R. M.; Burkholder, W. E.; Khorramshahi, A. Dominicalure 1 and 2 Components of the Aggregation Pheromone from the Male Lesser Grain Borer *Rhyzopertha dominica. J. Chem. Ecol.* **1981,** *7,* 759–781.

CHAPTER 11

MANAGEMENT OF STORED GRAIN AND SEED INSECTS BY PLANT PRODUCTS

CONTENTS

ABSTRACT

To find out the alternative of chemical fumigants in the plant kingdom several plants have been evaluated for their fumigant, repellent, attractant, and deterrent potential, and most interestingly, many plants have been identified which are equipped with fumigant properties against insect pests of stored grains. Now it has been proved beyond doubt under laboratory conditions the plant components can suppress feeding and breeding of stored grain insects or even kill them completely within a few hours just like chemical fumigants. Essential oil from more than 75 plant species belonging to different families, such as Anacardiaceae, Apiaceae (Umbeliferae), Araceae, Asteraceae (Compositae), Brassicaceae (Cruciferae), Chenopodiaceae, Cupressaceae, Graminaceae, Lamiaceae (Labiatae), Lauraceae, Liliaceae, Myrtaceae, Pinaceae, Rutaceae, and Zingiberaceae have been studied for fumigant toxicity against insect pests of storage grains. There have been extensive studies on the utilization of plant extracts, powder, ash, essential oil, sulary, or whole plant materials for insect management under storage condition, but few are used on a commercial scale but there are some problems with plant-based insecticides are lack of consistency, safety concerns, rate of application, and sometimes odor. The application of plant materials to protect stored products against insect attack has a long history. Several plant species concerned have also been used in traditional medicine by local communities and have been collected from the field or specifically cultivated for these purposes. Leaves, roots, twigs, ash, essential oils, and flowers have been admixed, as grain or seed protectants, fumigant, repellent, oviposition deterrent with various stored products in different parts of the world, particularly in different developing countries. Many commercially available spices and herbs, turmeric, basil, marjoram, anise, cumin, and coriander, are for management of stored grain insects but they are also able to suppress fungal and mite growth under storage condition.

11.1   INTRODUCTION

In India, only aluminum phosphide and methyl bromide are available for fumigation of food grains. The use of aluminum phosphide is restricted by law, while methyl bromide needs special infrastructures for its use. Due to injudicious use, aluminum phosphide is poisoning our environment, causing many fatal diseases in exposed persons, in addition to facilitating many to end the life. Its improper use is also developing resistance in the insect pests of stored grain, and in many parts of world including India it has become useless for fumigation. Now a time has reached when we have to think whether it is a boon or a bane. Scientists all over the world realize that presently it is impossible to survive without its use. But they also understand that it is necessary to minimize its use at least at the farmer's level by searching for other viable alternatives. For the last four decades, serious attempts have been made to find its alternative in the plant kingdom and hundreds of plants have been screened to study their fumigation potential. And most interestingly, many plants have been identified which are equipped with fumigant properties against insect pests of stored grain. Now it has been proved beyond doubt that under laboratory conditions the plant components can suppress feeding and breeding of stored grain insects or even kill them completely within a few hours just like chemical fumigants. Essential oil from more than 75 plant species belonging to different families, such as Anacardiaceae, Apiaceae (Umbeliferae), Araceae, Asteraceae (Compositae), Brassicaceae (Cruciferae), Chenopodiaceae, Cupressaceae, Graminaceae, Lamiaceae (Labiatae), Lauraceae, Liliaceae, Myrtaceae, Pinaceae, Rutaceae, and Zingiberaceae have been studied for fumigant toxicity against insect pests of storage grain (Rajendran & Sriranjini, 2008).

There have been extensive studies on the utilization of plant extracts, powder, ash, essential oil, sulary, or whole plant materials for insect management under storage condition, but few are used on a commercial scale (Rajendran & Sriranjini, 2008). Farmers often use locally available or naturally occurring plant materials for insect control in developing countries. The major problems with plant-based insecticides are lack of consistency, safety concerns, rate of application, and sometimes odor. The application of plant product for management of stored grain insects was first reported by Golob and Webley (1980) in the form of bibliography. Grainge and Ahmed (1988) reported the 2400 plant species which contain the insecticidal activities against stored grain insects. In addition to bibliography reported by Golob and Webley (1980), the alternative methods to synthetic insecticides for management of stored grain insects were reported by Rees et al. (1993) in the form of a database of 1100 plants; these methods include plant materials, extract, and essential oils. In some part of the world, vegetable oil and black pepper are used as grain or seed protectant of beans (Baler & Webster, 1992). In all over the world commercial application of plant extracts as insecticides began in 1850 with the introduction of nicotine from *Nicotiana tabacum,* rotenone from *Lonchocarpus* sp, derris dust from *Derris elliptica,* and pyrethrum from the flower heads of *Chrysanthemum cinerariaefolium,* and many insecticides have been developed from their constituents too.

It is often falsely assumed that because a plant material is used as a food flavoring or medicine that extracts from the material will be safe for human consumption. Various extracts from the neem tree, *Azadirachta indica*, collectively referred to as the insecticide Neem, are commercially available botanical insecticides, and local formulations have been widely used in some parts of the world for stored-product insect management (Kaul et al., 1990). However, commercial formulations show only moderate levels of efficacy (Abate, 2001; Kavallieratos et al., 2007). Crude pea flour and the protein-rich fraction of field peas, *Pisum* species, as well as that of other food legumes (e.g., species of *Pissum*, *Phaseolus*, and *Vignia*), are toxic and repellent to stored-product insects (Bodnaryk et al., 1999; Fields, 2006). Direct application of protein-enriched pea flour to bulk grain at 0.1% by weight resulted in substantial reductions in stored-product Coleoptera populations (Hou & Field, 2003), and broad scale application of pea flour to the inside of mills reportedly resulted in insect control, but such control was not at commercially acceptable levels like those achieved with synthetic fumigants. Pyrethrum, a commercial mixture of compounds derived from *Chrysanthemum cinerariifolium*, is perhaps the most successful botanical insecticide throughout all modern pest control, and this is certainly the case for stored products. The active ingredients from pyrethrum are called pyrethrins. Synergized pyrethrum commonly contains the synergist piperonyl butoxide, which suppresses metabolic degradation of pyrethrins in the insect. Synergized pyrethrum is commonly used as an aerosol in flour mills (Toews et al., 2006) and is usually combined with another insecticide that has longer residual activity because the pyrethrum achieves only quick knockdown of insect pests at best, while the other insecticide with which it is combined provides longer activity (Arthur, 2008). Organically compliant pyrethrum, which lacks any synthetic synergist, is extracted from chrysanthemum flowers. Table 11.1 elaborate that the different types of plants and their activity against major stored grain insect pests. Utilization of locally available plant materials for protection against stored grain insects will open the door for biopesticides which is biodegradable, eco-friendly, and sustainable, because they do not produce any adverse effect on grain or seeds and break down resistency against chemical insecticides. There are several non-cultivated plants that have been used traditionally in grain and safe to non-target organisms. Although commonly regarded as safe, many plants contain secondary metabolites, which are directly admixed into grain.

TABLE 11.1 Name of the Plant, their Secondary Metabolites and its Activity against Major Stored Grain Insect Pests.

Scientific name	Vernacular name	Family	Plant part	Formulation	Major secondary metabolites	Stored grain pests	Type of activity	References
Acorus calamus L.	Sweet flag	Araceae	Rhizome Leaves Seed	Oil, extract powder	B- asarone, acorin	*R. dominica, S. oryzae, C. chinensis T. castaneum*	Fumigant toxicity Protectant	Chander et al., 1999; Schmidt et al., 1991
			Seed	Powder		*S. oryzae, R. dominica*	Repellant	Chellappa & Chelliah, 1976
			Rhizome	Extract		*S. oryzae, C. chinensis*	Fumigants	Kim et al., 2003
			Rhizome	Oil		*C. chinensis*	Repellant	Agarwal et al., 1973
			Rhizome	Oil		*R. dominica*	Repellant	Agarwal et al., 1973
			Rhizome	Oil		*R. dominica*	Repellant Antifeedant	Jilani & Saxena, 1990
			Rhizome	Oil		*S. granary*	Ovipozition deterrent	Risha et al., 1990
			Rhizome	Oil		*S. oryzae*	Ovipozition deterrent	Risha et al., 1990
			Rhizome	Oil		*S. oryzae*	Ovipozition deterrent	Su, 1991
			Rhizome	Oil		*T. castaneum*	Ovipozition deterrent	Risha et al., 1990
			Rhizome	Oil		*C. chinensis*	Ovipozition deterrent	Yadav, 1971
			Rhizome	Oil		*C. maculatus*	Protectant	Su, 1991
			Rhizome	Oil		*C. maculatus*	Protectant	Su, 1991
			Rhizome	Oil		*C. chinensis*	Ovipozition deterrent	Risha et al., 1990
			Rhizome	Oil		*T. castaneum*	Protectant	Su, 1991
			Rhizome	Oil		*S. oryzae*	Protectant	Su, 1991
			Rhizome	Oil		*C. chinensis*	Fumigant	Bhonde et al., 2001

TABLE 11.1 *(Continued)*

Scientific name	Vernacular name	Family	Plant part	Formulation	Major secondary metabolites	Stored grain pests	Type of activity	References
			Rhizome	Powder		*C. chinensis*	Protectant	Deshpande et al., 2004
			Rhizome	Powder		*R. dominica*	Protectant	Jilani, 1984
			Rhizome	Powder		*S. oryzae*	Protectant	Jilani, 1984
			Rhizome	Powder		*C. chinensis*	Protectant	Khan, 1986
			Rhizome	Powder		*C. chinensis*	Ovipozition deterrent	Chiranjeevi, 1991
			Rhizome	Powder		*C. chinensis*	Repellant	Ignatowicz & Wesolowska, 1996
			Rhizome	Powder		*C. chinensis*	Repellant	Trehan &Pingle, 1947
			Rhizome	Powder		*C. chinensis*	Protectant	Chander & Ahmed, 1986
			Rhizome	Powder		*S. cerealella*	Protectant	Trehan &Pingle, 1947
			Rhizome	Powder		*S. granary*	Protectant	Ignatowicz & Wesolowska, 1996
			Rhizome	Powder		*S. granary*	Protectant	Paneru et al., 1997
			Rhizome	Powder		*S. granary*	Protectant	Paneru et al., 1997
			Rhizome	Powder		*S. oryzae*	Protectant	Paneru et al., 1993
			Rhizome	Powder		*S. oryzae*	Repellant	Ignatowicz & Wesolowska, 1996
			Rhizome	Powder		*S. oryzae*	Protectant	Paneru et al., 1997
			Rhizome	Powder		*S. oryzae*	Protectant	Paneru et al., 1997
			Rhizome	Powder		*S. oryzae*	Oviposition deterrent	Chander et al., 1990
			Rhizome	Powder		*T. castaneum*	Oviposition deterrent	Chander et al., 1990
			Rhizome	Powder		*S. oryzae*	Oviposition deterrent	Chander & Ahmed, 1983
			Seed	Oil		*T. castaneum*	Fumigant	Srivastava et al., 1965

TABLE 11.1 *(Continued)*

Scientific name	Vernacular name	Family	Plant part	Formulation	Major secondary metabolites	Stored grain pests	Type of activity	References
			Leaves	Powder		*R. dominica*	Protectant	Prakash et al., 1993
			Leaves	Powder		*S. cerealella*	Protectant	Prakash et al., 1993
			Leaves	Powder		*S. oryzae*	Protectant	Prakash et al., 1993
			Leaves	Whole (dried)		*S. cerealella*	Protectant	Prakash et al., 1983
			Leaves	Whole (dried)		*S. oryzae*	Protectant	Prakash et al., 1983
			Leaves	Whole (dried)		*R. dominica*	Protectant	Prakash et al., 1983
			Leaves	Whole (dried)		*S. oryzae*	Protectant	Prakash et al., 1983
			Leaves	Whole (dried)		*R. dominica*	Protectant	Prakash et al., 1983
			Leaves	Powder		*S. cerealella*	Protectant	Prakash et al., 1993
			Leaves	Powder		*S. oryzae*	Protectant	Prakash et al., 1993
			Leaves	Powder		*R. dominica*	Protectant	Prakash et al., 1993
			Leaves	Powder		*R. dominica*	Protectant	Rao & Prakash, 2002
			Leaves	Powder		*S. oryzae*	Protectant	Deka, 2003
			Leaves	Powder		*T. granarium*	Protectant	Deka, 2003
			Whole plant	Extract		*T. castaneum*	Protectant	Atal et. al., 1978
Azadirachta indica	Neem	Meliaceae	Leaves Seed, Bark	Oil, extract powder	Azadirechtinene	*C. chinensis, T. castaneum*	Repellent	Pathak & Krishna, 1991 Raguraman & Singh, 1997
			Leaves	Extract		*S. oryzae*	Oviposition deterrent	Vishweshwariah et al., 1971
			Leaves	Extract		*S. oryzae T. castaneum*	Repellent	Qadri, 1973
			Leaves	Extract		*S. oryzae*	Repellent	Qadri & Rao, 1977
			Seed	Extract		*S. oryzae*	Repellent	Qadri & Hasan, 1978
			Seed	Powder		*C. chinensis*	Protectant	Ohsawa et al., 1990
			Seed	Powder		*C. analis*	Protectant	Juneja & Patel, 2002
			Seed	Powder		*C. maculatus*	Protectant	Ecija, 2000

TABLE 11.1 (Continued)

Scientific name	Vernacular name	Family	Plant part	Formulation	Major secondary metabolites	Stored grain pests	Type of activity	References
			Seed	Oil		C. chinensis, C. maculatus	Protectant	Doharey et al., 1983
			Seed	Oil		C. maculatus	Protectant	Doharey et al., 1988
			Seed	Oil		C. chinensis	Protectant	Reddy et al., 1994
			Seed	Oil		C. chinensis	Protectant	Choudhary & Pathak, 1989
			Seed	Oil		C. chinensis	Protectant	Modgil & Mehta, 1997
			Seed	Oil		C. maculatus	Protectant	Onolememhen & Oigiangbe, 1991
			Seed	Oil		S. oryzae	Protectant	Ivbijaro, 1984
			Seed	Oil		S. oryzae	Protectant	Ivbijaro et al., 1985
			Seed	Oil		S. oryzae	Oviposition deterrent	Salas, 1985
			Seed	Oil		C. maculatus	Fumigant	Raghvani & Kapadia, 2003
			Seed	Oil		R. dominica	Protectant	Trivedi, 1987
			Seed	Oil		R. dominica, S. oryzae	Protectant	Shukla et al., 1992
			Seed	Oil		T. castaneum	Protectant	Trivedi, 1987
				Extract		T. castaneum	Protectant	Liu & Ho,1999
				Extract		T. castaneum	Protectant	Sendi et al., 2003
				Oil		T. castaneum	Repellant	Sendi et al., 2003
				Oil		C. maculatus, R. dominica, S. oryzae	Repellant	Agrawal et al., 2001
			Leaves	Extract		S. oryzae	Oviposition deterrent	Reddy & Singh, 1998
			Leaves	Extract		C. analis, C. chinensis, C. maculatus	Oviposition deterrent	Yadav, 1985

TABLE 11.1 (Continued)

Scientific name	Vernacular name	Family	Plant part	Formulation	Major secondary metabolites	Stored grain pests	Type of activity	References
			Leaves	Extract		R. dominica	Protectant	Sharma et al., 1989
			Leaves	Extract		R. dominica	Repellant	Suss et al., 1997
			Leaves	Extract		T. castaneum	Protectant	Rahman et al., 1997
			Leaves	Extract		C. chinensis	Protectant	Rayasekaran & Kumar-swami, 1985
			Leaves	Extract		R. dominica, S. oryzae	Protectant	Dakshinamurthy & Goel, 1992
			Leaves	Extract		C. chinensis	Oviposition deterrent, Protectant	Yadav & Bhargava, 2002
			Seed	Oil		R. dominica	Repellant, Antifeedant	Ketkar, 1976
			Seed	Oil		C. chinensis	Repellant, Antifeedant	Rani et al., 2000
			Seed	Oil		C. chinensis	Oviposition deterrent, Protectant	Sunarti, 2003
			Seed	Oil		C. chinensis	Oviposition deterrent, Protectant	Ba-Angood &Al-Suraidy, 2003
			Seed	Oil		C. chinensis	Oviposition deterrent, Protectant	Ba-Angood & Al-Suraidy, 2003
			Seed	Oil		C. chinensis	Oviposition deterrent, Protectant	Boeke et al., 2004b
			Seed	Oil		C. maculatus	Oviposition deterrent, Protectant	Raja & Ignacimuthu, 2001
			Seed	Oil		C. chinensis	Oviposition deterrent, Protectant	Ketkar, 1976
			Seed	Oil		C. chinensis	Oviposition deterrent, Protectant	Das, 1986

TABLE 11.1 *(Continued)*

Scientific name	Vernacular name	Family	Plant part	Formulation	Major secondary metabolites	Stored grain pests	Type of activity	References
			Seed	Oil		*C. chinensis*	Oviposition deterrent, Protectant	Das & Karim, 1986
			Seed	Oil		*C. chinensis*	Protectant	Das, 1987
			Seed	Oil		*C. chinensis*	Protectant	Choudhary & Pathak, 1989
			Seed	Oil		*C. chinensis*	Protectant	Das, 1989
			Seed	Oil		*C. chinensis*	Protectant	Jacob & Sheila, 1990
			Seed	Oil		*C. chinensis*	Protectant	Reddy et al., 1994
			Seed	Oil		*C. chinensis*	Protectant	Reddy et al., 1994
			Seed	Oil		*C. chinensis*	Protectant	Haque et al., 2002
			Seed	Oil		*C. chinensis*	Protectant	Singh & Sharma, 2003
			Seed	Oil		*C. chinensis*	Protectant	Singh & Yadav, 2003
			Seed	Oil		*C. chinensis*	Protectant	Ketkar, 1987
			Seed	Oil		*C. maculatus*	Protectant	Ketkar, 1987
			Seed	Oil		*C. chinensis*	Oviposition deterrent	Pathak &Krishna, 1991
			Seed	Oil		*C. maculatus*	Oviposition deterrent	Raghvani & Kapadia, 2003
			Seed	Oil		*C. chinensis*	Protectant	Khaire et al., 1992
			Seed	Oil		*C. chinensis*	Repellant	Khaire et al., 1992
			Seed	Oil		*R. dominica*	Repellant	Kathirvelu & Ezhilkumar, 2003
			Seed	Oil		*S. cerealella, S. oryzae*	Protectant	Rahman et al., 1997
			Seed	Oil		*S. oryzae, S. zeamais*	Growth inhibitor	Zhang & Zhao, 1983
			Seed	Oil		*S. oryzae*	Protectant	Dayal et al., 2003
			Seed	Oil		*S. granrary*	Fumigant	Tadesse & Basedow, 2004

TABLE 11.1 *(Continued)*

Scientific name	Vernacular name	Family	Plant part	Formulation	Major secondary metabolites	Stored grain pests	Type of activity	References
			Seed	Oil		C. chinensis	Protectant	Mansour et al., 1997
			Seed	Oil		C. maculatus	Protectant	Ivbijaro, 1990
			Seed	Powder		R. dominica	Protectant	Ketkar 1976.
			Seed	Powder		C. analis	Protectant	Juneja & Patel, 2002
			Seed	Powder		S. zeamais	Protectant	Tadesse & Basedow, 2005
			Flower	Powder		T. castaneum	Protectant	Mostafa, 1988
			Fruit	Powder		T. castaneum	Protectant	Mostafa, 1988
			Leaves	Dust		R. dominica	Protectant	Kathirvelu & Ezhilkumar, 2003
			Leaves	Powder		C. chinensis	Protectant	Chiranjeevi, 1991
			Leaves	Powder		C. maculatus	Protectant	Dhakshinamoorthy & Selvanarayanan, 2002
			Leaves	Powder		C. chinensis	Toxicant	Singh, 2003
			Leaves	Powder		C. maculatus	Protectant	Iloba & Ekrakene, 2006
			Leaves	Powder		S. zeamais	Protectant	Iloba & Ekrakene, 2006
			Leaves	Powder		C. chinensis	Protectant	Patel & Patel, 2002
			Leaves	Powder		S. oryzae	Protectant	Niber, 1994
			Leaves	Powder		C. chinensis	Protectant	Saradamma et al., 1977
			Fruit	Extract		S. oryzae	Protectant	Azmi et al., 1993
			Seed	Extract		R. dominica	Protectant	Chellappa & Chelliah, 1976
			Seed	Extract		S. cerealella	Protectant	Jotwani & Sircar, 1965
			Seed	Extract		C. chinensis	Oviposition deterrent	Lale & Kabeh, 2004
			Leaves	Extract		R. dominica	Oviposition deterrent	Rahim, 1998

TABLE 11.1 (Continued)

Scientific name	Vernacular name	Family	Plant part	Formulation	Major secondary metabolites	Stored grain pests	Type of activity	References
			Leaves	Extract		S. oryzae	Repellant	Suss et al., 1997
			Seed	Cake		S. cerealella	Protectant	Verma et al., 1983
			Seed	Extract		T. castaneum	Oviposition deterrent	Jilani & Malik, 1973
			Seed	Extract		S. oryzae	Growth inhibitor	Mohapatra et al., 1996
			Seed	Oil		S. oryzae	Protectant	Sharma, 1999
			Seed	Oil		R. dominica	Protectant	Jaipal et al., 1984
			Seed	Oil		T. castaneum	Repellant	Jilani et al., 1988
			Seed	Oil		C. chinensis	Oviposition deterrent	Lale & Kabeh, 2004
			Seed	Oil		C. maculatus	Oviposition deterrent	Lale & Kabeh, 2004
			Seed	Oil		C. maculatus	Oviposition deterrent	Maina & Lale, 2004
			Seed	Oil		C. chinensis	Oviposition deterrent	Das, 1987
			Seed	Oil		R. dominica	Repellant	Mohinddin et al., 1993
			Seed	Oil		T. castaneum	Repellant	Mohinddin et al., 1993
			Seed	Powder		C. chinensis	Protectant	Deshpande et al., 2004
			Seed	Powder		R. dominica	Protectant	Sharma et al., 1989
			Seed	Powder		R. dominica	Protectant	Ketkar, 1976
			Seed	Powder		S. cerealella	Protectant	Ketkar, 1976
			Seed	Powder		S. oryzae	Protectant	Ketkar, 1976
			Seed	Powder		S. oryzae	Protectant	Sharma et al., 1989
			Seed	Powder		C. chinensis	Protectant	Borikar & Pawar, 1995
			Seed	Powder		C. maculatus	Protectant	Ecija, 2000

TABLE 11.1 (Continued)

Scientific name	Vernacular name	Family	Plant part	Formulation	Major secondary metabolites	Stored grain pests	Type of activity	References
			Seed	Powder		S. zeamais	Oviposition deterrent	Silva et al., 2003
			Seed	Powder		R. dominica, S. cerealella	Oviposition deterrent	Ketkar, 1976
			Seed	Powder		C. chinensis	Protectant	Chiranjeevi, 1991
			Seed	Powder		R. dominica, S. oryzae, T. granarium	Protectant	Deka, 2003
			Seed	Powder		R. dominica	Protectant	Girish & Jain, 1974
			Seed	Powder		R. dominica, S. oryzae	Protectant	Jotwani & Sircar, 1965
			Fruit	Extract		S. zeamais	Protectant	Ho et al., 1997a
			Seed	Oil		S. zeamais, T. castaneum, R. dominica	Fumigant	Xu et al.,1993
			Root bark	Powder		S. oryza	Oviposition deterrent	Chander & Ahmed, 1983, Zhang & Zhao, 1983
			Leaves	Oil		S. oryzae, T. castaneum	Fumigant, Repellant	Mishra et al., 2002; Tapondjou et al., 2003
Acacia arabica	Babbol	Leguminosae	Wood	Ash		C. chinensis	Ovipozition deterrent	Chiranjeevi, 1991
Acacia concinna	Shikakai	Leguminosae	Fruits	Extract		S. oryzae	Repellant	Qadri, 1973
Achyranthes aspera	Apamarge	Amaranthaceae	Leaves	Powder		C. chinensis	Ovipozition deterrent	Chiranjeevi, 1991
Acorus gramineus	Grassy leaf sweet flag	Aeraceae	Rhizome	Extract		S. oryzae, C. chinensis	Fumigant	Kim et al., 2003
Adhatoda vasica	Malabar nut	Acanthaceae	Leaves Bark	Powder Extract	Adhatodine, vasicinine	Callasobruchs chinensis, T. castaneum	Protactant	Saxena et al., 1986
Aegle marmelos	Bengal quince	Rutaceae	Leaves	Powder	Aegeline, marmeline	S. cerealella, R. dominica, S. oryzae	Protectant	Umoetok, 2004

TABLE 11.1 (Continued)

Scientific name	Vernacular name	Family	Plant part	Formulation	Major secondary metabolites	Stored grain pests	Type of activity	References
Aempfera galanga	Kachoura	Zingiberaceae	Leaves	Powder		*T. castaneum*	Repellent	Gundurao & Majumdar, 1966
Aframomum melegueta	Melegueta pepper	Zingiberaceae	Leaves	Powder	Paradol	*S. zeamais, S. cerealella*	Protactant	Prakash et al., 1983
Agastache scrophulariae-folia	Purple giant hyssop	Agastache	Leaves	Oil		*C. maculatus*	Fumigant	Lognay et al., 2002
Agave americana	Sisal	Agavaceae	Leaves	Powder		*S. oryzae*	Protectant	Bin, 2003
Ageratum conyzoides	Goatweed	Asteraceae	Leaves	Oil Extract	Chromenes, precocene11	*C. chinensis*	Repellent	Deka, 2003
Ageratum houstonianum	Bedding plant	Asteraceae	Leaves	Extract		*T. castaneum*	Repellant	Qureshi et al., 1988
Ailanthus altissimma	Tree of heaven	Simaroubaceae	Leaves	Extract		*T. castaneum*	Fumigant	Pascual & Robledo, 1998
Ajuja iva	Ajuva	Labiateae	Leaves	Extract		*T. castaneum*	Oviposition deterrent	Jbilou et al., 2006
Allium sativum	Garlic	Liliaceae	Clove	Powder Extract	Allicine, diallyl sulphide	*T. castaneum*	Repellent	Ho et al., 1997b
Alpinia calcarata	Snap ginger	Zingiberaceae	Leaves	Oil		*C. maculatus*	Oviposition deterrent	Paranagama et al., 2003
Alpinia galangal	Greater galangal	Zingiberaceae	Leaves Rhizome	Oil Powder	Kaempferia, galangin	*S. oryzae*	Protactant	
Ammi visnaga	Tooth-pick plant	Apiaceae	Seeds	Extract		*S. graneries*	Oviposition deterrent	Abdel-Latif, 2004
Amomum subulatum	Large cardamom	Zingiberaceae	Seeds	Oil		*S. oryzae*	Fumigant	Singh et al., 1989
Anacardium occidentale	Cashew nut	Anacardiaceae	Leaves, Shell	Powder Extract	Anacardic acid, cardanol	*C. maculatus*	Protactant	

TABLE 11.1 *(Continued)*

Scientific name	Vernacular name	Family	Plant part	Formulation	Major secondary metabolites	Stored grain pests	Type of activity	References
Androgphis paniculata	Creat	Acanthaceae	Leaves	Oil		*S. oryzae*	Oviposition deterrent	Prakash et al., 1993
Anethum graveolens	Dill, Sowa	Apiaceae	Leaves	Extract		*S. oryzae*	Repellant	Su, 1985a, 1985b; Su, 1989
Anethum graveolens	Dill	Apiaceae	Seed	Powder Extract	Limonene, carvone	*S. oryzae, T. castaneum*	Repellant	
Anethum sowa	Indian Dill	Apiaceae	Leaves	Extract		*C. maculatus*	Oviposition deterrent	Dwivedi & Sharma, 2003
Annona reticulata	Bullocks heart	Annonaceae	Leaves	Powder Extract	Anonaine, roemerine	*C. chinensis*	Deterrent	
Annona reticulate	Custard apple	Annonaceae	Leaves	Extract		*T. castaneum*	Oviposition deterrent	Atal et al., 1978
Annona squomosa	Sugar apple	Annonaceae	Leaves Seed	Powder Extract	Anonaine, roemerine	*S. zeamais C. maculatus*	Oviposition deterrent Protactant	Haryadi & Yuniarti, 2003
Arachis hypogaea	Ground nut	Leguminasae	Seed	Oil		*C. chinensis*	Protectant	Bhargava & Meena, 2002
Aristolochia baetica	Dutchman's pipe	Aristolochia-ceae	Leaves	Extract		*T. castaneum*	Protectant Chemosterilant	Jbilou et al., 2006 Saxena et al., 1979
Artemicia annua	Worm wood	Asteraceae	Leaves	Extract		*C. maculatus, R. dominica, S. oryzae S. zeamais, T. castaneum*	Protectant Fumigant	Liu et al., 2006 Tripathi et al., 2001
Artemicia argyi	Buffelo herb	Asteraceae	Leaves	Extract		*S. zeamais, T. castaneum*	Protectant	Liu et al., 2007
Artemisia sieberi	Worm wood	Compositae	Leaves	Oil		*S. oryzae, C. chinensis T. castaneum*	Fumigant	Negahban et al., 2007
Artemisia vulgaris	Mug wort	Compositae	Leaves	Oil	Cineole	*T. castaneum*	Fumigant	Wang et al., 2 006
Artemisia maritima	Sea wormwood	Compositae	Leaves	Oil		*S. oryzae*	Fumigant	Singh et al., 1989

TABLE 11.1 (Continued)

Scientific name	Vernacular name	Family	Plant part	Formulation	Major secondary metabolites	Stored grain pests	Type of activity	References
Bassia latifolia	Mahua	*Sapotaceae*	Seed	Extract		*R. dominica, S. oryzae*	Oviposition deterrent	Bowery et al., 1984
Begonia picta	Begonia	*Begoniaceae*	Leaves	Powder		*R. dominica, S. oryzae, S. cerealella*	Protectant	Prakash et al., 1982
Boscia senegalensis	Hemmet	*Capparaceae*	Flower	Extract		*C. maculatus, S. cerealella s*	Oviposition deterrent, Toxicant	Seck et al., 1993
Boscia senegalensis	Seck	*Capparidaceae*	Leaves	Powder	Glucocapparin	*C. maculates*	Protactant toxicity	
Bougainvillea sp.	Bougainvillea	*Yctaginaceae*	Flower	Extract		*S. oryzae*	Toxicant	Rao, 1955
Brassica campestris	Yellow sarson	*Cruciferae*	Seed	Oil		*S. oryzae, C. chinensis*	Protectant	Ran-Pal et al., 1988; Sharma et al., 2002
Brassica juncea	Mustard	*Cruciferae*	Seed	Oil		*C. maculatus, C. chinensis*	Protectant	Doharey et al., 1983; Das, 1986; Singh & Yadav, 2003; Kim et al., 2003
Brassica napus	Rape seed	*Cruciferae*	Seed	Oil		*C. maculatus, C. chinensis*	Protectant	Sharma & Singh, 1993; Haque et al., 2002; Raghvani & Kapadia, 2003
Caesalpiniapul cherrima	Peacock	*Caesalpiniaceae*	Flower	Extract		*S. oryzae*	Protectant	Rao, 1955
Callicarpa marcrophylla	Beauty berry	*Verbenaceae*	Leaves	Oil		*S. oryzae*	Fumigant	Singh et al., 1989
Callistepus chinensis	China aster	*Asteraceae*	Flower	Extract		*S. oryzae*	Protectant	Rao, 1955
Callotropis gigantea	Akund fibre	*Ascelpidaceae*	Flower	Extract		*S. oryzae*	Protectant	Rao, 1955 and 1957
Callotropis procera	Aak	*Ascelepiadaceae*	Leaves, Flower	Powder		*R. dominica*	Protectant	Diarisso & Pendleton, 2005 Sharma, 1985

TABLE 11.1 (Continued)

Scientific name	Vernacular name	Family	Plant part	Formulation	Major secondary metabolites	Stored grain pests	Type of activity	References
Calophyllum inophyllgum	Laurel wood	Clusiaceae	Leaves	Oil	Friedelin, canophyllol	*S. oryzae*	Protectant	
Canaga odorata	Ylang-ylang	Annonaceae	Leaves	Oil		*S. oryzae*	Fumigant	Huang et al., 1999
Cannabis sativa	Hemp	Cannabidaceae	Leaves	Extract		*S. oryzae*	Protectant	Khare, 1972
Capsicum annum	Pepper	Solanaceae	Fruit Leaves	Powder Extract	Capsicine, capsaicine	*S. oryzae, S. cerealella*	Protectant	Debkirtaniya et al., 1980, Srinivasan and Nadara, 2004
Capsicum frutescens	Chilli, red pepper	Solanaceae	Fruit	Extract Powder	Capsicine	*T. granarium C. maculatus*	Protectant	Al-moajel, 2004
Carica papaya	Papaya	Caricaceae	Leaves	Slurry		*C. maculatus*	Repellant	Boeke et al., 2004b
Carthamum tinctorius	Safflower	Compositeae	Seed	Oil		*C. maculatus, C. chinensis*	Protectant	Doharey et al., 1983
Carum carvi	Caraway	Apiaceae	Fruit	Powder Extract	Carvone, saponins	*S. oryzae, R. dominica*	Protactant	
Carum roxburghianum	Bishop's weed	Umbelliferae	Seed	Powder		*S. oryzae*	Protectant	Chander & Ahmed, 1983
Cassia alata	Gua java	Caesalpiniaceae	Leaves	Extract		*S. cerealella*	Protectant	Srinivasan and Nadara, 2004
Cassia fistula	Golden-shower	Caesalpiniaceae	Flower	Extract		*S. oryzae*	Protectant	Rao, 1955 and Prakash et al., 1993
Cassia nigricans	Tanner's cassia	Leguminocae	Leaves	Powder		*R. dominica*	Protectant	Diarisso & Pendleton, 2005
Cassia occidentalis	Coffee senna	Caesalpiniaceae	Leaves	Extract		*T. granarium*	Oviposition deterrent	Dwivedi et al., 1998
Cassia sophera	Seena sophera	Caesalpiniaceae	Leaves	Powder, Extract		*S. oryzae, C. maculatus*	Protectant, Oviposition deterrent	Kestenholz et al., 2007
Casurina indica	Saru	Casuarinaceae	Wood	Ash		*C. chinensis*	Oviposition deterrent	Chiranjeevi, 1991

TABLE 11.1 *(Continued)*

Scientific name	Vernacular name	Family	Plant part	Formulation	Major secondary metabolites	Stored grain pests	Type of activity	References
Cedrus deodara	Himalayan cidar	Pinaceae	Wood	Oil		*S. oryzae*	Fumigant	Singh et al., 1989 Raguraman & Singh, 1997
Celastrus angulatus	Bitter tree	Celastraceae	Seed	Oil Powder	Maytansine, celengalin	*C. chinensis, S. oryzae*	Protectant	Singh & Sharma, 2003
Centratherum anthelminticum	Kalijiri	Asteraceae	Seed	Extract		*C. chinensis*	Protectant	Ansari et al, 2003
Cestrum nocturnum	Night Jasmine	Solanaceae	Leaves	Powder extract		*S. oryzae, T. castaneum*	Protectant	Tapondjou et al., 2003, Kemabonta & Okogbue, 2002
Chamaecyparis obtusa	Hinoki cypress	Cupressaceae	Leaves	Oil		*S. oryzae, C. chinensis*	Fumigant	Park et al., 2003
Chenopodium ambrosioides	Indian wormseed	Chenopodiaceae	Leaves Fruits	Powder, Oil Extract	Limonene, ascaridol	*T. castaneum, R. dominica*	Anti feedant Fumigant toxicity	
Chondrilla juncea	Skeleton weed	Compositae	Leaves	Extract		*T. castaneum*	Protectant	Pascual & Robledo, 1998
Chromolaea odorata	Siam weed	Asteraceae	leaves	Powder		*S. oryzae*	Protectant	Niber, 1994
Chrysanthemum cinerariifolium	Pyrethrum	Compositae	Flower	Extract		*S. oryzae*	Protectant	Rao, 1955
Cichorium intybus	Chicory	Compositeae	Leaves	Extract		*T. castaneum*	Protectant	Pascual & Robledo, 1998
Cinnamomum sieboldii	Sassafras	Lauraceae	Bark	Extract		*S. oryzae, C. chinensis*	Fumigant	Kim et al.,2003
Cinnamomum aromaticum	Cinnamon	Lauraceae	Leaves Bark	Extract	Cinnamic aldehyde	*S. zeamais T. castaneum*	Repellant Fumigant	Ho et al., 1997a Huang & Ho, 1998
Cinnamomum camphora	Camphor	Lauraceae	Leaves	Oil Powder		*S. oryzae, C. chinensis T. castaneum*	Fumigant Repellant	Huang et al., 1999 Al-jabr, 2003
Cinnamomum cassia	Chinese cinnamon	Lauraceae	Bark	Extract Oil		*S. oryzae*	Repellant Fumigant	Su, 1985 Kim et al.,2003

TABLE 11.1 *(Continued)*

Scientific name	Vernacular name	Family	Plant part	Formulation	Major secondary metabolites	Stored grain pests	Type of activity	References
Cinmamomum micranthum	Chinese sassafras	Lauraceae	Bark	oil		*T. castaneum*	Repellant	Xu et al., 1996
Cinnamomum zeylanicum	Ceylon cinnamom	Lauraceae	leaves	Oil		*S. cerealella C. maculatus*	Fumigant Oviposition deterrent	Paranagama et al., 2003
Cissampelos awariensis	Pareira brava	Menispermaceae	Leaves Root	Extract Powder	Bis benzylisoquinoline	*S. oryzae*	Protectant	Niber, 1994
Citrus aurantifolia	Lime	Rutaceae	Peel	Oil Powder	Limonene	*C. chinensis*	Oviposition deterrent Fumigant Toxicity Protectant	Bhargava et al., 2003
Citrus aurantium	Sour orange	Limonenae	Leaves	Oil	Limonene	*S. zeamais T. castaneum*	Fumigant	Haubruge et al., 1989
Citrus limon	Limon	Rutaceae	Leaves	Powder	Limonene, β-pinene	*T. castaneum*	Protectant	
Citrus paradisa	Grapefruit	Rutaceae	Peel	Oil Powder	Limonene	*S. granaries*	Fumigant toxicity Protectant	
Citrus reticulara	Mandrin	Rutaceae	Peel	Oil	Limonene	*S. oryzae, R. dominica, T. castaneum, S. cerealella, C. cephalonica, C. chinensis*	Fumigant	Kumar, 2015 Kumar, 2016
Citrus sinensis	Sweet orange	Rutaceae	Peel Leaves	Oil Powder	Citracridone, citbrasine	*C. maculatus*	Fumigant Toxicity Protectant	Don-Pedro, 1996
Cleome monophylla	Daziel	Capparidaceae	Leaves	Oil	Terpenolene, terpeneol	*S. zeamais*	Repellent	Lee et al., 2002
Clerodendrum inerme	Kashmir bouquet	Verbenaceae	Leaves	Extract		*S. cerealella*	Oviposition deterrent	El-Lakwah et al., 1997a
Cocholeria aroracia	Horse raddish	Cruciferae	Seed	Extract		*S. oryzae, C. chinensis*	Protectant	Kim et al.,2003
Cocos nucifera	Coconut	Arecaceae	Fruit	Oil	Trimyristin, triaurin	*C. chinensis*	Protectant	Sharma & Singh, 1993

TABLE 11.1 *(Continued)*

Scientific name	Vernacular name	Family	Plant part	Formulation	Major secondary metabolites	Stored grain pests	Type of activity	References
Coleus amboinicus	Indian borage	Lamiaceae	Leaves	Oil	Ethereal	*C. chinensis*	Fumigant toxicity	
Convolvulus arvensis	Field bind weed	Convolvulaceae	Leaves Seed	Extract	Convolvuline	*T. castaneum S. oryzae*	Protectant	Singh et al., 1976 and Peterson et al., 1989
Coriandrum sativum	Coriander	Apiaceae	Seed	Oil		*C. maculatus, S. oryzae, T. castaneum*	Protectant, Repellant	Pascual-Villalobos, 2003, Su, 1986
Crinum defixum	Kerasachottu	Amaryllidaceae	Leaves	Powder		*C. chinensis*	Oviposition deterrent	Chiranjeevi, 1991
Croton sparsiflorus	Croton	Euphorbiaceae	Leaves	Extract		*S. cerealella*	Protectant	Srinivasan & Nadara, 2004
Cucumis melo	Muskmelon	Cucurbitaceae	Seed	Oil		*S. oryzae, C. chinensis T. castaneum*	Fumigant	Wang et al., 2000
Cuminum cyminum	Cumin	Apiaceae	Seed	Oil		*S. oryzae, C. chinensis T. castaneum*	Fumigant	Tunc et al., 2000
Curcuma longa	Turmeric	Zingiberaceae	Leaves Rhizome	Oil, Extract Powder	Curcumin, turmerone	*C. maculatus, S. oryzae*	Fumigant Toxicity Protactant	Tripathi et al., 2002
Cymbopogon citrates	Lemon grass	Graminae	Leaves	Oil		*S. cerealella, C. maculatus*	Oviposition deterrent Fumigant	Paranagama et al., 2003 Gbolade & Adebayo, 1993
Cymbopogon martini	Palmarosa	Gramineae	Leaves	Oil	Geraniol	*C. chinensis*	Fumigant toxicity	
Cymbopogon nardus	Citronella grass	Poaceae	Leaves	Oil		*S. oryzae, C. chinensis T. castaneum C. maculatus*	Fumigant Oviposition deterrent	Raja et al., 2001 Ketoh et al., 2005 Boeke et al., 2004b
Cymbopogon schocnanthus	Camel grass	Poaceae	Leaves	Oil		*S. oryzae, C. chinensis T. castaneum*	Fumigant	Ketoh et al., 2005
Cymbopogon winterianus	Java citronella	Poaceae	Leaves	Oil		*S. oryzae*	Fumigant	Singh et al., 1989

TABLE 11.1 (Continued)

Scientific name	Vernacular name	Family	Plant part	Formulation	Major secondary metabolites	Stored grain pests	Type of activity	References
Delonix regia	Gulmohur	Leguminosae	Flower	Extract		*T. castaneum*	Oviposition deterrent	Saxena & Yadav, 1984
Derris elliptica	Derris	Fabaceae	Root	Powder		*C. chinensis S. oryzae*	Toxicant Protectant	Fukami et al., 1959; Kardinan et al., 1997
Dictamnus dascarpus	Densefruit	Rutaceae	Root	Extract		*T. castaneum, S. zeamais*	Protectant	Liu et al.,2007
Eclipta alba	Eclipta	Asteraceae	Root	Extract		*S. cerealella*	Protectant	Prakash et al, 1979
Eichhornia crassipes	Water hyacinth	Pontederia-ceae	Leaves	Extract		*C. chinensis S. oryzae, T. castaneum*	Protectant	Jamil et al., 1984
Elacis guineensis	Palm	Arecaceae	Seed	Oil		*C. chinensis S. oryzae, R. dominica*	Protectant	Jacob & Sheila, 1990; Ukonronkwo, 1991; Nualvatna et al., 2000
Elacis guineensis	African oil palm	Palmaceae	Palm	Oil	Trimyristin, triaurin	*C. chinensis*	Oviposition deterrent	
Embelia ribes	Baberang	Myrsinaceae	Beery	Extract	Embelin, quercitol	*S. Cerealella, S. oryzae, R. dominica, T. castaneum*	Protectant	Chander & Ahmed, 1987
Ertyrophleum suaveolans	Ordeal tree	Caesalpini-aceae	Leaves	Powder		*S. oryzae*	Protectant	Niber, 1994
Eruca vesicaric	Taramira	Crucifareae	Seed	Oil	Erueic acid	*C. chinensis*	Protectant Oviposition deterrent	Doharey et al., 1983, Singh et al., 1994
Erythrophleum suaveolens	Ordeal tree	Caesalpini-aceae	Bark	Extract Powder	Cassaidine cassaine	*S. zeamais*	Protactant	
Eucalyptus blakelyi	Blakely's red gum	Myrtaceae	Leaves	Oil		*S. oryzae*	Fumigant	Lee et al., 2004
Eucalyptus camaldulensis	Murray gum	Myrtaceae	Leaves	Oil		*S. oryzae, C. chinensis T. castaneum*	Fumigant	Tunc et al., 2000 Chairat et al., 2002
Eucalyptus camaldulensis	Murray red gum	Myrtaceae	Leaves	Oil		*T. castaneum, S. oryzae*	Fumigant	Sacac & Tunc, 1995

TABLE 11.1 *(Continued)*

Scientific name	Vernacular name	Family	Plant part	Formulation	Major secondary metabolites	Stored grain pests	Type of activity	References
Eucalyptus citriodora	Lemon scented gums	Myrtaceae	Leaves	Powder Oil	Citronellal	*C. chinensis*	Oviposition deterrent Fumigant	Subramanya et al., 1994
Eucalyptus globulus	Blue gums	Myrtaceae	Leaves	Oil Powder	Cineole-β-Phell	*C. chinensis S. oryzae, R. dominica, T. castaneum*	Fumigant Protectant	Pathak & Krishna, 1991 Patel & Patel, 2002; Rao & Prakash, 2002
Eucalyptus pauciflora	Alpine snow gums	Myrtaceae	Leaves	Oil		*S. oryzae, C. chinensis T. castaneum, R. dominica*	Fumigant	Shukla et al., 2002
Eucalyptus saligna	Sydney blue gums	Myrtaceae	Leaves	Oil Powder		*S. oryzae, R. dominica, T. castaneum C. chinensis*	Fumigant Repellant Oviposition deterrent	Tapondjou et al., 2003, 2005, 2007 Nukenine et al., 2007
Eugenia caryophyllata	Clove	Myrtaceae	Leaves Clove	Oil Powder		*C. maculatus*	Fumigant Repellant	Rajapakse et al., 2002
Eugenia uniflora	Surinam cherry	Myrtaceae	Leaves	Oil		*C. chinensis*	Fumigant	Gbolade & Adebayo, 1993
Eupatorium odoratum	Jack in the bush	Asteraceae	Leaves	Powder		*S. oryzae, R. dominica, T. castaneum*	Protectant	Deka, 2003
Evodia rutaecarpa	Spindle tree	Rutaceae	Leaves	Extract		*S. zeamais, T. castaneum*	Protectant	Liu et al., 2007
Ficus exasperate	Pipal	Moraceae	Leaves	Powder		*C. maculatus*	Protectant	Boeke et al., 2004a
Foeniculum vulgare	Sweet fennel	Umbelliferae	Seed	Extract		*S. oryzae, C. cephalonica*	Protectant	Kim et al., 2003
Fumaria indica	Fumitory	Papaveraceae	Leaves	Powder		*S. oryzae*	Protectant	Chander & Ahmed, 1983
Ganoderma lucidum	Reishi mushroom	Polyporaceae	Flower	Extract		*S. oryzae*	Protectant	Rao, 1955,1957
Gardenia fosbergii	Gardenia	Rubiaceae	Leaves bud	Extract		*S. oryzae*	Repellant	Kestenholz & Stevenson, 1998
Gaultheria procumbens	Boxberry	Ericaceae	Leaves	Oil	Methyl salicylate	*C. chinensis*	Fumigant toxicity	

TABLE 11.1 (Continued)

Scientific name	Vernacular name	Family	Plant part	Formulation	Major secondary metabolites	Stored grain pests	Type of activity	References
Glycine max	Soybean	Fabaceae	Seed	Oil	Lipoxy oxides	*C. chinensis, S. oryzae, R. dominica*	Oviposition deterrent Protectant	Singh & Singh, 1989; Das, 1986; Nualvatna et al., 2000; Kassis & Shemais, 2002
Gossypium sps	Cotton	Malvaceae	Seed	Oil	Furfurol, quercetin	*C. chinensis, S. oryzae, R. dominica*	Protectant	Sharma & Singh 1993; Rao &Prakash, 2002
Gynandropsis gynandra	Jakhiya	Capparaceae	Seed	Extract		*C. cephalonica*	Protectant	Ansari et al., 2003
Helianthus annus	Sun flower	Asteraceae	Seed	Oil	Linoleic acid	*C. chinensis, C. maculatus*	Protectant	Doharey et al., 1983
Helleborus foetidus	Stinkende Nieswurz	Ranunculaceae	Leaves	Extract		*T. castaneum*	Protectant	Pascual & Robledo, 1998
Hibiscus chinensis	Rose mallow	Malvaceae	Flower	Extract		*S. oryzae*	Protectant	Rao, 1955,1957
Hyptis spicigera	Black sesame	Labiatae	Leaves	Powder		*R. dominica*	Protectant	Ajayi et al., 1987
Hyptis suaveolens	Pignut	Labiatae	Leaves Seed	Powder Extract Powder	Menthol	*S. oryzae, R. dominica, S. cerealella*	Oviposition deterrent Protectant	Rao & Prakash, 2002; Iloba & Ekrakene, 2006; Prakash & Rao, 2006
Illium verum	Star anise	Illiciaceae	Leaves Seed	Powder Oil		*S. oryzae, R. dominica, T. castaneum*	Protectant Fumigant	Xu et al., 1993 Kim et al., 2003
Intsia bijuga	Ipil, Merbau	Leguminosae	Seed	Oil		*T. castaneum*	Repellent	Mohiuddin, 1987
Ipomoea palmata	Cairo morning glory	Convolvulaceae	Leaves	Extract		*C. chinensis*	Oviposition deterrent Protectant	El-Ghar et al., 1987
Itrus sinensis	Sweet Orange	Rutaceae	Peel	Powder Oil		*C. analis*	Protectant	Juneja & Patel, 2002; Bhargava et al., 2003
Jatropha curcas	Physic nut	Euphorbiaceae	Fruit	Oil	Toxalbumin, curcin	*C. maculates*	Protectant	
Juniperus virginiana	Cedar wood	Coniferae	Wood	Oil		*S. oryzae*	Repellent	Sighamony et al., 1986

TABLE 11.1 *(Continued)*

Scientific name	Vernacular name	Family	Plant part	Formulation	Major secondary metabolites	Stored grain pests	Type of activity	References
Justicia adhatoda	Malabar nut tree	Acanthaceae	Leaves	Extract		*T. castaneum*	Protectant	Srivastava & Awasthi, 1958
Justicia betonica	Squirrel's tail	Acanthaceae	Leaves	Powder		*C. cephalonica, S. oryzae*	Oviposition deterrent	Chander & Ahmed, 1983, 1986
Lantana camara	Lantana	Verbenaceae	Leaves	Oil Powder Extract	A-amyrine, β-sistosterol	*S. oryzae, R. dominica, T. castaneum*	Fumigant Prptectant Repellent	Kumar & Tiwari, 2015 Srinivasan &Nadara, 2004; Ba-Angood & Al-Suraidy, 2003
Laurus nobilis	Bay laurel	Lauraceae	Leaves	oil	1,8-cineole, borneol and thymol cineol, sabinene	*S. oryzae, T. castaneum, R. dominica*	Fumigants	Rozman et al., 2006
Lavendula angustifolia	Lavender	Lamiaceae	Leaves	oil	1,8-cineole, borneol and thymol Linalyl, linalool	*S. oryzae, T. castaneum, R. dominica*	Fumigants	Rozman et al., 2006
Lawsonia inermis	Henna	Lythraceae	Seed	Oil		*C. chinensis*	Protectant	Singh & Sharma, 2003; Singh & Yadav, 2003
Ledum palustre	Labrador tea	Ericaceae	Twig	Powder		*C. chinensis, S. oryzae*	Repellent	Ignatowicz & Wesolowhka, 1996
Lepidium sativa	Alfalfa	Brassicaceae	Leaves	Powder		*T. castaneum*	Protectant	Al-jabr, 2003
Limonia acidissima	Bergapten	Rutaceae	Bark	Extract	Bergapten	*C. chinensis*	Oviposition deterent	
Linum usitatissimum	Linseed	Linaceae	Seed cake	Powder Oil		*S. oryzae*	Repellent	Bowry et al., 1984
Lippia alba	Bushy matgrass	Verbinaceae	Leaves	Oil	Linalool 1,8-cineole	*S. oryzae, C. maculatus, T. castaneum, R. dominica*	Fumigans	Verma et al, 2001
Lippia germinata	Wild sage	Verbenaceae	Leaves	Powder		*S. zeamais, R. dominica*	Repellant	Prakash & Rao, 1984; Nukenine et al., 2007
Lupinus termis	Egyptian lupin	Papilionaceae	Leaves	Powder		*T. castaneum*	Repellant	Al-jabr, 2003

TABLE 11.1 *(Continued)*

Scientific name	Vernacular name	Family	Plant part	Formulation	Major secondary metabolites	Stored grain pests	Type of activity	References
Lycopersicon esculentum N.	Tomato	Solanaceae	Leaves	Extract	Coumaroylputrescine	*S. oryzae*	Protactant	Raja & Ignacimuthu, 2001
Madhuca longifolia	Indian butter tree	Sapotaceae	Seed	Oil		*C. maculatus*	Protectant	
Marjorana syriaca	Syrian marjoram	Lamiaceae	Leaves	Oil		*S. oryzae, T. casta-neum, R. dominica*	Fumigants	Shaaya et al., 1990
Melaleuca fulgens	Scarlet Honey Myrtle	Myrtaceae	Leaves	Oil		*S. oryzae*	Fumigants	Lee et al., 2004
Melia azadarach	China berry	Meliaceae	Leaves	Powder	Azadirechtinene	*S. zeamais*	Oviposition deterrent	Haryadi & Yuniarti, 2003
Mentha arvensis	Japenese mint	Lamiaceae	Leaves	Oil	Menthol, men-thone, menthyl acetate	*S. oryzae, C. maculatus T. castaneum, R. dominica*	Fumigants	Lee et al., 2001 Raja et al., 2001
Mentha citrata	Bergamot mint	Lamiaceae	Leaves	Oil	Menthol, menthone	*S. oryzae, C. maculatus R. dominica*	Fumigants	Singh et al., 1989 Tripathi et al., 2000
Mentha longifolia	Horse mint	Lamiaceae	Leaves	Oil	Menthol, men-thone, piperitone	*S. oryzae*	Fumigant	Malkani et al., 1990
Mentha piperita	Peppermint	Lamiaceae	Leaves	Oil	Menthol, men-thone, Menthyl acetate	*S. oryzae, C. maculatus R. dominica*	Fumigants	Raja et al., 2001; Lee et al., 2002
Mentha spicata	Spearmint; Pudina	Lamiaceae	Leaves	Oil	Carvone	*S. oryzae, C. maculatus R. dominica*	Fumigants	Tripathi et al., 2000; Khanuja et al., 2001; Rao & Prakash, 2002
Mikania cordata	Hempvine	Compositae	Leaves	Powder		*S. cerealella*	Oviposition deterrent	Prakash & Rao, 2006
Millettia ferruginea	Mimosa	Papilionaceae	Leaves	Powder		*C. chinensis*	Protectant	Damte & Chichaybelu, 2002
Momordica charantia	Bitter gourd	Cucurbitaceae	Leaves	Powder	Charantin, momordicin	*T. castaneum C. maculatus*	Protectant	Boeke et al., 2004a

TABLE 11.1　(Continued)

Scientific name	Vernacular name	Family	Plant part	Formulation	Major secondary metabolites	Stored grain pests	Type of activity	References
Monodora myristica	Jamaican nutmeg	Annonaceae	Seed	Powder Extract	Linoleic acid, oleic acid	C. maculates	Deterrent	
Murraya koenigii	Curry leaf	Rutaceae	Leaves	Oil		S. oryzae, C. chinensis T. castaneum, C. cephalonica, S. cerealella R. dominica	Fumigants	Pathak et al., 1997. Kumar et al., 2015
Myristica fragrans	Nutmeg	Myristicaceae	Seeds	Oil	A-pinene, sabinene	S. oryzae, T. castaneum, R. dominica	Fumigants	Huang et al., 1997
Narcissus tazetta	French daffodil	Amaryllidaceae	Leaves	Extract		T. castaneum	Protectant	Liu et al.,2007
Nardostachys jatamansi	Indian spikenard	Valerianaceae	Leaves	Oil		S. oryzae, R. dominica	Fumigants	Wang et al., 2000
Nerium oleander	Oleander	Apocynaceae	Whole plant	Powder	Cardiac glycosides, oleandrin	C. chinensis	Protactant	
Nicotiana tabacum	Tobacco	Solanaceae	Leaves	Powder	Nicotine, normicotine	C. chinensis	Protectant	Damte & Chichaybelu, 2002
Nigella sativa	Black cumin	Ranunculaceae	Seed	Powder	Melanthin	T. castaneum	Protectant	Al-jabr, 2003
Ocimum americanum	Basil	Lamiaceae	Leaves	Powder		R. dominica	Protectant	Juneja & Patel, 2002
Ocimum bacilicum	Sweet basil	Labiatae	Leaves	Oil	Linalool	S. oryzae, T. castaneum, R. dominica	Fumigants	Ketoh et al., 2005
Ocimum canum	Camphor basil	Lamiaceae	Leaves	Oil Extract	Linalool	S. oryzae	Fumigant toxicity	
Ocimum gratissimum	African basil	Lamiaceae	Leaves	Powder Oil	Camphin	S. zeamais	Fumigant Protectant	Ngamo et al., 2001 Iloba & Ekrakene, 2006
Ocimum kilimandscharicum	Jembere basil	Lamiaceae	Leaves	Oil		S. oryzae, R. dominica	Fumigant toxicity	

TABLE 11.1 (Continued)

Scientific name	Vernacular name	Family	Plant part	Formulation	Major secondary metabolites	Stored grain pests	Type of activity	References
Ocimum sanctum	Sacred basil	Lamiaceae	Leaves	Powder		*C. maculatus*	Oviposition deterrent	Rajapakse et al., 2002
Ocimum suave	Wild basil	Lamiaceae	Leaves	Oil	Eugenol	*S. zeamais*	Repellent	
Olea europaea	Olive tree	Oleaceae	Fruit	Oil	Cinchonidine, cinchonine	*S. oryzae*	Contact Toxicity	
Oreganum syriacum	Oregano	Lamiaceae	Leaves	Oil		*S. oryzae, T. castaneum, R. dominica*	Fumigants	Tunc et al., 2000
Origanum vulgare	Oregano	Lamiaceae	Leaves	Oil	Carvacrol, thymol	*S. oryzae, R. dominica*	Fumigant toxicity	
Oryza sativa	Rice	Graminae	Bran	Oil		*C. chinensis*	Protectant	Singh & Sharma, 2003; Nualvatna et al., 2000
Peganum harmala	Hermala	Zygaphyl-laceae	Leaves	Extract		*T. castaneum*	Oviposition deterrent	Jbilou et al., 2006
Pelargonium graveolens	Geranium	Geraniaceae	Leaves	Oil		*S. oryzae, T. castaneum*	Fumigants	Wang et al., 2000
Pimentha racemosa	Bayrum tree	Myrtaceae	Leaves	Oil		*T. castaneum*	Fumigants	Lee et al., 2002
Pimpinella anisum	Anise	Umbelliferae	Seed Leaves	Oil Extract	Pinene, phellandrene	*S. oryzae, T. castaneum, R. dominica*	Fumigants Protectant	Tunc & Erler, 2000; Al-jabr, 2003
Pinus longifolia	Chir	Pinaceae	Exudation	Oil		*S. oryzae*	Fumigants	Singh et al., 1989
Piper aduncum	Spikenard pepper	Piperaceae	Seed	Oil		*S. oryzae*	Fumigants	Estrela et al., 2006
Piper cubeda	Java long pepper	Piperaceae	Fruit	Oil, Powder Extract	Piperine	*S. oryzae, C. maculatus*	Fumigant toxicity Protectant	Su, 1990
Piper guineense	Ashanti Black pepper	Piperaceae	Fruit	Oil, Powder Extract	Piperine, sylvatine	*S. zeamais, C. maculatus*	Fumigant toxicity Protectant	Umoetok, 2004
Piper hispidinervum	Long pepper	Piperaceae	Seed	Oil	Piperine, sylvatine	*S. oryzae*	Fumigants	Estrela et al., 2006

TABLE 11.1 *(Continued)*

Scientific name	Vernacular name	Family	Plant part	Formulation	Major secondary metabolites	Stored grain pests	Type of activity	References
Piper nigrum	Black pepper	Piperaceae	Seed	Oil	Piperine	*S. oryzae, R. dominica, T. castaneum*	Fumigant Repellant	Kumar & Tiwari 2015, Kumar 2016 Tripathi et al., 1997
Pittosporum toriba	Japanese cheesewood	Pittosporaceae	Leaves	Extract		*S. oryzae, C. maculatus*	Protectant	Salama et al., 2004
Pogostemon patchouli	Patchouli	Labiatae	Leaves	Oil		*S. oryzae, T. casta-neum, R. dominica*	Fumigants	Wang et al., 2000
Pongamia glabra	Karanj	Fabaceae	Leaves	Oil	Karanjin (pongamol)	*S. oryzae, R. dominica T. castaneum*	Fumigant Oviposition deterrent Protectant	Kumarswami, 1985; Reddy et al, 1994; Negi et al, 1997; Singh & Sharma, 2003
Psidium guajava	Guava	Myrtaceae	Leaves	Powder	Cineol, tannins	*S. oryzae*	Protectant	Sharaby, 1989
Psoralea coylifolia	Forsk	Fabaceae	Seed	Extract	Oleoresin, psoraline	*C. chinensis*	Protectant	
Raphanus sativus	Radish	Brassicaceae	Seed	Oil	Spirobrassinin	*S. oryzae*	Fumigant toxicity	
Ricinus communis	Castor	Euphorbiacease	Seed	Cake Oil	Ricinine	*S. oryzae, R. dominica, T. castaneum, C. chinensis*	Oviposition deterrent Protectant	Verma et al., 1983, Bhargava & Meena, 2002; Singh & Sharma, 2003
Rosemarinus officinalis	Rosemary	Lamiaceae	Leaves	Oil	Camphor, borneol	*S. oryzae, C. maculatus T. castaneum, R. dominica*	Fumigants	Tunc et al., 2000, Isikber et al., 2006; Rozman et al., 2006
Salvia bracteata	Salvia	Lamiaceae	Leaves	Oil		*S. oryzae, C. maculatus T. castaneum, R. dominica*	Fumigants	Shakarami et al., 2005
Salvia officinalis	Salvia	Lamiaceae	Leaves	Oil	Linalyl acetae, terpene	*S. oryzae, C. maculatus T. castaneum, R. dominica*	Fumigants	Shaaya et al., 1990
Sapindus emarginatus	Soap nut	Sapindaceae	Seed	Extract		*S. cerealella*	Protectant	Srinivasan & Nadara, 2004

TABLE 11.1 (Continued)

Scientific name	Vernacular name	Family	Plant part	Formulation	Major secondary metabolites	Stored grain pests	Type of activity	References
Sassafras albidum	Sassafras	Lauraceae	Leaves	Oil		*S. zeamaize, T. castaneum*	Fumigants	Huang et al., 1999
Saussurea lappa	Costus	Asteraceae	Rhizome	Extract	Saussurine	*T. castaneum*	Repellent	
Sesamum indicum	Til	Pedaliaceae	Seed	Oil		*C. chinensis*	Oviposition deterrent	Srinivasan & Nadara, 2004
Solanum nigrum	Wonderberry	Solanaceae	Leaves	Extract	Solanine, saponin	*S. oryzae*	Contact toxicity	
Sophora flavescens	Henritte's	Fabaceae	Seed	Extract		*S. zeamais, T. castaneum*	Protectant	Liu et al.,2007
Sphaeranthus indicus	Mundi	Asteraceae	Leaves	Extract		*S. cerealella*	Protectant	Srinivasan & Nadara, 2004
Steganotaenia areliacea	Carrot tree	Apiaceae	Leaves	Powder		*S. zeamais*	Protectant	Nukenine et al., 2007
Stemona sessilifolia	Stemona	Stemonaceae	Leaves	Extract		*S. zeamais, T. castaneum*	Protectant	Liu et al.,2007
Tagetes erecta	Marigold	Asteraceae	Leaves	Oil		*S. oryzae, C. maculatus T. castaneum, R. dominica*	Fumigants	Krishina et al., 2005
Tagetes minuta	Maxican Marigold	Asteraceae	Leaves	Oil		*S. oryzae, C. maculatus T. castaneum, R. dominica*	Fumigants	Krishina et al., 2005
Tagetes patula	French Marigold	Asteraceae	Leaves	Oil		*S. oryzae, C. maculatus T. castaneum, R. dominica*	Fumigants	Krishina et al., 2005
Tagetus indica	Marigold	Asteraceae	Leaves	Powder Extract		*C. cephalonica*	Oviposition deterrent	Ansari et al., 2003
Tephrosia purpurea	Wild indigo	Papilionaceae	Leaves	Extract		*S. cerealella*	Protectant	Srinivasan & Nadara, 2004
Tetradenia riparia	Umuruvumba	Lamiaceae	Leaves	Powder	Linalool, diterpenol	*S. oryzae, R. dominica*	Oviposition deterrent	

TABLE 11.1 *(Continued)*

Scientific name	Vernacular name	Family	Plant part	Formulation	Major secondary metabolites	Stored grain pests	Type of activity	References
Thevetia nereifolia	Kaner, Nerium	Apocynaceae	Seed	Extract		*T. castaneum, C. chinensis*	Protectant	Malek et al., 1996; Ba-Angood & Al-Sunaidy, 2003
Thymus vulgaris L.	Garden thyme	Lamiaceae	Leaves	Oil	p-cymene, thymol	*R. dominica*	Fumigant toxicity	Rozman et al., 2006; Shaaya et al., 1990
Tithonia diversifolia	Tree marigold	Asteraceae	Leaves	Extract		*C. maculatus*	Protectant	Adedire & Akinneye, 2004
Torreya grandis	Chinese nut meg	Taxaceae	Leaves	Extract		*S. zeamais, T. castaneum*	Protectant	Liu et al., 2007
Trepterygium wilfordii	Thunder godwine	Calastraceae	Flower	Extract		*S. zeamais, T. castaneum*	Protectant	Liu et al., 2007
Trigonella froenum-graecum	Fenugreek	Fabaceae	Seed	Oil Powder	Trigonelline, sapogenin	*R. dominica, T. castaneum*	Repellant	Mohiuddin et al., 1993 Al-jabr, 2003
Valeriana wallichi	Valeriana	Valerianaceae	Leaves	Extract		*T. castaneum*	Repellant	Qureshi et al., 1988
Vitex negundo	Lagundi	Verbenaceaea	Leaves	Oil	Glucononitol	*S. cerealella, R. dominica, S. oryzae*	Fumigant Toxicity	
Withania somnifera	Winter cherry	Solanaceae	Root	Extract	Withanine	*C. chinensis*	Protectant	
Zanthoxylum alatum	Tejbal	Rutaceae	Pericarp Leaves	Extract Oil	A-phellandrene, linalool	*T. castaneum S. oryzae*	Repellent Fumigants	Singh et al., 1989
Zea mays	Maize	Gramineae	Corn	Oil		*C. chinensis*	Oviposition deterent	
Zingiber officinale	Ginger	Zingiberaceae	Rhizome Leaves	Powder Oil	Zingiberene, zingeiberol	*C. maculatus, S. oryzae T. castaneum*	Protactant Fumigant Repellant	Sunarti, 2003

11.2 UTILIZATION OF PLANT PRODUCTS

The application of plant materials to protect stored products against insect attack has a long history. Several plant species concerned have also been used in traditional medicine by local communities and have been collected from the field or specifically cultivated for these purposes. Leaves, roots, twigs, ash, essential oils, and flowers have been admixed, as grain or seed protectants, fumigant, repellent, oviposition deterrent with various stored products in different parts of the world, particularly in different developing countries. Many commercially available spices and herbs, turmeric, basil, marjoram, anise, cumin, and coriander, are for management of stored grain insects but they are also able to fungal growth under storage condition (Hito-koto et al., 1980). According to a survey in northern, semi-arid regions of Ghana, only 16 plants were identified as being used as grain protectants (Brice et al., 1996). Apart from neem, none of the plants featured in the list of stored-product protectants were used in the Ashanti Region. Two of the plants, *Chamaecrista nigricans* and *C. kirkii* (both known locally as "lodel"), said to be the most effective, have not been recorded elsewhere, worldwide, as having this property and, consequently, do not feature in any research programs devoted to plant protectants of stored products; these particular plants, like many of the others used in Northern Ghana, are weeds and serve no other useful purpose. A survey report of plants used as domestic insecticides in the Ashanti Region in Central Ghana indicates that 26 different plant species were found to be used as grain storage protectants, the most common being *Chromoleana odorata*, *Azadirachta indica*, and *Capsicum annum*. (FAO, 1996). It is clear that there is a lack of information concerning the actual use of plants by farmers at small scale grain protection. If local production of plant protectants is to be encouraged then it will generate more information related to unidentified plants also. Several methods have been applied to determine the effectiveness of plant and plant products for the management of stored grain insects, but almost all trials were laboratory based and of short duration. Although laboratory experiments may provide useful information regarding effectiveness of plant products, there are various parameters for evaluating plant products as bio pesticides reported from all over the world. The plant products can be mixed with stored commodities directly and their contact toxicity can be assessed through progeny production and indirectly measuring the percent infestation and percent weight loss. Laudani and Swank (1954) developed the method for repellency test of pyrethrum in stored maize. The plant products may be used as extracts depending on the solubility of the active components. The most common solvents are chloroform, petroleum ether, hexane, methanol, ethanol, and acetone. Essential and vegetable oils may be extracted from appropriate plant species in the laboratory or alternative commercial preparations may be examined. Vegetable oils are applied at 5–10 ml/kg commodity although higher values, in excess of 50 ml/kg, have been recorded (Don-Pedro, 1989). Extract of plant products applied as antifeedant for management of stored grain insects.

11.2.1 ESSENTIAL OILS AS STORAGE INSECTICIDES

The essential oils are complex mixture of volatile organic compounds produced as the secondary metabolites whose functions are other than nutrition. The essential oils of several botanical origins are known to have repellent and insecticidal activities against stored grain insects (Tripathi et al., 2000; Verma et al., 2000). Essential oils are found in glandular hairs or secretary cavities of plant cell wall and are present as droplets of fluid in the leaves, stems, bark, flowers, root, and fruits of different plants. Generally the essential oils are liquid at normal room temperature but it easily transforms into gaseous form after some time without any degradation.

Plant oils have been reported to possess repellent and insecticidal (Shaaya et al., 1997; Kostyukovsky et al., 2002; Papachristos & Stamopoulos, 2002), nematicidal (Oka et al., 2000), antifungal (Paster et al., 1995), antibacterial (Matasyoh et al., 2007), virucidal (Schuhmacher et al., 2003), and antifeedant, reproduction inhibitory activity (Raja et al., 2001; Papachristos & Stamopoulos., 2002). Exploration of fumigantional potential of essential oils may lead to the development of non-synthetic, ecologically safe, and easily degradable seed as well as grain protactant. Plants contain many compounds that are responsible for the ovicidal, repellent, antifeedant, chemosterilant, and toxic effects in insects (Nawrot & Harmatha, 1994; Isman, 2006). There are many reviews dealing with the use of plant products against insect pests of stored products (Lale, 1995; Golob & Gudrups, 1999; Adler et al., 2000; Weaver & Subramanyam, 2000; Isman, 2006), specifically on essential oils (Saxena & Koul, 1978; Singh & Upadhyay, 1993; Regnault-Roger, 1997) and on monoterpenoids (Coats et al., 1991). Steam distillation of aromatic plant yields essential oil, long used as fragrances and flavorings in the perfume and food industries, respectively, and more recently for aromatherapy and as herbal medicines (Buckle, 2003; Coppen, 1995; Isman, 2006). The essential oil of a plant may contain hundreds of different constituents but certain components are present in larger quantities.

For example, 1,8-cineole is predominant in the essential oil of *Eucalyptus* spp., linalool in *Ocimum* spp., eugenol in clove oil (*Syzygium aromaticum*), thymol in garden thyme (*Thymus vulgaris*), and menthol in various species of mint (*Mentha* spp.), limonene in *Citrus* spp., myrcene in *Curcuma longa*, carvone in *Caruma carvi*, asarone in *Acorus calamus*, and glucosinolates in plants belonging to Brassicaceae, cyanohydrins in *Manihot esculenta*, thiosulphinates in *Allium* spp., methyl salicylate in *Securidaca longepeduncu-lata*, and carvacrol as well as β-thujoplicine in *T. dolabrata* (Isman, 1999; Rajendran & Sriranjini, 2008). A number of the source plants have been traditionally used for protection of stored commodities, especially in the Mediterranean region and in southern Asia, but interest in the essential oils was renewed with emerging demonstration of their fumigant and contact toxicity activities to a wide range of pests in the last decade (Isman, 2000). Gu-Yanfang et al. (1997) reported that *Allium Sativum* (Garlic) oil showed 88% repellency after 72 h against *T. castaneum* at 0.2 ml/culture dish. In another study conducted by Ho et al.(1996), complete mortality of eggs of *T. castaneum* was achieved using the filter paper impregnation bioassay at 4.4 mg/cm². Adults of *T. castaneum* were more susceptible than *S. zeamais* with LD_{50} value of 1.32 mg/cm² and 7.65 mg/cm², respectively. Eggs laid on wheat and rice treated with the garlic oil failed to produce F1 progeny at >2000 ppm in rice for *T. castaneum* and 5000 ppm in wheat for *S. zeamais*. Huang et al. (2000) evaluated garlic oil against *S. zeamais* and *T. castaneum* and reported that of the two major constituents, diallyl trisulfide was more potent contact toxicant, fumigant, and feeding deterrent than methyl allyl disulfide. Diallyl trisulfide completely suppressed egg hatching at 0.32 mg/cm², larval and adult emergence at 0.08 mg/cm². The ovicidal activity of vapor of essential oil of garlic was evaluated against eggs of *T. confusum*. The essential oil of garlic showed strong ovicidal activity varying from 42.2 to 100% (Karci & Isikber, 2007).

Essential oil of Sweet Annie, *Artemisia annua* (Compositae), evaluated by Tripathi et al. (2000) against *T. castaneum* and *C. maculatus* at 1% v/v, was proved adult repellent and the study revealed negative correlation between larval survival, pupal survival, and adult emergence of *T. casta-neum*. Tripathi et al. (2001) observed the effect of oil against *T. castaneum*. The major compound of the Artemisia oil is 1,8 cineole. Adults were more susceptible than larvas. The LC_{50} value was 108.4 μg/mg body weight (adult) in contact assay and 1.52 mg/l alr for (adult) in fumigant assay. In filter paper assay this oil at 3.22–16.10 mg/cm² reduced egg hatchability significantly. This oil also showed feeding deterrence of 81.9% at 121.9 mg/g food by adults and 68.8% by larva (Goel et al., 2007). The oil of *Artemisia aucheri* was tested against *C. maculatus, T. castaneum, S. oryzae,* and *S. granaries* for its fumigant toxicity and repellency at various concentrations (0.037, 0.185, 0.370, 0.556, 0.741, and 0.926 micro ml/cm^3). The result indicated that mortality of adult insects increased as the concentration of oil increased resulting in 84.41, 85.41, 84.70, and 83.04% against *C. maculatus, T. castaneum, S. oryzae,* and *S. granaries,* respectively. The adults of *S. granarus* and *C. maculatus* were more susceptible to the tested oil (Shakarami et al., 2004).

In laboratory tests, essential oil of *Callistemon lancelets* was found effective against *C. maculatus.* The LC_{50} was lower and it decreased with the increase in exposure duration (Mishra et al., 1989). In another study conducted by Ansari and Mishra (1990) LC_{50} decreased with the time, being 0.16 and 0.18 ml after 24 and 96 h, respectively.

Lee et al. (2004) evaluated fumigant toxicity of essential oil of *C. sieberi* (Myrtaceae) against *S. oryzae, T. castaneum,* and *R. dominica.* The oil exhibited potent fumigant toxicity at the LD_{50} and LD_{95} of the essential oil against *S. oryzae* adults which was between 19.0 and 30.6 and between 43.6 and 56.0 µl/l air, respectively. The oil was also approximately twice as toxic to *T. castaneum* and *R. dominica* at the $LD_{95.}$

Toxicity, repellency, and oviposition inhibitory activity of essential oil isolated from commonly used spice *Carum copticum* (Ajwain) was evaluated against *C. chinensis* (Upadhyay et al., 2007). It Inhibited oviposition and resulted in suppression of reproductive potential and showed good repellent activity and inhibited feeding activity at low concentration. Data confirmed toxicity with LC50 value of 1.05 µl. It also inhibited egg hatching at all concentrations (0.00, 0.10, 0.20, 0.50, 1.00, and 2.00%) with minimum LC50 value of 0.156 ± 0.0321 (Bhargava & Meena, 2001). Complete mortality at concentration higher than 185.2 µl/l was obtained with *C. copticum* seed oil and it was observed that *S. oryzae* was more susceptible than *T. castaneum* with LC50 value of 0.91 µl/l and 33.14 µl/l, respectively (Sahaf et al., 2007). In another study, Sahaf and Moharramipour (2009) determined high feeding deterrent activity of the *C. copticum* seed oil against *T. castaneum* at 100–1500 ppm.

An essential oil of *Chenopodium ambrosioides* (Chenopodium) was evaluated for its toxicity and repellency to adults of *C. maculatus, S. oryzae, L. serricorne,* and *T. confusum* under laboratory conditions (Su, 1991). The result indicated that the oil was effective in reducing insect (*S. oryzae* and *C. maculatus*) infestation when applied to wheat or cowpea seed at dosages of 2000 and 1000 ppm, respectively. It was also highly toxic to *C. maculatus* and caused 100% mortality at 40 µg/insect and moderately toxic to *S. oryzae* and caused 52.5% mortality and was slightly toxic to *T. confusum* at 50 µg/insect. The oil also showed strong insect repellent activity against *S. oryzae*, *T. castaneum* at 300 ppm (Mishra et al., 2002). Leaf oil of *C. ambrosioides* was further investigated by Tapondjou et al. (2002) as post-harvest grain protectant against six stored-product beetles. The result indicated 80–100% mortality of all the test insects (*C. chinensis, A. obtectus, S. granarius*, and *Prostephanus truncatus*) within 24 h of treatment except *C. maculatus* and *S. zeamais*, where this dosage resulted in 20% and 5% mortality, respectively. It was further reported to have insecticidal and repellent properties against *C. maculatus* and recorded more toxic to insects with DL_{50} value of 0.17 micro l/g grains in treated cowpea seeds (Tapondjou et al., 2003).

Efficacy of leaf oil of Dalchini, *Cinnamomum zeylanicum* (Lauraceae) against the eggs of *C. cephalonica* was studied by Bhargava and Meena (2001) at 0.00, 0.10, 0.20, 0.50, 1.00, and 2.00% concentrations. The oil resulted in inhibition of egg hatching which increased with the increase in the concentration of the oil. The minimum LC_{50} value for cinnamon leaf oil recorded was 0.3773 ± 0.0368. Paranagama et al. (2003) tested this oil on stored rice in wooden boxes lined with aluminium foil against *S. cerealella* and found that the population of test insect was significantly lower in oil-treated rice and the milling quality of rice was not affected, but it enhanced flavor and stickiness of cooked rice. The oil also showed a significant inhibition of oviposition and F1 progeny emergence compared to the control during no-choice test when tested against *C. maculatus* on cowpea seed. Percent of eggs laid was also reported to be zero in samples treated with 160 mg of *C. zeylanicum*. Further oil obtained from its bark exhibited repellency to *C. chinensis* (Tiwari et al., 1997).

Don-Pedro (1996) conducted laboratory experiments to investigate the fumigant toxicity of citrus peel essential oil against *C. maculatus* or *S. zeamais* (>10 ml/kg) and *C. maculatus* (>20 ml/kg). Reduction in oviposition and larval emergence through parental adult mortality was recorded, but no residual activity on the eggs and larvas produced by survivors was noticed. Lime oil extracted from citrus species (Rutaceae) exhibited potent fumigant toxicity against *T. castaneum* with LD_{50} value of 17.9 µl/l air (Lee et al., 2002). All the citrus oils were almost completely effective in preventing the emergence of adults of the F1 progeny when applied at 0.75% against *C. maculatus* in cowpea and *S.* spp. in wheat grain (El-Sayed et al., 1989). A study was undertaken by (Bhargava et al., 2003) to explore the possibility of some citrus oils, that is, sweet lime (*C. aurantiifolia*), mandarin (*C. reticulata*) and orange (*C. sinensis*) as grain protectant against *C. cephalonica*. For this purpose twenty freshly laid eggs (0–24 h old) were treated with different concentrations (0.0, 0.5, 1.0, 2.0, 3.0, and 5.0%) of the citrus oils through dipping method. The result showed that the oils were highly superior in reducing the adult emergence over the control and reduction in adult emergence increased with increase in concentration of citrus oil. The reduction in adult emergence at 1.0 ml/100 g seed was 46.91, 50.79, and 53.76% in mandarin orange and sweet lime oil, respectively, as compared to the control. Orange (*C. sinensis*) oil was found to be most potent repellent (89%) against *T. castaneum* (0.2 ml/culture dish) (Gu-Yanfang et al., 1997), These oils were also effective against *T. castaneum* with LC_{50} value of 19.47 µl/l (Mohamed & Abdelgaleil, 2008), highly toxic to *C. maculatus* (Su, 1972) and to have better fumigation efficacy to *S. zeamais* at dose of 40 µl/l (Zang et al., 2005). Essential oil of *Citrus limon* (Lemon) exhibited potent fumigant toxicity to *T. castaneum* with LD_{50} value of 16.2 µl/l air (Lee et al., 2002).

Tunc et al. (2000) tested fumigant toxicity of essential oil from cumin (*Cuminum cyminum*) against eggs of two stored-product insects, *T. confusum* and *Ephestia kuehniella*. It caused 100% mortality of the eggs and also inhibited egg hatching at all concentration tested (0.00, 0.10, 0.20, 0.50, 1.00, and 2.00%) with minimum LC_{50} value of 0.2015 ± 0.0393 (Bhargava & Meena, 2001). Cumin oil also showed repellent, oviposition deterrent and development inhibitory effect against *S. oryzae* and *T. castaneum* (Chaubey, 2007; Upadhyay et al., 2007; Chaubey, 2008). This oil was reported most active against *A. obtectus* with LC_{50} value of 28.01 µl/l (Karakoc et al., 2006). Cumin oil was reported ineffective in fumigant toxicity test against *T. castaneum* (Mohiuddin et al., 1987).

Essential oil isolated from commonly used spices, viz., clove (*Eugenia aromaticum*) evaluated for insecticidal, oviposition inhibitory, and repellent activity against pulse beetle, *C. chinensis* showed good repellent activity at 0.4–4.0 µl/l concentration (Upadhyay et al., 2007).

Huang et al. (1997) investigated antifeedant effects of essential oil extracted from nutmeg seed (*Myristica fragrans*) against *T. castaneum*. Nutritional studies showed that nutmeg oil significantly ($p < 0.05$) affected the growth rate and food consumption of the insect species. At 20ml nutmeg oil/100ml, the feeding deterrence index of *T. castaneum* was about 7%. (Karci & Isikber, 2007) tested the ovicidal activity of vapour of essential oil from nutmeg (*Myristica fragrans*) against eggs of *T. confusum*. Eggs of *T. confusum* were exposed to a dose of 100 µ l/l air for a periods of 24, 48, and 72 h. Vapour of nutmeg essential oil exhibited low ovicidal toxicity against eggs of *T. confusum* at all exposure times by <20% of corrected mortality.

Chemical composition and insecticidal properties of the *Jatropha curcas* seed oil was investigated by (Adebowale & Adedire, 2006). Anti-oviposition and ovicidal effects were observed against *C. maculatus* in cowpea at 0–2% (v/w) concentration at five days interval. Susceptibility of *C. maculatus* and *D. basalis* to Jatropha seed oil was evaluated under laboratory condition (Boateng & Kusi, 2008). Eggs and adults both were susceptible. The oil was ineffective against pupae and had less persistency. (Jadhav & Jadhav, 1984) concluded from their studies that *J. curcas* oil at 0.1, 0.2, and 0.3% (v/w) inhibited emergence of adult beetles of *C. maculatus*.

Oil obtained from Spanich flag, *Lantana camara* (Verbinaceae) was found to be unsuitable for fumigation purpose on cowpea against *C. maculatus* at 5–50 µl/ 9.9 g (Globade & Adebayo, 1993). In another study its leaf oil at 0.5–30 µl doses was found effective to cause mortality in adult of *C. maculatus* (Adebayo & Globade, 1994). Good insecticidal activity was displayed by oil against *S. oryzae* with LC_{50} value of 29.47 µl/l in the fumigant assay test (Mohamed & Abdelgaleil, 2008).

Bio- efficacy of leaf oil of Curry leaf plant *Murraya koenigii* (Rutaceae) was evaluated against *C. chinensis* (Paranagama et al., 2002). Oil attracted insects at 25 mg dose and repelled at 300 mg dose in the dual choice repellency test. In both contact (0.125 mg/cm^2) and fumigant (22.5 mg/ml) toxicity test 100% mortality was observed with LC_{50} value of 0.08 mg/cm^2 and 22.5 mg/ml, respectivily. (Pathak et al., 1997) investigated toxicity and repellent activity of *Murraya koenigii* against *C. chinensis* in stored green gram, chickpea at 340 ppm. Oil showed toxic effect and ovicidal properties. They concluded that the oil can be used for small level protection of legumes. In another experiment, Paranagama et al. (2002) tested the same oil at 7.5 ml/l for their toxicity and repellency against *C. maculatus* in stored cowpea. The result indicated 100% bruchid mortality, reduced oviposition, and F1 progeny emergence during fumigant toxicity bioassay while lowest LC_{50} value of 0.240 ml/l was revealed during contact toxicity bioassay. Only 7.0% bruchids settled in at the dosage of 160 mg in choice chamber bioassay for curry leaf oil.

Mohiuddin et al. (1993) evaluated twelve vegetable oils for their toxicity against *T. castaneum* and *R. dominica* for a period of eight weeks. *Piper nigrum* oil exhibited the greatest toxicity at 0.25% w/w by surface treatment of wheat grains. Chaubey (2007) tested essential oil of *P. nigrum* against *T. castaneum* for its repellent, toxic, and developmental inhibitory activities. The essential oil was found to repel the adults of *T. castaneum* at 0.2% concentration (v/v) with LC_{50} value of 14.02 for larval stage and 15.26 µl/l for adults. The oil also caused inhibition of development and reduction in oviposition potential of the test insect. Insecticidal, ovipositional, egg hatching and developmental inhibitory activities of the oil against *C. chinensis* was also determined (Chaubey, 2008). In another study oil obtained from *P. aduncum* (Spikenard pepper) and *P. hispidinervum* (Long pepper) belonging to family Piperaceae possessed insecticidal properties against *S. zeamais* (Estrela et al., 2006).

Bachrouch et al. (2010) investigated the chemical constituents and fumigant toxicity of *Pistacia lentiscus* (Anacardiaceae) essential oil against carob moth *Ectomyelois ceratoniae* and Mediterranean flour moth *Ephestia kuehniella*. Result showed that *P. lentiscus* essential oil contains terpinene-4-ol (23.32%), alpha terpineol (7.12%), and beta caryophyllene (22.62%) as major compounds. Fumigant toxicity test showed that *P. lentiscus* oil was more toxic to *E. kuehniella* (LC_{50} = 1.84 µl/l, LC_{95} = 5.14 µl/l) than *E. ceratoniae* (LC_{50} = 3.29 µl/l, LC_{95} = 14.24 µl/l). The fecundity and hatching rate of both insects decreased with increase in concentration of exposure time to the oil. At 136 µl/l air, fecundity and hatching rate were respectively 35 eggs/female and 42.86% for *E ceratoniae* and 78 eggs/female & 29.49% for *E. kuehniella*.

The fumigant toxicity of eucalyptus oil and 1,8-cineol, the major constituents of eucalyptus oil was reported against a Chlorpyrifos-methyl resistant strain and pure strain of the saw-toothed grain beetle, *Oryzaephilus surinanensis*. The resistant strain showed 1.9 and 2.2 fold higher tolerance against essential oil and 1,8-cineol in fumigant toxicity, respectively, relative to pure strain (Lee et al., 2000).

Cosimi et al. (2009) reported that essential oils extracted from Bay Laurel (*Laurus nobilis*), Bergamot (*Citrus bergamia*), Fennel (*Foeniculum vulgare*) and Lavandin (*Lavendula hybrida*) were tested for repellency against *Sitophilus zeamais*, *Cryptolestes ferrugineus* adults and *Tenebrio moliator* larvas. The essential oil of *L. nobilis* contain limonene followed by linalyl acetate and gama terpinene. Linalool and linalyl acetate were the main component of fennel. In filter paper bioassays, essential oil of fennel showed different activity on *S. zeamais* after 3 h of exposure. Bergamot oil and lavandin oil showed highest repellency against *S. zeamais*, *C. ferrugineus* and *T. malitor* after 24 h of exposure. Triterpenes from *Junellia aspera* (Verbenaceae) and chemical derivatives were evaluated for their antifeedant and toxic effects against adults of *S. oryzae*. The compounds maslinic acid, daucasterol and 3 beta- hydroxyl-12 alpha bromine oxide showed highest toxic and antifeedatnt activities against test insect (Pungitore et al., 2005). Cymol, one of the major compounds of *Eucalyptus saligna* and *Cupressus sempervirens* leaves oil evaluated against *Sitophilus zeamais* and *Tribolium confusum*. Filter paper bioassay showed that *Eucalyptus* oil was more toxic than *Cupressus* oil to both test insect (LD_{50} = 0.36 µl/cm² for *S. zeamais* and 0.48 µl/cm² for *T. confusum*). Both oil considerably reduced the F1 progeny production and grain weight loss. Moreover, both *curd* oil extracts produced a stronger repellent activity against the test insects (Tapondjou et al., 2005).

Liu et al. (2007) evaluated 40 species of Chinese medicinal herbs from 32 different botanical families for their contact, fumigant and feeding deterent activities against two stored grain coleopterans *S. zeamais* and *T. castaneum*. Thirty Chinese medicinal herbs exhibited insecticidal or feeding deterent activities against test insects. The oil of *Artemisia argyi, Dictamnus dasycarpus, Evodia rutaecarpa, Lietsea cubeba, Narcissus tazetta* var *chinensis, Polygonum aviculare, Rhododendrum molle, Sophora flavescens, Stemona sessilifolia, Tripteryginni wilfordii,* and *Torreya grandis* were most effective against both test insects.

The uses of some cooking oils viz., Noug oil, Soybean oil, Sunflower oil, Corn oil, and Olive oil and the reputedly non toxic botanical "Triplex" were evaluated against *S. zeamais* in stored maize grain in local Ethiopian storage condition. All the cooking oils were toxic to *S. zeamais*. The oil treatments showed higher mortality of adult weevils at each sampling period compared with untreated grains. These oils also reduced the weight loss and grain damage as compared to untreated grains. The "Triplex" caused higher mortality in adult weevils, low percent grain damage, and low percent weight loss (Demissie et al., 2008).

Rajapakse et al. (1997) reported that four vegetable oils corn groundnut, sunflower and sesame significantly reduced the oviposition of three bruchid species *C. maculatus, C. chinensis* and *C. rhodesianus* at 10 ml/kg and also reduced the longevity of adults of *C. maculatus* and *C. chinensis* at this dose.

Papachristos et al. (2002) evaluated thirteen essential oils *Apium graveolens, Citrus sinnensis, Eucalyptus globules, Junipers oxycedrus, Laurus nobilis, Lavandula hybrida, Mentha microphylla, Mentha viridis, Ocimum basilicum, Origanum vulgare, Pistacia terebinthus, Rosmarinus officinalis,* and *Thuja orientalis* in their vapor form against *Acathoscelides obtectus*. The most of oils showed repellent action, reduced fecundity, decreased egg hatchability, increased neonate larval mortality and adversely influenced offspring emergence. Furthermore, some oils were strongly toxic to *A. obtectus*. Males appeared more susceptible than females.

The adults of *S. zeamais* and *T. castaneum* were equally susceptible to the contact toxicity of the Cardamom (*Ellataria cardamomum*) oil at LD_{50} level with LD_{50} values of 56 and 52 µg/mg insect respectively. However *S. zeamais* was more susceptible than *T. castaneum* at the LD_{95} level. For fumigant toxicity *S. zeamais* adults were more than twice as susceptible as *T. castaneum* adults at both LD $_{50}$ and LD$_{95}$ levels. Furthermore, 12 days larvas of *T. castaneum* were more tolerant than the adults to the contact toxicity of the oil but 14 and 16 days larvas had the same susceptibility as the adults, the susceptibility of the larvas to contact toxicity increased with age. In contrast, all the larvas (12–16 days old) of *T. castaneum* were much more tolerant than the adults to the fumigant action, and larvas of different ages had similar susceptibility. Cardamom oil also decreased the egg hatchability of *T. castaneum* at 1.04–2.34 mg/cm concentration and reduced survival rate but could not show any growth inhibitory or feeding deterrence activity against either adults or larvas of *T. castaneum* (Huang et al., 2000).

The fumigant toxicity of essential oil of *Cinnamomum vulgare, Cochlearia aroracia* and *Brassica juncea* was investigated against *L. serricorne*. Results showed that all oils caused 90% mortality at 3.5 mg/cm^2. However cinnamon, horseradish and mustard oil were highly toxic at 0.7 mg/cm^2 against the test insects (Kim et al., 2003).

Papachristos and Stamopoulos (2004) reported the fumigant toxicity of the essential oil of *Lavandula hybrida, Rosmarinus officinalis* and *Eucalyptus globlus* against eggs of *A. obtectus*. The essential oil vapors were toxic to eggs with LC$_{50}$ values ranging between 1.3 and 35.1 µl/l air depending upon egg age and the essential oil. In all cases the young eggs (≤3 day old) were more tolerant to essential oil vapors than older ones (≥ 4 day old). Apart from the inhibition and hatching, the exposure of eggs to essential oil vapors significantly increased the subsequent mortality of hatched larvas.

The combined effect of Niger seed oil and Malathion 5% dust was evaluated against *S. zeamais* to determine the minimum effective rates of the combinations that can provide adequate protection to maize seed against *S. zeamais*. Niger seed oil at 0, 10, 20, 30, 40, 50, and 100% of the recommended application rate 5 ml/kg was combined with malathion at the respective rates of 100, 50, 40, 30, 20, and 0% of the recommended rate 0.5 g/kg. All combination provides complete protection to maize seed from the *S. zeamais* up to 90 days of treatments. To determine the residual effects of the treatments, weevil was reintroduced to the grain that had been treated 90 days previously. In addition to 100% malthion, 10% niger seed oil + 50% malathion and 20% niger seed oil + 40% malathion were fully effective in controlling *S. zeamais* for further 156 days after re- infestation (Yuya et al., 2009).

The essential oil of *L. cubeba* strongly repelled *S. zeamais* and *T. castaneum* even at low concentrations, but its repellency was more marked toward *T. castaneum*. This oil was also highly effective against both test insects in fumigant toxicity and contact toxicity study (Waraporn et al., 2009).

Ilboudo et al. (2010) reported that essential oil of *Ocimum americanum* was very toxic against *C. maculates* adults (LC_{61} = 2.23 µl/l) while oil from *Hyptis suaveolens, Hyptis spicigera* and *Lippia multiflora* exhibited higher LC_{61} values (1.30 µl/l: 5.53 µl/l and 6.44 µl/l, respectively). The persistency of the biological activities of four oil was variable and that from *O. americanum* was most persistent after 14 days of post application.

Yang et al. (2010) evaluated the efficacy of garlic oil applied alone or with diatomaceous earth (DE) against adult of *S. oryzae* and *T. castaneum*. Combination treatments were significantly more effective than either treatment alone. In addition, the results also showed that the simultaneous application of essential oil plus DE significantly reduced the concentration of essential oil alone required for an effective treatment and the application rate of DE can be reduced when combined with essential oil. Moreover, the activity of the combination treatment lasted longer than that of essential oil alone and the survival of eggs or larvas to adult stage was significantly inhibited in the combined treatments against both test insects as compared to essential oil alone.

Raghvani et al. (2003) evaluated eight vegetable oils at 2.5, 5.0, and 10.0 ml/kg against *C. maculatus*. The neem and coconut oil at 10 ml/kg seed provided complete control of *C. maculatus* for six months, followed by karanj, mustard, and castor oil at 10 ml/kg of seed. Sesamum and groundnut oil at 10 ml/kg of seed and karanj oil at 5 ml/kg of seed gave more than 94% protection up to four months of storage. The germination of pigeon pea seeds was not impaired due to different oil treatments.

The biological activities of essential oil of *Ocimum basilicum* and *Salvia officinalis* against *S. oryzae* indicated that the oil of both plant caused significantly mortality, repellency and anti reproduction effects at 2% (Popovic et al., 2006).

Contact and fumigant toxicity of three essential oils Cardamom (*Elleteria cardamomum*), Cinnamon (*Cinnamomum aromaticum*) and Clove (*Syzygium aromaticum*) against *T. castaneum* larvas and adults revealed that three days old adults and 10 days old larvas were equally susceptible to the contact toxicity of cinnamon oil with LD_{50} values of 0.074 and 0.196 mg/cm^2, respectively. Cardamom oil provided more toxicity to 14–18 days old larvas having LD_{50} value of 0.10 mg/cm^2. In fumigation bioassay cinnamon oil provided the highest toxic to adult and 10, 14, and 18 days old larvas with LD_{50} values of 0.03, 0.05, 0.008, and 0.09 mg/cm^2, respectively. In contact and fumigation toxicity adults and all stage of larvas were more resistant to clove oil (Mondal et al., 2006).

The repellent, toxic and inhibitory activities of three common species *Trachyspermum ammi, Anethum gravolens* and *Nigella sativa* spp was investigated against *Tribolium castaneum*. All essential oils repelled the adults of *T. castaneum* at low concentration. The death of larvas and adults of *T. castaneum* was caused by fumigation with these oils. The LD_{50} of *T. ammi, A. graveolens* and *N. sativa* essential oil against larval and adult stage of the insect were 11.62, 14.78, and 9.46 µl/l and 13.48, 16.66, and 10. 87 µl/l, respectively (Caubey, 2007).

Behal (1998) investigated that *Acorus calamus* (Sweet flag) oil effectively repelled the larvas of *C. cephalonica* at 0.1, 0.3, and 0.5% concentration. Schmidt et al. (1991) in another experiment tested the efficacy of rhizome oil of *A. calamus* against *S. granarius, S. oryzae* and *C. chinensis* and reported progeny reduction by the vapour of oil with not any significantly difference noticed in treatment and control in case of *T. castaneum*.

Essential oil from the leaves of Purple giant hyssop *(Agastache scrophulariaefolia)* showed fumigant toxicity at 0.8% concentration against *C. maculatus*. The active compounds in this oil were limonene, menthone, isomenthone, and pulegone (Lognay et al., 2002).

Oil of Large cardamom *Amomum subulatum* (Zingiberaceae) protected the wheat grain for 30 days at 1000 ppm against *S. oryzae* (Singh et al., 1989).

Devaraj and Srilatha (1993) reported greatest antifeedent activity of essential oil of Neem, *Azadirachta indica* (Meliaceae) against *C. cephalonica*. About 80–100% repellent activity in oil of *A. indica* against *T. castaneum* and *R. dominica* was also observed by Mohiuddin et al. (1993). The oil at 1.0 ml/100 g concentration resulted in longest total life cycle (57.8 days), highest reduction in adult emergence (85.7%) and egg viability (65.3%), lowest number of egg/ female, and shortest longevity for males (3.3 days) and females (4.8 days) against same insect on sorghum (Yadav & Bhargava, 2002). The oil was also reported to have adverse effect on oviposition of *C. chinensis* at 0.05–0.03%v/w. Gradual decline in egg laying with rise in concentration of oil was also noticed by (Ghosal et al., 2008). The oil was ineffective as fumigant against *C. cephalonica* since post embryonic development and adult emergence was not adversely affected in presence of volatile substances (Pathak & Krishna, 1991). In chickpea treated with seed oil, 98.3, 118.7, and 57 eggs/5 female were observed at 3, 2, and 1%, respectively (Raguraman & Singh, 1997). Seed oil was further evaluated by Arivudainambi et al. (2003) for its efficacy against *T. granarium* at different concentration (5, 10, 25, 50, 100, and 200 ml/l). The observations revealed 100% mortality at a dose of 20–200 ml in adults and grubs. Not any ovicidal activity was found at lower doses. The order of susceptibility was: adults > grubs > eggs.

Singh et al. (1989) reported that oil of Beauty berry *Callicarpa marcrophylla* (Verbenaceae) was ineffective against *S. oryzae* at 1000 ppm in acetone as progeny increased more in treated than control.

Singh et al. (1989) reported that wheat grain treated with essential oil of Himalayan cidar, *Cedrus deodara* (Pinaceae) could be protected from the attack of *S. oryzae* at 1000 ppm in acetone. Raguraman and Singh (1997) evaluated wood oil of *C. deodara* against *C. chinensis* on chickpea at different concentration and reported 134.3, 114, and 112.3 eggs/5 female at 3, 2, and 1% concentration of the oil, respectively.

Park et al. (2003) evaluated the essential oil of *Chamaecyparis obtusa* (Hinoki cypress) family Cupressaceae against *C. chinensis* and *S. oryzae*. Filter paper impregnated with leaf oil was used for the study. Bornyl acetate and alpha-phellandrene caused 97% mortality of *C. chinensis* at 0.1 mg/cm^2 dose while terpetene caused 93% mortality at 0.26 mg/cm^2 dose in case of *S. oryzae*.

Wanyika et al. (2009) determined the contact toxicity of stabilized natural pyrethrum (*Chrysanthemum cinerarifolium*) against adults of maize weevils *S. zeamais*. Result indicated that the natural pyrethrum extract blended with cotton seed oil was the most potent toxicant against *S. zeamais* and this potency was concentration- time dependent.

Globade & Adebayo (1993) reported that the essential oil of Triffid weed, *Chromolaena odorata* (Asteraceae) was unsuitable for fumigation when applied at the rate of 5–50 ml/9.9 g of cowpea seed and proved ineffective against *C. maculatus*.

Safrole, an effective component of essential oil of *Cinnamomum camphora* (Camphor) family Lauraceae showed fumigant toxicity to *S. zeamais* and *T. castaneum* in flour disc test (Huang et al., 1999). In another experiment the oil was evaluated under laboratory condition against adults of *O. surinamensis* and *T. castaneum*. Five concentrations each of essential oil (0.125, 0.25, 0.5, 0.75, and 1.0%) were tested. Complete mortality of *O. surinamensis* was achieved by *C. camphora* at concentration more than 0.5% (Al- Jabr, 2006). The oil was strong repellent against *R. dominica* (Al Jabr, 2006).

Toxicity of Citrus clean (composed of citronella oil, pine oil and natural oils from lemongrass and marigold) against *C. cephalonica* was evaluated by Dwivedi et al. (2000). Oil formulation of Citrus clean inhibited 100% oviposition and 89.5% mortality of larvas. Its efficacy against *C. chinensis* in cowpea reduced the oviposition of pulse beetle and caused 66.65% egg mortality and 100% repellent action. It also resulted in 100% eggs mortality of *C. cephalonica* at concentrations of 50, 75, and 100%. The lowest concentration, that is, 25% gave 91% mortality (Dwivedi et al., 2000).

Wang et al. (2000) reported high fumigant activity in essential oil of Muskmelon, *Cucumis melo* (Cucurbitaceae) against *S. zeamais*, T. *confusum*, *L. serricorne* and *Falsogastrallus sauteri*.

Jilani et al. (1988) evaluated oil of Turmeric, *Curcuma longa* (Zingiberaceae) against *T. castaneum* on rice and concluded that fewer adults settled in rice grain treated with 100, 500, or 1000 ppm of the oil. Repellency increased with increasing concentration. About 40–60% repellency was reported by Mohiuddin et al. (1993) against *T. castaneum* and *R. dominica*. Yalamanchilli and Punukollu (2000) observed that the oil obtained from the leaves *C. longa* could effectively protect the seeds at a low concentration of 2% (w/w) under the experimental conditions. Turmeric oil was also found effective in deterring the attack of stored grain pest, *C. chinensis* on four pulses and wheat grains. The oil when investigated by Tripathi et al. (2002) proved effective when applied on rice, wheat, wheat flour to control *R. dominica, S. oryzae* and *T. castaneum*. Oil was much toxic in both contact and fumigant assay. The adults of *R. dominica* were highly susceptible in contact action with LD_{50} value of 36.71 µg/ml where as *S. oryzae* adults were susceptible in fumigant assay with LC_{50} value of 11.36 mg/l of air. At 5.2 mg/cm^2 dose oviposition and egg hatchability was reduced by 72 and 80% in *T. castaneum* and showed >81% antifeedant activity against *R. dominica, S. oryzae and T. castaneum* at 40.5 mg/g food dose. Karci and Isikber (2007) reported low ovicidal toxicity of turmeric oil to eggs of *T. confusum* at all exposure times (24, 48, and 72 h) by <20% of corrected mortality at 100 µl/l air.

Essential oil of Lemon grass *Cymbopogon citratus* (Poaceae) exhibited fumigant toxicity against *C. maculatus*, when applied on cowpea at dose 5–50 µg/9.9 g of seed (Globade & Adebayo, 1993). This oil caused 100% mortality at 0.15 g/l with lowest LC_{50} value of 0.026 g/l in contact toxicity bioassay and the number of eggs laid was zero in stored cowpea. It also reduced oviposition and F1 progeny emergence (Paranagama et al., 2002). Oil showed a significant inhibition of oviposition and F1 progeny emergence compared to the control during no-choice tests on stored cowpea (Paranagama et al., 2003). In another study conducted on *S. cerealella* on rice, the population of the test insect was significantly less with lower grain damage. However, the treatment reduced seed germination and enhanced flavor and stickiness of cooked rice (Paranagama et al., 2003). Owolabi et al. (2009) concluded that activity of the lemongrass oil was concentration dependent and strongly toxic with LD_{50} value of 0.560 for *S. zeamais* on maize. Chahal et al. (2007) tested the bio-efficacy of lemon grass oil and its fractions against the adults of *T. castaneum* at 50–2000 μg/g and reported very effective at all concentration. Polar fraction was more active than non polar fraction. Bhargava et al. (2005) investigated effect of essential oil (*C. flexuosus*) on food utilization of *C. cephalonica* on sorghum seeds at 1.0 ml/100 g seeds for 72 h. The essential oils had feeding deterrent activity. The weight of food ingested and digested as well as the weight gained by the test insect was reduced when fed on treated seeds. The efficiency of conversion of ingested and digested food to body substances also decreased. Yadav and Bhargava (2002) tested the oil for its efficacy against *C. cephalonica*. The oil had no adverse effect on the germination of sorghum seeds when treated with 0.1, 0.5, 1.0 ml/100 g concentration. Michaelraj et al. (2006) evaluated the ovicidal activity of the essential oil on stored maize at 0, 25, 50, 75, 100, 150, 200, or 250 ppm against 0–24 h old eggs of *C. cephalonica* by the contact method. The oil was found effective against 10 days old larvas. It recorded LD_{50} value of 0.1033 μl/cm². Maximum inhibition of hatching was observed at 250 ppm while 96.6% mortality was obtained at 2.5 ml/kg. Similarly in case of *R. dominica* lemongrass (*C. flexuosus*) oil caused nearly 1/4th mortality at the same dose (Michaelraj et al., 2007). In case of eggs of *C. cephalonica* complete inhibition in hatching was obtained at all the doses of lemongrass (50, 100, and 150 μl) after 48 h of exposure (Michaelraj et al., 2008). Tewari and Tiwari (2008) evaluated toxicity of the leaf oil of *C. flexuosus* to three species of insect pests viz., *S. oryzae, T. castaneum* and *R. dominica* and concluded that only *S. oryzae* and *R. dominica* adults were susceptible to the tested oil as it resulted in 90% progeny inhibition at 0.1–0.4% for *R. dominica* and 1.0% for *S. oryzae*. The adults of *T. castaneum* were least affected. Effect of *C. flexuosus* on *C. maculatus* was studied by Raja and William (2008) who revealed that highest mortality and ovicidal activity (92 and 45.25%) was observed at 96 h exposure period.

Saraswathi and Rao (1987) studied the repellent effect of citronella oil *Cymbopogon nardus* (Poaceae) against *T. castaneum, C. chinensis* and *P. americana*, under laboratory conditions. Repellency of this oil was noted up to 52 h after treatment. Lale (1991) determined the bio efficacy of essential oil on *C. maculatus* on cowpea and significant mortality was observed in all bioassay tests. Similarly *C. nardus* oil was found effective at higher concentration (0.5%) to 10 days old *C. cephalonica* larvas (Behal, 1998). But Raja et al. (2001) reported that oil was not much effective in protecting stored cowpea. On the other hand high dose of oil slightly affected the survival of the 4th instar or pupa of *C. maculatus* (Ketoh et al., 2002). Deterrent effect of essential oil on oviposition and progeny production of the cowpea bruchid,

C. maculatus was investigated by Paranagama et al. (2003). It showed a significant inhibition of oviposition and F1 progeny emergence compared to the control during no-choice tests. With the increasing concentrations of the oil ranging from 10 to 160 mg the percent of eggs deposited decreased. The percent of eggs laid was recorded zero in sample treated with 40 mg of *C. nardus*. In similar study conducted on *S. cerealella*, the population of test insect and percent grain damage was lower in oil-treated rice (Paranagama et al., 2003). In another experiment it was found that the treated seeds did not affect reproduction of *C. maculatus* (Ketoh et al., 2005). Biswas and Biswas (2005) evaluated *C. nardus* oil against *C. chinensis* on gram and concluded that the oil effectively controlled its population by reducing oviposition rate at 2.5 and 5.0 ml/kg seeds. It also recorded least seed damage, weight loss and highest percent seed germination. The oil at same concentration was found most effective in reducing *S. oryzae* infestation in rice (Biswas & Biswas, 2006). In case of *S. zeamais* the oil proved to be promising material for prevention of post harvest pests (Nakahara et al., 2006). A combination of oils of citronella and *Vitex negundo* at a ratio of 1:1 was evaluated by Krishnarajah et al. (1985) who reported it most effective by causing knockdown and inactivation of *S. cerealella*. Its component p-Cymene was the most effective terpenes tested in relation to knockdown, inactivation and mortality while β-pinene was the most effective repellent.

The bio-efficacy of essential oil of Camel grass *Cymbopogon schoenanthus* (Poaceae) was investigated by Ketoh et al. (2005) to evaluate the residual effects of essential oil on *C. maculatus* survival and female reproduction and cowpea seed germination. The result indicated that the treated seeds did not affect reproduction of *C. maculatus*. Singh et al. (1989) reported very low grain damage in wheat grain treated with 1000 ppm concentration of Java citronella *Cymbopogon winterianus* (Poaceae) under laboratory conditions. The effect of *C. winterianus* oil on food utilization of *C. cephalonica* was studied by Bhargava et al. (2005). The Oil had feeding deterrent activity on fifth-instar larvas fed on broken sorghum seeds treated with oil at 1.0 ml/100 g seeds for 72 h. The weight of food ingested and digested as well as the weight gained by the test insect (fifth instars larvas was reduced. The oil was also found effective in causing adult mortality of *C. chinensis* (Ghosal et al., 2005). The toxicity of vapour of essential oil obtained from *C. winterianus* was determined against eggs of *T. confusum* (Karci & Isikber, 2007). Oil had low ovicidal toxicity to eggs when exposed to a dose of 100 µl/l air at all exposure times (24, 48, and 72 h) by <20% of corrected mortality. Ghosal et al. (2008) tested the efficacy of non-edible oil (*C. winterianus*) as seed protectant against adult *C. chinensis*. Gradual decline in egg laying was observed with the rise in concentration (0.05–0.03%v/w). The oil also proved more effective by causing death of the beetles, and adverse effect on oviposition. The oil also showed knockdown action against *S. cerealella* (Krishnarajah et al., 1985).

Baskaran and Janarthanan (2000) reported that dust formulation of Palmarosa, *Cymbopogon martinii* were more effective than the plant oils alone against *C. chinensis*. Less adult emergence (79.0/100 g seed) was recorded when treated with Palmarosa oil (PO) + Neem oil (NO) + Iluppai oil (IO) 10% D (1:2:2 v/v/v) at 2%. It further caused reduction in percent bored grains (12.6%) and wt. loss (5.0%). Palmarosa Oil was less effective at 10% D (2%). Only 3.0 eggs/50 g was observed for cowpea treated with Palmarosa Oil + Neem oil 10% D (2%) and 6.5 eggs/ 50 g seeds at 1%, respectively. Bhargava et al. (2005) evaluated feeding deterrent activity of *C. martinii* oil against *C. cephalonica* in sorghum at 1.0 ml/100 g of seed and found it effective. There was reduction in the weight of food ingested and digested and weight gained by the test insect when fed on treated seeds. The efficiency of conversion of ingested and digested food also decreased.

Essential oil of Blakely's red gum *Eucalyptus blakelyi* (Myrtaceae) was found effective against *S. oryzae* at 100 μl/l air. The adults were most susceptible followed by larvas then pupae and eggs (Lee et al., 2004). Oil of *Eucalyptus camaldulensis* (Murray gum) showed 45% mortality of eggs of *T. confusum* and *E. kuhniella* (Tunc et al., 2000). In another experiment this oil exhibited fumigant toxicity against *T. castaneum, S. zeamais,* and *C. maculatus*. Active compounds were monoterpenes. The efficacy of wet season oil was more as compare to dry season oil (Chairat et al., 2002). Essential oil of *E. camaldulensis* was also tested to determine its potential insecticidal activity against *S. oryzae* and *T. castaneum* (Mohamed & Abdelgaleil, 2008). Oil showed highest toxicity against *T. castaneum* in contact bioassay test where as good toxicity was exhibited against *S. oryzae* in fumigation bioassay test. *Eucalyptus globulus* oil when applied against *C. cephalonica* adversely affected post-embryonic development and adult emergence in presence of volatile oil. It also showed detrimental effect on egg hatchability (Pathak & Krishna, 1991). The oil also showed promising toxicant and repellent when screened against *Musca domestica, Stegobium paniceum, S. oryzae,* and *C. chinensis* (Ahmed & Eapen, 1986). Srivastava et al. (1988) concluded that essential oil of *E. globulus* at 0.40% can be used for control of *C. chinensis* on gram. Toxicity and repellent activity of this essential oil was studied by Padin et al. (2000) on *S. oryzae* and *T. castaneum* who reported that it had no repellent effect on the test insects. Rao and Prakash (2002) conducted experiment under controlled condition to determine the efficacy of eucalyptus (*E. globulus*) oil as paddy grain protectant against *R. dominica* and concluded that the multiplication of the pest was absolutely controlled by the oil for 180 days. In an attempt to find natural and cheap source of paddy grain protectant. Prakash and Rao (2006) evaluated *E. globulus* oil against *S. cerealella* as prophylactic treatment and post-phylactic treatment under controlled conditions. When admixed with the seed at 0.50 and 1.00% v/w it gave absolute protection up to 90 days against both the insects. Karci and Isikber (2007) reported low ovicidal toxicity of this oil to eggs of *T. confusum* at all exposure times by <20% of corrected mortality when exposed to a dose of 100 μl/l air 100 μg/l air. The effectiveness of eucalyptus oil in protecting faba beans against *C. chinensis* was also studied by Richa et al. (1993) which showed some effect on oviposition but was non toxic to adults. Abd-El-Aziz (2001) also investigated persistence of *E. globulus* oil against *C. maculatus* during storage. The oil showed promising oviposition deterrence and toxic effect and protected cowpea seeds for four months in addition to impairing the fecundity of the beetles. In further study the efficacy of *E. citridora* (Nilgiri) oil on the olfactory response of the larvas of rice-moth, *C. cephalonica* was observed by Behal (1998). Eucalyptus oil was effective at the higher concentrations when 10-day-old larvas were exposed to 0.1, 0.3, or 0.5% concentrations. The evaluation of insecticidal activity of *E. citridora* was carried out by Ngamo et al. (2004) in another experiment. The oil exhibited significant insecticidal activity on *S. zeamais* from the first day of application and after two or four days this activity decreased. The essential oil of *E. citridora* was reported to have feeding deterrent activity to fifth-instar larvas of *C. cephalonica* which were fed on broken sorghum seeds treated at the rate 1.0 ml/100 g of seed. Reduction in the weight of food ingested and digested as well as the weight gained by the test insect was observed during the experiment (Bhargava et al., 2005). *Eucalyptus citridora* oil isolated from the plant was also found repellent when analyzed and evaluated for its insecticidal activity at doses between 0.063 and 0.503 μl/cm^2 against *S. zeamais* using area preference method (Nerio et al., 2009). The comparative efficacy of four essential oils against *C. maculatus* and *S. oryzae* was studied by Gakuru and Foua (1995). Not any effect on *S. oryzae* was found but it proved potent toxicant against *C. maculatus* with LD_{50} of 1.26 and 1.49 ml in acetone (2% dilution). Coitinho et al. (2006) conducted experiment to determine the toxicity of the essential oil of *E. citridora* to adults of *S. zeamais* in stored maize grain. The LC_{50} of the oil tested was 44.69 μg/20 g of maize. The effect of oil obtained from the leaves of Alpine snow gums, *Eucalyptus pauciflora* was also assessed. The result revealed mortality of insects in 40–60 min after application during contact toxicity test when applied at the rate of 5% on wheat and groundnut commodity infested by *R. dominica, S. oryzae, T. granarium, C. cephalonica* and *E. cautella*. During fumigant toxicity test it caused mortality of *R. dominica* and *S. oryzae* in 7 h, *T. granarium* in 5 h and *C. cephalonica* and *E. cautella* in 5 h, respectively (Shukla et al., 2002). The essential oil of *Eucalyptus tereticornis* had noeffect on *S. oryzae* and *C. maculatus* (Gakuru & Foua, 1995). Bioactivity of the same essential oil was studied by Hou et al. (2002) against *S. oryzae*. High fumigant and antifeedant activity was exhibited to third instar larvas and it caused 100% mortality of *S. zeamais* at 0.4%.

Globade & Adebayo (1993) reported that oil of *Eugenia uniflora* (Surinam cherry) family Myrtaceae exhibited fumigant toxicity at 5–50 µg/9.9 g of seed against *C. maculatus* in stored cowpea. Oil was also effective against *C. chinensis* by causing mortality of beetles (Ghosal et al., 2005; Ghosal et al., 2008).

Essential oil of *Evodia rutaecarpa* (Rutaceae) was evaluated for fumigant and contact toxicity against *S. zeamais* and *T. castaneum*. In fumigant toxicity assay *S. zeamais* was less susceptible than *T. castaneum* where as in contact toxicity assay *S. zeamais* adults were more susceptible than *T. castaneum*. In repellency test weaker feeding deterrence was observed against *T. castaneum* than *S. zeamais* (Liu & Ho, 1999).

Ghosal et al. (2008) evaluated the essential oil of *Hydnocarpus kurzii* (Chaulmoogra) against the adult of *C. chinensis*. It was evident that non-edible oil was more effective to cause death of the beetles. The persistency of toxicity was longest for 120 days after treatment. It also caused mortality of adult *C. chinensis* beetle (Ghosal et al., 2005).

Essential oil of Bay laurel, *Laurus nobilis* (Lauraceae) was investigated against *R. dominica, S. oryzae and T. castaneum* (0.1–100 µl). At lowest dose the active compound 1,8-cineole, borneol and thymol were highly effective against *S. oryzae*. Camphor and linalool gave 100% mortality of *R. dominica* and less than 20% mortality at highest dose in case of *T. castaneum* (Rozman et al., 2007). Monoterpenes in essential oil of *L. nobilis* were effective against the maize pest, *S. zeamais* (Rozman et al., 2007). In another study the oil was found most effective against *R. dominica* at 1000 ppm (Shaaya et al., 1991) and vapors of laurel were found to have a low ovicidal toxicity at exposure times of 24, 48, and 72 h, respectively, by <20% of corrected mortality (Karci & Isikber, 2007).

Wang et al. (2000) determined the toxicity of geranium oil against museum insect pests viz., *L. serricorne, F. sauteri, S. zeamais* and *T. confusum*. High fumigation insecticidal activity was exhibited. They also concluded that plant essential oils are a promising prospect for development as potential botanical pesticides. In another experiment conducted by Richa et al. (1993) geranium oil exhibited greatest insecticidal activity in protecting faba beans against *C. chinensis*. In experiment conducted by Michaelraj et al. (2007) adults of *S. oryzae* and *R. dominica* were exposed to essential oil of geranium *Geranium viscosissimum* in the fumigation chamber. It was observed that geranium oil caused nearly 1/4th mortality at the same dose. In a similar study, Michaelraj et al. (2008) reported complete inhibition of hatching of eggs of *C. cephalonica* at a dose of 150 µl of geranium after 48 h of exposure. Michaelraj and Sharma (2006) evaluated the toxicity of geranium essential oil against 10-day-old larvas of *C. cephalonica* on maize (cv. Basi local) grain using the film method. It was most effective against 10-day-old larvas, as it recorded the lowest LD_{50} value of 0.0891 µl/cm². On treatment of maize grain with different concentration of the oil 95% mortality with no grain damage was recorded at 2.5 ml/kg grains. Feeding deterrent activity of geranium *Pelargonium graveolens* oil at 1.0 ml/100 g of seed against *C. cephalonica* on treated sorghum seed was also recorded (Bhargava et al., 2005).

Essential oil of Bay laurel, *Laurus nobilis* (Lauraceae) was tested against *R. dominica, S. oryzae and T. castaneum* at 0.1–100 µl/720 ml volume. The active compound 1,8-cineole, borneol and thymol were highly effective against *S. oryzae* at lowest dose. Camphor and linalool gave 100% mortality against *R. dominica*. Less than 20% mortality at highest dose for *T. castaneum* was observed (Rozman et al., 2006). In another study the oil was found very effective against *R. dominica* at 1000 ppm (Shaaya et al., 1990) and it was toxic to all stages of *T. confusum* (Isikber et al ., 2006).

Shaaya et al. (1990) investigated the bioefficacy of *Lavendula angustifolia* (Lavender) essential oil against *R. dominica, S. oryzae, O. surinamensis* and *T. castaneum* which was found most effective against *R. dominica*. Kalinovic et al. (1997) studied insecticidal activity of 5 aromatic plants on wheat grains and *Phaseolus vulgaris* against *S. granarius* and *Acanthoscelides obtectus* under laboratory condition. Oil extract of *L. angustifolia* was recorded most efficient in controlling *A. obtectus* on *P. vulgaris* with no negative effect on seed germination. Maga et al. (2000) in an experiment concluded that 1,8-cineole an active component of oil was responsible for the insecticidal properties against *T. castaneum*. Novokmet et al. (2002) evaluated essential oil of *L. angustifolia* for its insecticidal activity against two species of storage pests (*Plodia interpunctella* and *T. castaneum*). The oil was applied at 0.5 and 1 ml/kg on three kinds of bird food. Under winter storage conditions, 100% mortality of *P. interpunctella* and

T. castaneum was observed after 40 and 44 days of exposure, respectively, in the bird food treated with the essential oil at 1 mg/kg. Under summer storage conditions, 100% mortality was obtained after 68 days of exposure to the bird food treated with the same concentration of the essential oil. The same level of storage insect control at 0.5 ml/kg concentration was obtained by prolonging the exposure period for 4–8 days. Further study was carried out by Rozman et al. (2006) to investigate the fumigant toxicity of naturally occurring compounds isolated from essential oil of lavender (*L. angustifolia*) oil on stored wheat. Insect showed very high susceptibility to the compound 1,8-cineole (eucalyptol) with average mortality of 96.5–100% when tested at 0.1 µl/720 ml and had no negative effect on dough properties. These compounds were again evaluated by Rozman et al. (2007) for fumigant activity against adults of *S. oryzae*, *R. dominica* and *T. castaneum*. 1,8-cineole, borneol and thymol were highly effective against *S. oryzae* at lowest dose (0.1 µl/720 ml volume). Camphor and linalool gave 100% mortality against *R. dominica* while less than 20% mortality was recorded at higher dose (100 µl/720 ml volume) against T. castaneum. Monoterpenes in the essential oils were effective against the maize pest *S. zeamais* (Rozman et al., 2007).

Essential oil of Bushy matgrass, *Lippia alba* (Verbinaceae) was tested against *R. dominica, S. oryzae, T.* castaneum and C. *maculatus* and the oil was more effective in contact toxicity than fumigant toxicity. The adults of *C. maculatus* were more susceptible as compared to other insects in fumigant toxicity test and showed 50% reduction in oviposition and 70% reduction in case of *C. maculatus* and *T. castaneum*. The major compounds were linalool 65.2% and 1,8-cineole (eucalyptol) 6.6% (Verma et al., 2001). Insecticidal and oviposition deterrent activity of oil against *C. chinensis* on stored cowpeas was also assessed by Dubey et al. (2004). Oil of *Lippia gracilis* showed strong insecticidal activity at higher dose. Apart from this, reduction in the number of eggs laid on the treated seeds was also reported. In another study conducted by Coitinho et al. (2006) the oil provided highest control of *S. zeamais* as LC_{50} of the oil tested was 11.45 µl/g of maize. Pereira et al. (2008) reported absolute mortality of *C. maculatus* in cowpea at all concentration (10, 20, 30, 40, and 50 µl/20 g) and the reduction in the number of viable eggs and emerged insects was approximately 100%. Oil resulted in absolute mortality of the same insect in treated cowpea at 50 µl/20 g. Apart from this, low residual effect of the tested oil was noticed (Pereira et al., 2009). Koumaglo et al. (1996) tested the efficacy of geranial and neral, major constituents of *Lippia multiflora* leaf oil for protection against C. *maculatus*. The essential oil exhibited insecticidal activity against *C. maculatus* in a dose-dependent manner. Kouninki et al. (2007) reported essential oil of *Lippia rugosa* (Lamiaceae) promising in contact and inhalation tests because of its efficacy on the life stages of *T. castaneum*. The oil was found effective with 100% mortality of larvas. Ngamo et al. (2004) reported significant insecticidal activity of the oil on the first day application. Another study showed that oil of *Lippia adoensis* (Kosseret) exhibited fumigant toxicity when tested against *C. maculatus* in stored cowpea at 5–50 µg/ 9.9 g of seed (Glabode & Adebayo, 1993). Odeyemi (1993) studied the insecticidal properties of certain indigenous plant oils against *S. zeamais* at the rate of doses 0.1, 0.5, 0.7, and 1.0 ml oil/50 g maize. A reduction in the number of F1 progeny was also observed following treatments, and this was significant ($p < 0.05$) for the higher dose of *L. adoensis*.

The fumigant toxicity of essential oil extracted from Syrian marjoram, *Marjorana syriaca* (Lamiaceae) was evaluated against adults of *R. dominica, O. surinamensis, T. castaneum and S. oryzae*. The oil was reported most effective against *O. surinamensis* (Shaaya et al., 1991).

Lee et al. (2004) studied efficacy of Scarlet Honey Myrtle *Melaleuca fulgens* essential oil against *S. oryzae* at 100 µl/l air. Adults were most susceptible followed by larvas then pupae and eggs.

Vapour toxicity and repellency of essential oil obtained from Japanese mint, *Mentha arvensis* (Lamiaceae) against insect pests (*Musca domestica, Stegobium paniceum S. oryzae* and *C. chinensis*) was reported by (Ahmed & Eapen, 1986). Oil showed strong toxicity and repellent action against *C. maculatus* and *T. castaneum* (Tripathi et al., 2000) where *C. maculatus* was more susceptible than *T. castaneum*. In protecting stored rice from *S. oryzae* oil proved significantly superior over untreated control at 0.5 ml/kg concentration based on cumulative percent mortality of adults (Dayal et al., 2003). Lee et al. (2001) concluded that *M. arvensis* oil was very potent fumigant against *S. oryzae* with LC_{50} value of 45.5 µl/l air. Singh et al. (1995) found this oil as an effective fumigant against *S. oryzae* at 166 µl/l air with no effect on percent seed germination and recommended it for seed storage rather than for food storage in case of sorghum. The oil was effective in preventing oviposition in red gram at 0.2% concentration against *C. cephalonica* (18.66 eggs/ adult 90 DAT) and gave complete control of oviposition (Srivastava et al., 1988). When it was tested for its fumigant properties against *C. chinensis* infesting stored pigeon pea, it reduced infestation and seed germination and had no adverse effect on nutritive value (Srivastava et al., 1989). However, in another it was very effective against *S. oryzae* (Ahmed & Eapen, 1986). Leaf oil of *M. arvensis* was most effective and recorded 100% mortality of test insects when applied at 1–2 g /25 g in rice against *S. oryzae* (Khajuria et al., 2003).

Tripathi et al. (2000) investigated the toxicity of essential oil of Bergamot mint, *Mentha citrata* (Lamiaceae) against *C. maculatus* and *T. castaneum* and recorded that *C. maculatus* was more susceptible than *T. castaneum*. Singh et al. (1989) tested leaf oil (1000 ppm in acetone) of *M. citrata* for reproduction retardant, fumigant toxicity, and grain protection capability against *S. oryzae* under laboratory conditions. Oil showed significant fumigant toxicity when applied to stored wheat grain. However, this effect lasted for only 30 days.

Raja et al. (2001) observed *Mentha piperita* (Peppermint) oil effective in protecting cowpea from the attack of *C. maculatus*. The experiment was conducted to investigate fumigant toxicity of essential oil of (1000 ppm) against *S. oryzae*, *R. dominica*, *O. surinamensis* and *T. castaneum* on wheat by Shaaya et al. (1990). Oil was observed active against *T. castaneum* and 1,8-cineole was the active compound. *Callosobruchus maculatus* was more susceptible than *T. castaneum* (Tripathi et al., 2000). Essential oil was further evaluated for its toxicology and repellent activity against *S. oryzae* and *T. castaneum* by (Padin et al., 2000). No repellent activity was observed. Potent fumigant toxicity was exhibited against *T. castaneum* at 25.8 µl/l air (Lee et al., 2002). The contact toxicity of the peppermint essential oil was investigated against 10-day-old larvas of *C. cephalonica* on stored maize grain using the film method. The level of inhibition of hatching ranged from 25.00 to 89.33%, at 250 ppm being the most effective. The greatest mortality (100%) with no grain damage was recorded at 2.5 ml/kg (Michaelraj et al., 2006). Peppermint oil at all the concentrations caused 100% mortality of adults against *S. oryzae* with no progeny production and grain damage. In the case of *R. dominica*, except at lower doses, 100% mortality was obtained for peppermint oil (Michaelraj et al., 2007). Complete mortality of adults of *S. oryzae* was recorded at 100 and 150 µl/250 ml after 48 h of exposure where as absolute mortality was observed in *R. dominica*, at all the doses (50, 100, 150, and 200 µl/250 ml). The adult of *C. cephalonica* was not sensitive to peppermint at 5 1/250 ml. In case of 0–24 h old eggs of *C. cephalonica*, complete inhibition in hatching was recorded at all the doses of peppermint (25, 50, 100 and 150 µl) after 48 h of exposure (Michaelraj et al., 2008). More than 90% mortality was observed against *S. cereallela* (Klingauf et al., 1983). The same was reported by essential oil at 100 µl/l air also exhibited low ovicidal activity against *T. confusum* (Karci &Isikber, 2007). Activation of the acetylcholinesterase, glutamic transaminase, and amylase in the body of oil-treated insect (*R. dominica*) was also reported (El-Sheikh et al., 2005).

In a study conducted by Tripathi et al. (2000) an essential oil extracted from Spearmint, *Mentha spicata* (Lamiaceae) was found to acted as strong toxicant, repellent and grain protectant against *C. maculatus* and *T. castaneum*. Study also revealed that *C. maculatus* was more susceptible than *T. castaneum*. Rao and Prakash (2002) evaluated *M. spicata* oil as paddy grain protectant against *R. dominica* and reached to the conclusion that the adult emergence was as low as 1.20 and 0.40 at 90 DAT (days after treatment) and 6.00 and 2.60 at 180 DAT at the rates of 0.5 and 1.0%, respectively. The oil was further evaluated in the laboratory for its effectiveness as a toxiciant and repellent against adults of *O. surinamensis* and T. *castaneum*. Five concentrations (0.125, 0.25, 0.5, 0.75, and 1.0%) were tested. Complete mortality of *O. surinamensis* at 0.5% and *T. castaneum* at 1.0% was obtained. *Tribolium castaneum* was recorded less susceptible as compared to *O. surinamensis* (Al-Jabr, 2006). Bioactivity of essential oil of *M. spicata* against *M. separata, H. armigera, P. xylostella*, and *S. zeamais* was studied by Hou et al. (2002). High fumigation effect of oil was the reason for the absolute mortality of *S. zeamais*. Its effectiveness against *C. maculatus* in stored cowpea was confirmed by Raja et al. (2001). Antimicrobial and insecticidal activity of the leaf oil of *M. spicata* was studied against *T. castaneum, C .maculatus, R. dominica*, and *S. oryzae* which was found effective against these insects (Khanuja et al., 2001). The fumigant toxicity of this essential oil was investigated against *S. zeamais* and *C. pusillus* at concentration 40 µl/l at different exposure times during which it show better fumigation efficacy (Zang et al., 2005). The oil also exhibited higher fumigant toxicity and 100% mortality in 24 h against three important stored-product insects, *A. obtectus, S. granarius and S. oryzae*, under laboratory conditions (Karakoc et al., 2006).

Oil of Indian spikenard *Nardostachys jatamansi* (Valerianaceae) was found to have high fumigation insecticidal activity against the museum pests, *L. serricorne, S. zeamais, T. confusum and F. sauteri* (Wang et al., 2000).

Essential oil of *O. basilicum* was evaluated by Padin et al. (2000) for their repellent activity against *S. oryzae* and *T. castaneum*. They reported absence of any repellent activity in tested essential oil. On the other hand, the 1,8-cineole, a major component of essential oil of *O. kenyense*, was found to have strong repellent action against *S. granaries* and *S. zeamais* but was moderately repellent to *T. castaneum* and *Prostephanus truncatus* (Obeng et al., 1996). In case of *C. maculatus* this oil showed 80% mortality at 25 µl/ vial in 12 h with LC$_{50}$ value of 65 µl/g. This oil also gave complete protection for three months at 400 µl (Keita et al., 2001). The treated seeds did not affect reproduction of *C. maculatus* in stored cowpea (Ketoh et al., 2005). Comparative efficacy of plant essential oil from *O. basilicum* was evaluated against *C. maculatus* and *S. oryzae* in the laboratory by Petri dish method. Not any significant effect was found against *S. oryzae*. The result exhibited its potency against *C. maculatus* with LD_{50} value of 1.26 and 1.49 µl in acetone at 2% dilution (Gakuru & Foua, 1995). In another experiment, Dubey et al. (2004) evaluated the *Ocimum* oil for its insecticidal and oviposition deterrent activity against *C. chinensis* on stored cowpea at different concentrations (0.1, 1, 10, and 100 µl). The result revealed that the oil exhibited strong insecticidal activity at higher doses and showed insecticidal activity above 50% even at 10 µl/l and decreased oviposition even at 1 µl/l concentration. It also reduced the number of eggs laid on the treated seeds. Sweet basil oil tested against *R. dominica*, *S. oryzae*, *O. surnamensis* and *T. castaneum* in stored wheat at 1000 ppm was found most effective against *O. surnamensis*. Linalool, alpha-terpineol and carvacrol are active compound of *O. bacilicum* (Shaaya et al., 1990). Percent of germination of stored beans was highest after treatment with the oil (Boeke et al., 2004a, 2004b). Toxic compounds in essential oil of basil were determined for their insecticidal activity against *S. oryzae*, *R. dominica* and *C. pusillus*. Methyl eugenol/estragole, estragole and estragole/linalool chemotypes were three major essential oil profiles present in the five varieties of *O. basilicum* analyzed. The abundance of component had a strong influence on the outcome of the bioassays (Lopez et al., 2008). A comparative study of the anti-bruchid activity and chemical composition of the essential oil of a germplasm collection of 18 *Ocimum basilicum* accessions was undertaken by Pascual and Ballesta (2003). Some accessions inhibited the oviposition and other caused mortality. Methyl chavicol and linalool were the active compound present in the oil. Keita et al. (2001) performed an experiment to determine the efficacy of African basil *Ocimum gratissimum* essential oil obtained by steam distillation as an effective fumigant and concluded that the oil was responsible for 70% mortality of *C. maculatus* at 25 µl/vial dose in 12 h with LC_{50} value of 116 µl/g and exhibited a significant effect both on the egg hatching rate and on the emergence of adults. Complete protection for three months at 400 µl was also achieved. Ngamo et al. (2001) appraised the potential of essential oil of *O. gratissimum* to protect stored maize against *S. zeamais*. The oil caused high mortality (74%) after four days of treatment. The oil also showed significant insecticidal activity on the first day of application (Ngamo et al., 2001). The same essential oil obtained from the different parts of the plant was evaluated for its toxicity and repellency against *T. castaneum* by Andronikashvili and Reichmuth (2003). The result indicated that at 60 µl/ml, 76 and 88% mortality was obtained for *Og* (1) and *Og* (2) oils, and fecundity was reduced. Toxicity was positively correlated with the concentration up to

50–80 µl/ml and negatively with concentration > 80–100µl/ml. Tinkeu et al. (2004) reported insecticidal activity of oil against *S. zeamais*. Essential oil of *O. gratissimum* recorded higher contact toxicity resulting in 98% mortality against *S. zeamais* (Kouninki et al., 2005). It had no contact and ingestion effect on adults of *T. castaneum* (Kouninki et al., 2007).

The effect of the essential oil of *O. grattissimum* against *S. zeamais* was evaluated under laboratory condition for repellency, mortality, progeny emergence, and maize damage. The oil was found to be moderately repellent to the maize weevil and induced high mortality in the weevils. In addition, grains treated with the essential oil showed significant reduction in the number of progeny. There was no observable feeding damage on grains treated with the higher dosage of the essential oil. Thymol (32.7%), para-cymene (25.4%), gamma -terpinene (10.8%), β-selinene (4.5%), phellandrene (3.9%), and β -myrcene (3.1%) were major constituents of essential oil (Asawalam et al., 2008). A laboratory experiment was conducted by Ousman et al. (2007) to investigate the insecticidal activities of essential oil of *O. grattissimum* (Lamiaceae) leaves against *S. zeamais* in stored maize. Results showed significant insect mortality. The oil also revealed effective insecticidal activity and contact action after 96 h. The mortality rate increased with increase in the concentration of essential oil. Ogendo et al. (2008) evaluated the bio-efficacy of *O. grattissimum* and its major constituents (β-Z-ocimene and eugenol) for fumigant toxicity and repellent effect against *S. oryzae, T. castaneum, O. surinamensis, R. dominica* and *C. chinensis* to find potential alternatives to synthetic fumigants in the treatment of durable agricultural product. The fumigant toxicity of oil and two of its constituents were assessed at four rates (0, 1, 5 and 10 µl/l air) in space fumigation, whereas repellency of the oil and eugenol in acetone was evaluated in choice bioassays at five rates (0, 1.0, 2.0, 3.0, and 4.0 µl oil/2 g grain). The result indicated that fumigant toxicity and repellency was significant ($p < 0.0001$). At 1 µl/l air the oil caused 98, 99, 100% mortality of *R. dominica, O. surinamensis* and *C. chinensis,* respectively. Eugenol achieved 79, 61, and 100% mortality of the same insects. β -Z - ocimene caused 8, 11, and 59% mortality of *R. dominica, O. surinamensis* and *C. chinensis*. LC_{50} values of 0.20–14, 0.01–17, and 0.80–23 µl/l air were recorded for *O. grattissimum* oil, eugenol & β Z - ocimene. The oil was more effective in case of *T. castaneum* at 0.05, 1 or 5% (Bhattacharyya & Bordoloi, 1986).

Bekele and Hassanali (2001) reported toxic action of *Ocimum kilimandscharicum* (basil shrub) oil against *R. dominica*. Leaf oil induced 100% mortality and caused significant reduction in the number of progeny and survival rate against *S. zeamais, R. dominica* and *S. cerealella* in maize and sorghum grain. It showed highest repellent action to *S. zeamais* when applied at 0.3 g/250 g of grain (Jembere et al., 1995). Camphor, a major component was highly toxic to all the insects (*S. granarius, S. zeamais, T. castaneum* and *Prostephanus truncatus*). It caused about 93–100% mortality in *S. granarius, S. zeamais* and *P. truncatus* and 70–100% in *T. castaneum*. The oil completely inhibited the epergnes of progeny. This oil showed 80–100% repellency of test insects (Obeng et al., 1998). Leaf oil of *Ocimum suave* (Holy basil) was highly repellent and toxic to S. *zeamais* and induced 100% mortality in *R. dominica*. The grain moth *S. cerealella* was most susceptible (Bekele et al., 1996). Eugenol, a major component of essential oil of *O. suave* was more effective on grain than on filter paper discs when applied at the rate of 1 µl/kg and also resulted in complete mortality and inhibition of test insects (*S. granarius, S. zeamais, T. castaneum* and *P. truncatus*). It was also highly repellent to the insects with overall repellency in the range of 80–100%.

Ovicidal activity of essential oil from oregano *Origanum syriacum* (Lamiaceae) was evaluated against two stored-product insects. Oregano achieved mortalities as high as 77–89% in *T. confusum* and *E. kuehniella*, respectively (Tunc et al., 2000). It was reported most active against *O. surinamensis* when tested for fumigant toxicity (Shaaya et al., 1991).

Lee et al. (2002) reported potent fumigant toxicity of essential oil of *Pimentha racemosa* (Myrtaceae) against *T. castaneum* with LD50 value of 17.8 µl/l air.

Tunc et al. (2000) evaluated the essential oil vapors distilled from *Pimpinella anisum* (anise) for fumigant toxicity against eggs of two stored-product insects. It caused 100% mortality of the eggs. The LT_{99} were 60.9 and 253.0 h at 98.5 µl anise essential oil/l air for *E. kuehniella* and *T. confusum*, respectively. It was also most effective against *T. castaneum* (Shaaya et al., 1991).

Essential oil of chir *Pinus longifolia* (Pinaceae) was evaluated by Singh et al. (1989) for reproduction retardant, fumigant toxicity, and grain protactant against *S. oryzae* (1000 ppm in acetone) under laboratory conditions in stored wheat. The result indicated 37.51, 75.21, and 86.82% reduction in the population of *S. oryzae* after 30, 60, and 90 days of application, respectively. Zang et al. (2005) reported better fumigation efficacy of essential oil of *P. tabulaeformis* against S. *zeamais* and *C. pusillus* at 40 µl/l concentration.

Wang et al. (2000) reported high fumigant activity of essential oil obtained from the leaves of Patchouli *Pogostemon patchouli* (Labiaceae) against *L. serricorne, S zeamais, T confusum,* and *F. sauteri*.

Lee et al. (2002) reported that essential oil of Rosemary *Rosemarinus officinalis* (Labiaceae) was a potent fumigant against *T. castaneum* with LD_{50}

of 7.4 µl/l air. When the oil was tested against *E. kuhniella* and *T .confusum* 65% mortality of eggs was observed (Tunc et al., 2000). It was reported toxic to all stages of *T. confusum* (Isikber et al., 2006). Treatment of rosemary oil against *R. dominica, S .oryzae* and *T. castaneum* in stored wheat showed that compound 1,8-cineole, borneol and thymol were highly effective against *S. oryzae* at lowest dose. Camphor and linalool gave 100% mortality against *R. dominica* while less than 20% mortality was recorded at highest dose for *T. castaneum* (Rozman et al., 2006). The oil was found most effective against *R. dominica* in stored wheat (Shaaya et al., 1991). It killed the *T. castaneum* effectively (Padin et al., 2000).

Shakarami et al. (2004) tested the efficacy of essential oil from the leaves of *Salvia bracteata* (Labiaceae) at six concentrations against first instar larvas, egg hatching, and oviposition rate of *C. maculatus*. At maximum concentration (0.56 µl/cm^3) the oil caused 91.66% mortality of eggs and 98.33% mortality of first instar larvas, respectively. Essential oil reduced the oviposition rate of adults significantly. The highest concentration (0.37 µl/cm^3) of essential oil led to 96.78% oviposition deterrence. Shakarami et al. (2005) tested the oil and reported it to be effective against *C. maculatus, T. castaneum, and S. oryzae* and *S. granaries*. At highest concentration of 0.926 µl/ml, the oil showed 76.52, 75.15, 76.26, and 78.0% mortality of *C. maculatus, T. castaneum, S. oryzae* and *S. granarius,* respectively, with LC$_{50}$ values of 0.188, 0.293, 0.252, and 0.231 µl/ml, respectively. Oil also showed significant repellency against test insects. In another experiment, oil of *S. officinalis* was found most effective against *R. dominica* in stored wheat at 1000 ppm (Shaaya et al., 1990).

Huang et al. (1999) reported fumigant toxicity of essential oil of Sassafras *Sassafras, albidum* (Lauraceae) to *S. zeamais* and *T. castaneum.*

Lale and Ajayi (2000) investigated the insecticidal, larvicidal, and antibiotic effects of *Syzygium aromaticum* oil in suppressing infestation of stored pearl millet. At the rate of 80–100 mg/5 g seed this oil caused complete suppression of the development of adult and larvas of *T. castaneum* and 96.3 and 65.7% adult and larval mortality, respectively. Sharma and Meshram (2006) evaluated essential oil of *S. aromaticum* against *S. oryzae* at 10 and 20 µl/l, 500 ppm. The oil caused inhibition of F1 progeny (50.42–72.5%) and decreased percent weight loss. Absolute protection of wheat from the insect was achieved at 10 and 20 µl/l. No infestation was noticed at 500 ppm. The oil was reported to be effective in both contact and fumigant toxicity test against *C. maculatus* (Mahfuz & Khalequzzaman, 2007). Eugenol, an active compound from clove oil significantly reduced longevity(1+ or –0.0 day), number of eggs laid (0.0+ or –0.0 eggs) and adult emergence rate(0.0+ or

–0.0 adult) of *C. maculatus* at 5 µl/50 g of seed(v/w) in chickpea (*Cicer arietinum*) (Kellouche & Soltani, 2004) . In case of *C. chinensis* the oil was effective at higher concentration (0.1, 0.3, and 0.5%; Behal, 1998). At 0.00, 0.10, 0.20, 0.50, 1.00, and 2.00% concentration the oil inhibited egg hatching of *C. cephalonica* with LC_{50} value of 0.2602 ± 0.0298 (Bhargava & Meena, 2001). It was significantly superior over the untreated control at 0.5 ml/ kg against *S. oryzae* (Dayal et al., 2003). In case of cowpea, the oil exhibited significant mortality in bioassay test and no appreciable mortality was noticed in fumigation test (Lale, 1991). The oil caused complete mortality of *S. oryzae* at 25–100 ppm with higher persistence and toxicity which was lower for *R. dominica* (Sighamony et al., 1986).

Krishna et al. (2005) investigated efficacy of essential oils obtained from different species of Marigold *Tagetes erecta* (Asteraceae) *Tagetes minuta* (Maxican Marigold) and *Tagetes patula* (French Marigold) against *S. oryzae*, *C. maculatus* and *T. castaneum* at a concentration of 50,000 ppm. In case of *T. erecta*, TE genotype 13 was effective toxicant whereas *T. minuta*, TM genotype I induced 100% mortality in three insects and in *T. patula*, TP genotype II caused 100% mortality of three insects. The efficacy of essential oils extracted from pyrethrum (*Tanacetum cinerariifolium*) was evaluated against adult of *S. oryzae* and *T. castaneum* and results indicated that former was more susceptible than latter. Toxicity increased with the number of days after application. Repellency was observed to be concentration dependent with higher concentrations (100, 50, and 25%) showing higher repellency than lower concentrations (12.5 and 6.25%). A study was carried out to investigate the insecticidal activity of naturally occurring compounds thymol and carvacral isolated from essential oils of thyme (*Thymus vulgaris*) each at 0.1 µl/720 ml on stored wheat against *S. granarius*. Very high susceptibility with average mortalities of 96.5–100% with no negative effect on dough properties was observed (Rozman et al., 2006). The compound 1,8-cineole, borneol and thymol were highly effective against S. oryzae when applied for 24 h at the lowest dose (0.1 µl/720 ml volume). For *R. dominica,* camphor, and linalool were highly effective and produced 100% mortality in the same conditions (Rozman et al., 2007). The oil was also most effective against *O. surinamensis* when assessed against adults of *R. dominica, O. surinamensis, T. castaneum* and *S. oryzae* for fumigant toxicity (Shaaya et al., 1991). Reproduction retardant and fumigant properties of the essential oil of Winged prickly ash *Zanthoxylum alatum* (Rutaceae) (1000 ppm in acetone) was studied by Singh et al. (1989) against *S. oryzae* who reported that oil was ineffective in preventing wheat grain from damage.

11.2.2   TOXICOLOGY OF PLANT MATERIALS

The secondary metabolite of plants is a complex mixture of several chemical compounds, which is not related to function of photosynthesis, growth or other aspects of plant physiology. The major secondary metabolites include alkaloids, terpenoids, phenolics, flavonoids, chromenes, and other minor chemicals Table 11.2. They can affect insects in several different ways: they may disrupt major metabolic pathways and cause rapid death, act as attractants, deterrents, phagostimulants, antifeedants, or modify oviposition. They may retard or accelerate development or interfere with the life cycle of the insect in other ways (Bell et al., 1990). Essential oils have the specific mode of action some of them act as good synergists, but most of essential oils act as fumigants and affect the respiratory system of stored grain insects.

TABLE 11.2   Chemical Constituents Commonly Known Plants.

S. no.	Commonly known plants	Chemical constituents
1	*Carum, Cinnamomum, Citrus, Lycopersicon, Mentha, Nicotiana, Pimpinella, Rosmarinus*	Acetaldehyde
2	*Gossypium spp.*	Acetic acid
3	*Coriandrum, Lycopersicon, Pogostemon*	Acetone
4	*Allium*	Allicin
5	*Mentha, Origanum*	Amyl alcohol
6	*Nicotiana spp.*	Anabasine
7	*Anethum, Ocimum, Pimpinella, Piper*	Anethole
8	*Apium, Piper*	Apiole scumus
9	*Acorus, Piper ssp.*	Asarone
10	*Chenopodium*	Ascaridole
11	Neem, melia	Azadirachtin
12	*Canaga, Canangium, Eugenia*	Benzyl alcohol
13	*Angelica, Blumea, Cinnamomum, Citrus, Coriandrum, Cymbopogon, Eucalyptus, Lavendula, Myristica, Ocimum, Rosmarinus, Salvia, Thymus, Valeriana, Zingiber*	Borneol
14	*Artemisia, Eucalyptus, Lavendula, Raphanus*	Butyraldehyde
15	*Eucalyptus, Gossypium, Sapindus*	Butyric acid
16	*Artemisia, Cinnamomum, Croton, Eucalyptus, Laurus, Lavendula*	iso-Butyric acid
17	*Acorus*	Calamus oil
18	*Rosmarinus, Zingiber*	d-Camphene

TABLE 11.2   *(Continued)*

S. no.	Commonly known plants	Chemical constituents
19	*Acorus, Alpina, Artemisia, Chenopodium, Chrysanthemum, Cinnamomum, Curcuma, Lavendula, Lippia, Ocimum, Rosmarinus, Salvia*	Camphor
20	*Chrysanthemum, Cinnamomum,* Cocos, *Laurus, Lavendula, Mentha*	Caproic acid
21	*Cinnamomum, Citrus,* Cocos, *Cymbopogon, Mentha, Myristica*	Caprylic acid
22	*Capsicum*	Capsaicin
23	*Carum*	Caraway oil
24	*Chenopodium, Citrus, Eucalpyrus, Piper*	Carene
25	*Anabasis, Carum, Cinnamomum, Mentha, Ocimum, Origanum, Thymus, Zea*	Carvacrol
26	*Anethum, Carum, Chrysanthemum, Citrus, Cymbppogon, Eucalpyrus, Lavendula, Lippia, Mentha, Syzygium*	Carvone
27	*Cinnamomum*	Cassia oil
28	*Acorus, Artemisia, Eupatorium, Glycine, Gossypium, Helianthus, Linum, Olea, Pimpinella, Trigonella, Valeriana*	Choline
29	*Alpina, Artemisia, Blumea, Cinnamomum, Curcuma, Eucalyptus, Eugenia, Laurus, Lavendula, Lippia, Mentha, Ocimum, Piper, Psidium, Rosmarinus, Salvia, Syzygium Zingiber*	Cineole (1–8)
30	*Cassia, Cinnamomum, Lavendula, Pogostemon*	Cinnamaldehyde
31	*Citrus, Cymbopogon, Eucalyptus, Lavendula, Lippia, Ocimum, Piper, Thymus, Zingiber*	Citral
32	*Cymbopogon*	Citronella oil
33	*Cinnamomum, Citrus, Cymbopogon, Lippia*	Citronellol
34	*Syzygium*	Clove oil
35	*Coriandrum*	Coriander oil
36	*Ageratum, Cinnamomum, Lavendula*	Coumarin
37	*Cuminum*	Cumene, Cumin oil
38	*Cinnamomum, Cuminum, Eucalyptus, Lavendula*	Cuminic alcohol
39	*Aegele, Artemisia, Cinnamomum, Cuminum,* ADI 0–0.1 mg *Lavendula, Zanthoxylum*	Cuminic aldehyde
40	*Curcuma longa*	Curcumin
41	*Aegele, Chenopodium, Cinnamomum, Citrus, Coriandrum, Croton, Cuminum, Cupressus, Eucalyptus, Eupatorium, Myristica, Origanum, Pimpinella, Salvia, Thymus*	p-Cymene

TABLE 11.2   *(Continued)*

S. no.	Commonly known plants	Chemical constituents
42	*Cassia, Cinnamomum, Citrus, Coriandrum, Lavendula*	Decanal
43	*Citrus, Nicotiana*	Ergosterol
44	*Antemisia, Ocimum, Pimpinella*	Esdragole (estragole)
45	*Citrus, Eucalyptus, Lycopersicon, Mentha, Nicotiana*	Ethyl alcohol
46	*Eucalpyrus*	Eucalyptol oil
47	*Acorus, Ageratum, Alpina,* ADI 0.1–2.5 mg/kg *Cinnamomum, Citrus, Cymbopogon, Eugenia, Lantana, Laurus, Lavendula, Myristica, Nicotiana, Ocimum, Pimpinella, Piper, Pogostemon, Syzygium*	Eugenol
48	*Canaga, Canangium, Cinnamomum, Myristica*	iso-Eugenol
48	*Artemisia, Cinnamomum, Citrus, Mentha,*	Formic acid
50	*Angelica, Carum, Cinnamomum, Citrus, Cupressus, Cymbopogon, Gossypium, Lavendula, Mentha, Myristica, Syzygium, Zea*	Furfural
51	*Gaultheria*	*Gaultheria* oil
52	*Cinnamomum, Citrus, Eucalyptus, Eugenia, Laurus, Lavendula, Lippia, Myristica, Thymus, Zingiber*	Geraniol
53	*Gossypium*	Gossypol
54	*Chenopodium, Citrus, Lycopersicon*	Histamine
55	*Ageratum, Allium, Annona, Boscia, Cassia, Chenopodium, Cinnamomum, Citrus, Combretum, Crotalaria, Cymbopogon, Erythrophleum, Eucalyptus, Euphorbia, Gliricidia, Glycine, Ipomaea, Jatropha, Lantana, Linum, Nerium, Nicotiana, Ocimum, Piper, Psidium, Ricinus, Tagetes, Tephrosia*	Hydrocyanic acid
56	*Cinnamomum, Citrus, Cymbopogon, Eucalyptus, Glycine, Lavendula, Mentha, Raphanus*	Isovaleraldehyde
57	*Cinnamomum, Citrus, Cocos, Elaois, Laurus*	Lauric acid
58	*Anethum, Apium, Carum, Chenopodium, Cinnamomum, Citrus, Coriandrum, Croton, Cuminum, Cymbopogon, Eucalyptus, Hyptis, Lavendula, Lippia, Mentha, Myristica, Nicotiana, Ocimum, Origanum, Pimpinella, Piper, Rosmarinus, Salvia, Syzygium, Valeriana*	Limonene
59	*Artemisia, Cinnamomum, Citrus, Coriandrum,* g *Cymbopogon, Eucalyptus, Laurus, Lavendula, Mentha, Myristica, Ocimum, Origanum, Rosmarinus, Salvia, Syzygium, Thymus, Zingiber*	Linalool
60	*Helianthus, Ricinus*	Menthol
61	*Artemisia, Capsicum, Coriandrum, Helianthus, Juniperus, Lycopersicon, Nicotiana, Ricinus*	Malic acid

TABLE 11.2   *(Continued)*

S. no.	Commonly known plants	Chemical constituents
62	*Mentha, Thymus*	Linoleic acid
63	*Carum, Gossypium, Lycospericon, Mentha, Syzygium*	Methanol
64	*Cananga, Canangium, Citrus*	Methyl anthranilate
65	*Artemisia, Ocimum, Pimpinella*	Methyl chavicol
66	*Syzygium*	Methyl cugenol
67	*Chenopodium, Erythroxylum, Eugenia, Gaultheria, Myristica, Syzygium, Xanthophyllum*	Methyl salicylate
68	*Cinnamomum, Citrus, Cocos, Croton, Elaeis, Helianthus, Myristica, Pimpinella*	Myristic acid
69	*Anethum, Carum, Cinnamomum, Myristica*	Myristicin
70	*Erythroxylum, Nicotiana*	Nicotine
71	*Nicotiana*	Nornicotine
72	*Lycopersicon*	Noscapine
73	*Myristica*	Nutmeg oil
74	Major component of oilseeds	Oleic acid
75	*Origanum*	*Origanum* oil
76	*Allium,* Anacardium, *Capsicum, Chenopodium, Citrus, Coriandrum, Glycine, Ipomaea, Juniperus, Lycopersicon, Mentha, Momordica, Nicotiana, Raphanus*	Oxalic acid
77	Common in oilseeds, *Apium, Gossypium, Ricinus*	Palmitic acid
78	*Mentha spp.*	Pennyroyal oil
79	*Aegele, Anethum, Angelica, Artemisia, Cinnamomum, Citrus, Coriandrum, Cuminum, Curcuma, Cymbopogon, Eucalyptus, Lantana, Laurus, Lavendula, Mentha, Pimpinella, Piper, Zingiber*	Phellandrene
80	*Coriandrum, Cuminum, Eucalyptus, Laurus, Myristica, Origanum, Pimpinella, Piper, Rosmarinus, Syzygium*	a-Pinene
81	Widespread, including *Cinnamomum, Lavendula, Pimpinella*	Propionic acid
82	*Bystropogon, Mentha, Origanum*	Pulegone
83	*Chrysanthemum spp.*	Pyrethrum
84	*Allium, Citrus, Eucalyptus, Psidium, Raphanus*	Pyrocatechol
85	*Carum*	Pyrogallol
86	*Allium, Euphoria, Euphorbia, Gossypium, Psidium, Shorea, Zea*	Quercitin
87	*Cassia, Ricinus*	Ricinoleic acid
88	*Derris, Lonchocarpus, Tephrosia*	Rotenone

TABLE 11.2 *(Continued)*

S. no.	Commonly known plants	Chemical constituents
89	Widespread, including *Artemisia, Calotropis*	Rutin
90	*Chenopodium, Cinnamomum, Myristica,*	Safrole
91	*Salvia*	Sage oil
92	*Cinnamomum, Gaultheria, Gossypium, Mentha*	Salicylic acid
93	*Sapindus, Zanthoxylum*	Sanguinarine
94	Very widespread, including *Acorus, Aegele, Allium, Anabasis, Bassia, Cassia, Chenopodium, Citrus, Eucalyptus, Eugenia, Euphorbia, Glycine, Helianthus, Lippia, Momordica, Nicotiana, Ocimum, Olea, Raphanus, Ricinus, Salvia, Sapindus, Thymus*	Saponin
95	*Ipomaea*	Sebacic acid
96	*Capsicum, Lycopersicon*	Solanine
97	*Artemisia, Gossypium, Helianthus*	Succinic acid
98	Widespread, including *Acorus, Artemisia, Catharanthus, Cinnamomum, Coriandrum, Cuminum, Eucalyptus, Eugenia, Eupatorium, Euphoria, Helianthus, Ipomaea, Lavendula, Melia, Myristica, Pimpinella, Rosmarinus, Salvia, Syzygium*	Tannic acid
99	*Artemisia, Cinnamomum, Citrus, Cuminum, Cupressus, Cymbopogon, Lantana, Laurus, Lippia, Myrisfica, Origanum, Pimpinella, Rosmarinus, Salvia, Thymus*	Terpinen-4-ol
100	*Artemisia, Lavendula, Lippia, Salvia*	Thujone
101	*Thymus*	Thyme oil
102	*Anabasis, Carum, Lavendula, Ocimum, Origanum, Thymus*	Thymol
103	*Lycopersicon*	Tomatine
104	*Acorus, Chenopodium, Gossypium, Nicotiana*	Trimethylamine
105	*Glycine, Helianthus, Ipomaea, Linum, Lycopersicon, Nicotiana*	Tryptothane
106	*Curcuma*	Turmurone
107	*Artemisia, Cocos, Thymus*	Undecanoic acid
108	*Cananga, Canangium, Croton, Laurus Lavendula, Mentha, Valeriana*	Valeric acid
109	*Artemisia, Calotropis, Cinnamomum, Citrus, Croton, Eucalyptus, Lavendula, Lippia, Mentha, Nicotiana, Rosmarinus, Valeriana*	iso-Valeric acid

Several organic solvents have been used in extraction of plant products However plant extracts are tested for insecticidal or non-insecticidal activities. Many of the plants which have been tested are found across a broad geographical range. It is well established in plant chemistry that differing climatic, soil and seasonal conditions can affect the type and quantity of the components isolated from extracts. For example, *Acorus calamus* is widely distributed in ponds and lakes in Asia, North America, and Europe. The rhizomes of the plant contains ß-asarone, a compound which shows toxic and sterilizing effects against insects. Three caryotypes of *A. calamus* have been found: the tetraploid caryotype ($4n = 48$) found in India, East Asia and Japan contains 70–96% ß-asarone whilst the diploid caryotype ($2n = 24$) found in North America and the triploid caryotype ($3n = 36$) present in Central Europe contain less than 15% of the active ingredient (Schmidt & Streloke, 1994).

KEYWORDS

- plant products
- essential oils
- toxicology
- storage insecticides
- secondary metabolites
- plant extract
- antifeedant
- repellant
- deterent

REFERENCES

Abate-Zeru, S. The Potential for Natural Products as Grain Protectants against Stored Product Insects. M.S. Thesis, Oklahoma State University Stillwater, Oklahoma, US, 2001, p 68.

Abd-El-Aziz, S. E. Persistence of Some Plant Oils against the Bruchid Beetle, *Callosobruchus maculatus* (F.) (Coleoptera: Bruchidae) during Storage. *J. Agril. Sci.* **2001,** *9* (1), 423–432.

Abdel Latif, A. M. Tooth-Pick Seed (*Ammi visnaga* L.) Extracts as Grain Protectant against the Granary Weevil (*Sitophilus granarius* L.). *Egyptian J. Agric. Res.* **2004,** *82* (4), 1599–1608.

Adebayo, T.; Gbolade, A. A. Protection of Stored Cowpea from *Callosobruchus maculatus* Using Plant Products. *Insect Sci. Appl.* **1994,** *15* (2), 185–189.

Adebowale, K. O; Adedire, C. O. Chemical Composition and Insecticidal Properties of the Underutilized Jatropha Curcas Seed Oil. *Afr. J. Biotechnol.* **2006,** *5* (10), 901–906.

Adedire, C. O.; Akinneye, J. O. Biological Activity of Tree Marigold, *Tithonia diversifolia,* on Cowpea Seed Bruchid, *Callosobruchus maculatus* (Coleoptera: Bruchidae), *Ann. Appl. Biol.* **2004,** *144* (2), 185–189.

Adler, C.; Ojimelukwe, P.; Tapondjou, A. L. In *Utilization of Phytochemicals against Stored Product Insects,* Proceeding of the Meeting of the OILB Working Group "Integrated Protection in Stored Products", Berlin, Germany, Aug 22–24, 1999; Adler, C., Schoeller, M. Eds.; IOBC/WPRS Bulletin No. 23; 2000; pp 169–175.

Agarwal, D. C.; Deshpande, R. S.; Tipnis, H. P. Insecticidal Activity of *Acorus calamus* on Stored Grain Insects. *Pesticides.* **1973,** *7,* 21.

Aggarwal, K. K.; Tripathi, A. K.; Prajapati, V.; Kumar, S. Toxicity of 1,8-Cineole towards Three Species of Stored Product Coleopterans, *Insect Sci. Appl.* **2001,** *21* (2), 155–160.

Ahmed, S. M.; Eapen, M. Vapour Toxicity and Repellency of Some Essential Oils to Insect Pests. *Indian Perfumer.* **1986,** *30* (1), 273–278.

Ajayi, O.; Arokoyo, J. T.; Nezan, J. T.; Olayinka, O. O.; Ndirmbula, B. M.; Kannike, O. A. Laboratory Assessment of the Efficacy of Some Local Plant Materials for the Control of Storage Insect Pests. *Samaru J. agric. Res.* **1987,** *5,* 81–86.

Al-Jabr, A. M. Toxicity and Repellency of Seven Plant Essential Oils to *Oryzaephilus surinamensis* (Coleoptera: Silvanidae) and *Tribolium castaneum* (Coleoptera: Tenebrioidae). *Sci. J. King Faisal Uni. Basic Appl. Sci.* **2006,** *7* (1), 49–60.

Al-Jabr, A. M. Insecticidal and Repellent Properties of Eight Botanical Powders to *Oryzaephilus surinamensis* (Coleoptera: Silvanidae) and *Tribolium castaneum* (Coleoptera: Tenebrioidae). *Alexandria J. Agric. Res.* **2003,** *48* (3), 163–171.

Al-Moajel, N. H. Testing Some Various Botanical Powders for Protection of Wheat Grain Against *Trogoderma granarium* Everts. *J. Biol. Sci.* **2004,** *4* (5), 592–597.

Andronikashvili, M.; Reichmuth, C. In *Repellency and Toxicity of Essential Oils from Ocimum gratissimum (Lamiaceae) and Laurus nobilis (Lauraceae) from Georgia Against the Rust-Red Flour Beetle (Tribolium castaneum Herbst) (Coleoptera: Tenebrionidae),* Proceedings of the 8th International Working Conference on Stored Product Protection, York, UK, Jul 22–26, 2002; 2003, pp 749–762.

Ansari, B. A.; Mishra, D. N. Toxicity of Some Essential Oils against the Pulse Beetle, *Callosobruchus maculatus* (F.) (Coleoptera: Bruchidae). *J. Inst. Agric. Anim. Sci.* **1990,** *11,* 95–98.

Ansari, K. K.; Prakash, S.; Pandey, P. N. Effect of Some Indigenous Plant Products on the Loss of Weight in Jowar Infested with *Corcyra cephalonica* (Stainton). *Flora Fauna Jhansi.* **2003,** *9* (2), 75–76.

Arthur, F. H. Aerosol Distribution and Efficacy in a Commercial Food Warehouse. *Insect Sci.* **2008,** *15,*133–140.

Asawalam, E. F.; Emosairue, S. O.; Hassanali, A. Essential Oil of *Ocimum grattissimum* (Labiatae) as *Sitophilus zeamais* (Coleoptera: Curculionidae) Protectant. *African J. Biotech.* **2008,** *7* (20), 3771–3776.

Atal, C. K.; Srivastava, J. B.; Wali, B. K.; Chakravarty, R. B.; Dhawan, B. N.; Rastogi, R. P. Screening of Indian Plants for Biological Activity; Part VIII. *Indian J. Exp. Biol.* **1978,** *16,* 330–349.

Azmi, M. A.; Neqvi, S. N. H.; Akhtar, K.; Jahan, M.; Habib, R.; Tabassum, R. Toxicological Studies of Solfac and Neem Formulation against *Sitophilus oryzae. Natl. Aca. Sci. Lett.* **1993,** *16,* 187–190.

Ba-Angood, S. A.; Al-Sunaidy, M. A. Effect of Neem Oil and Some Plant Powders on Egg Laying and Hatchability of the Cowpea Beetle *Callosobruchus chinensis* Eggs on Stored Cowpea Seeds. *J. Natural Appl. Sci.* **2003,** *7* (2), 195–202.

Bachrouch, O.; Jouda, M.; Aidi, W. W.; Thierry, T.; Brahim, M.; Manef, A. Composition And Insecticidal Activity of Essential Oil *from Pistacia lentiscus* L. against *Ectomyelois ceratonie* Z. and *Ephestia kuhniella* Z. (Lepidoptera : Pyralidae). *J. Stored Prod. Res.* **2010,** *46,* 242–247.

BAIER, A. H.; WEBSTER, B. D. Control of Acanthoscelides Obtectus Say (Coleoptera: Bruchidae) in *Phaseolus vulgaris* L. Seed Stored on Small Farms. I. Evaluation of Damage. *J. Stored Product Res.* **1992,** *28* (4), 289–293.

Baskaran, R. K. M.; Janarthanan, R. Effect of Dust Formulations of Certain Plant Oils against Important Pests of Paddy and Cowpea in Storage. *J. Entomol. Res.* **2000,** *24* (3), 271–278.

Behal, S. R. Effect of Some Plant Oils on the Olfactory Response of Rice Moth *Corcyra cephaloncia* (Stainton). *Ann. Plant Prot. Sci.* **1998,** *6* (2), 146–150.

Bekele, J.; Hassanali, A. Blend Effects in the Toxicity of the Essential Oil Constituents of *Ocimum kilimandscharicum* and *Ocimum kenyense* (Labiateae) on Two Post-Harvest Insect Pests. *Phytochemistry.* **2001,** *57* (3), 385–391.

Bekele, A. J.; Obeng-Ofori, D.; Hassanali, A. Evaluation of *Ocimum suave* (Willd) as A Source of Repellents, Toxicants and Protectants in Storage against Three Stored Product Insect Pests. *Int. J. Pest Mant.* **1996,** *42,* 139–142.

Bell, A. F.; Fellows, L. E.; Simmonds, S. J. Natural Products from Plants for the Control of Insect Pests. In *Safer Insecticide Development and Use;* Hodgson, E., Kuhr, R. J., Eds.; Marcel Dekker: USA, 1990; p 91.

Bhargava, M. C.; Meena, B. L. Effect of Some Spice Oils on the Eggs of *Corcyra cephalonica* Stainton. *Insect Environ.* **2001,** *7* (1), 43–44.

Bhargava, M. C.; Choudhary, R. K.; Hussain, A. Effect of Different Essential Oils on Food Utilization of *Corcyra cephalonica* (Stainton) (Lepidoptera: Pyralidae). *Insect Environ.* **2005,** *11* (1), 27–29.

Bhargava, M. C.; Meena, B. L. Efficacy of Some Vegetable Oils against Pulse Beetle, *Callosobruchus chinensis* (Linn.) on cowpea, *Vigna unguiculata* (L.), *Indian J. Plant Prot.* **2002,** *30* (1), 46–50.

Bhargava, M. C.; Pareek, B. L.; Yadav, J. P. Preliminary Screening of Citrus Oils on *Corcyra cephaloncia* (Stainton), *Insect Environ.* **2003,** *9* (1), 36–37.

Bhattacharyya, P. R.; Bordoloi, D. N. Insect Growth Retardant Activity of Some Essential Oil Bearing Plants. *Indian Perfumer.* **1986,** *30* (2–3), 361–371.

Bhonde, S. B.; Deshpande, S. G.; Sharma, R. N. Toxicity of Some Plant Products to Selected Insect-Pests, and its Enhancement by Combinations. *Pest. Res. J.* **2001,** *13* (2), 228–234.

Bin, M. T. M. The Utilization of Botanical Dusts in the Control of Foodstuff Storage Insect Pests in Kivu (Democratic Republic of Congo). *Cah. Agric.* **2003,** *12* (1), 23–31.

Biswas, N. P.; Biswas, A. K. Effect of Some Non-Edible Oils against Pulse Bettle, *Callosobruchus chinensis* in Stored Gram. *J. Interacademicia.* **2005,** *9* (3), 448–450.

Biswas, N. P.; Biswas, A. K. Use of Non-Edible Oils as Grain Protectant against Rice Weevil *(Sitophilus oryzae* L.) and their Subsequent Effect on Germination. *Adv. Plant Sci.* **2006,** *19* (2), 653–656.

Boateng, B. A.; Kusi, F. Toxicity of Jatropha Seed Oil to *Callosobruchus maculatus* (Coleoptera: Bruchidae) and its parasitoid, *Dinarmus basalis* (Hymenoptera: Pteromalidae). *J. Appl. Sci. Res.* **2008,** *8,* 945–951.

Bodnaryk, R. P.; Fields, P. G.; Xie, Y.; Fulcher, K. A.In Secticidal Factor from Peas. U.S. Patent 5,955,082, 1999.

Boeke, S.; Barnaud, C.; Loon, J. J. A. V.; Kossou, D. K.; Huis, A. V.; Dicke, M. Efficacy of Plant Extracts against the Cowpea Beetle, *Callosobruchus maculates*. *Int. J. Pest Mang.* **2004a,** *50* (4), 251–258.

Boeke, S.; Kossou, D. K.; Huis, A. V.; Loon, J. J. A. V.; Dicke, M. Field Trials with Plant Products to Protect Stored Cowpea against Insect Damage. *Int. J. Pest Manag.* **2004b,** *50* (1), 1–9.

Borikar, P. S.; Pawar, V. M. Relative Efficacy of Some Grain Protectants against *Callosobruchus chinensis* (Lin). *Pest. Res. J.* **1995,** *7,* 125–127.

Bowery, S. K.; Pandey, N. D.; Tripathi, R. A. Evaluation of Certain Oil Seed Cake Powders as Grain Protectant against *Sitophilus oryzae* Linn. *Indian J. Ent.* **1984,** *16,* 196–200.

Buckle, J. *Clinical Aromatherapy: Essential Oils in Practice;* Churchill Livingstone: Edinburgh, Scotland, 2003; p 416.

Chaubey, M. K. Insecticidal Activity of *Trachyspermum ammi* (Umbelliferae), *Anethum graveolens* (Umbelliferae) and *Nigella sativa* (Ranunculaceae) Essential Oils against Stored Product Beetle *Tribolium castaneum* H (Coleoptera: Tenebrionidae). *African J. Agril. Res.* **2007,** *2* (11), 596–600.

Chaubey, M. K. Fumigant Toxicity of Essential Oils from Some Common Spices against Pulse Beetle, *Callosobruchus chinensis* (Coleoptera: Bruchidae). *J. Oleo Sci.* **2008,** *57* (3), 171–179.

Chahal, K. K.; Kumari, M.; Joia, B. S.; Chhabra, B. R. Chemistry and Potential of Lemon Grass Oil as Stored Grain Protectant. *Pestic. Res. J.* **2007,** *19* (2), 141–144.

Chairat, C.; Siripan, T.; Suratwadee, J.; Somchai, I. Fumigant Toxicity of Eucalyptus Oil against Three Stored Product Beetles. *Thai. J. Agric. Sci.* **2002,** *35* (3), 265–272.

Chander, H.; Ahmed, S. M. Effect of Some Plant Materials on the Development of Rice Moth, *Corcyra cephalonica* (Staint.). *Entomon.* **1986,** *11,* 273–276.

Chander, H.; Ahmed, S. M. Laboratory Evaluation of Natural Embelin as a Grain Protectant against Some Insect-Pests of Wheat in Storage. *J. Stored Prod. Res.* **1987,** *23,* 41–46.

Chander, H.; Ahmed, S. M. Potential of Some New Plant Products as Grain Protectants against Insect Infestation. *Bull. Grain Tech.* **1983,** *21,* 179–188.

Chander, H.; Kulkarni, S. G.; Berry, S. K. *Acorus calamus* Rhizomes as a Protectant of Milled Rice against *Sitophilus oryzae* and *Tribolium castaneum. J. Food. Sci. Tech.* **1990,** *27,* 171–174.

Chander, H.; Nagender, A.; Ahuja, D. K.; Berry, S. K. Laboratory Evaluation of Plant Extracts as Repellents to the Rust Red Flour Beetle, *Tribolium castaneum* (Herbst), on Jute Fabric. *Int. Pest Control.* **1999,** *41,* 18–20.

Chellappa, K.; Chelliah, S. Studies on the Efficiency of Malathion and Certain Plant Products in the Control of *Sitotroga cerealella* and *Rhyzopertha dominica* Infesting Rice Grains. *Madras agric. J.* **1976,** *63,* 190–192.

Chiranjeevi, C. Relative Efficacy of Some Indigenous Plant Material and Ashes on the Egg Laying Capacity, Adult Emergence and Developmental Period of Pulse Beetle in Green Gram. *Andhra Agric. J.* **1991,** *38,* 285–287.

Choudhary, B. S.; Pathak, S. C. Efficacy of Organic Materials for the Control of *Callosobruchus chinensis* Linn. *Indian J. Plant Prot.* **1989,** *17,* 47–51.

Coats, J. R.; Kar, L. L.; Drewes, C. D. Toxicity and Neurotoxic Effects of Monoterpenoids in Insects and Earthworms. *In Naturally Occurring Pest Bioregulators;* Hedin, P. A., Ed.;

ACS Symposium Series No. 449. American Chemical Society: Washington, DC. 1991; pp 305–316.

Coitinho, R. L.; B. de. C.; Oliveira, J. V.; De.; Gondim-Junior, M. G. C.; Camara, C. A. G. da. Toxicity of Vegetable Oils to Adults of *Sitophilus zeamais* Mots. (Coleoptera: Curculionidae) in Stored Maize Grain. *Rev. Bras. Armazena.* **2006,** *31* (1), 29–34.

Coppen, J. J. W. *Flavours and Fragrances of Plant Origin;* Food and Agricultural Organisation: Rome, 1995, p 101.

Cosimi, S.; Rossi, E.; Coioni, P. L.; Canale, A. Bioactivity and Qualitative Analysis of some Essential Oil from Mediterranean Plants against Stored Product Pests: Evaluation of Repellency against *Sitophilus zeamais, Cryptolestes ferrugineus* and *Tenebrio molitor. J. Stored Prod. Res.* **2009,** *45,* 125–132.

Dakshinamurthy, A.; Goel, S. C. In *Insect Management in Grain and Seed Storage of Wheat Using Non-Hazardous Materials,* Proceedings of the National Symposium on Growth, Development and Control Technology of Insect Pests, 1992; pp 265–268.

Damte, T.; Chichaybelu, M. The Efficacy of Some Botanicals in Controlling Adzuki Bean Beetle, *Callosobruchus chinensis* in Stored Chickpea. *Trop. Sci.* **2002,** *42* (4), 192–195.

Das, G. P.; Karim, M. A. Effectiveness of Neem Seed Kernel Oil as Surface Protectant against the Pulse Beetle, *Callosobruchus chinensis* L. (Bruchidae: Coleoptera) *Trop. Grain Legume Bull.* **1986,** *33,* 30–33.

Das, G. P.; Effect of Different Concentrations of Neem Oil on the Adult Mortality and Oviposition of *Callosobruchus chinensis* L. (Bruchidae: Coleoptera). *Indian J. Agric. Sci.* **1986,** *10,* 743–744.

Das, G. P. Effect of the Duration of Storing Chickpea Seeds Treated with Neem Oil on the Oviposition of the Brucid *Callosobruchus chinensis* L. (Bruchidae: Coleoptera). *Bangladesh J. Zool.* **1989,** *17,* 199–201.

Das, G. P.; Efficacy of Neem Oil on the Egg and Grub Mortality of *Callosobruchus chinensis* L. (Bruchidae: Coleoptera). *Trop. Grain Legume Bull.* **1987,** *34,* 14–15.

Dayal, R.; Tripathi, R. A.; Renu. Comparative Efficacy of Some Botanicals as Protectant against *Sitophilus oryzae* in Rice and its Palatability. *Ann. Plant Prot. Sci.* **2003,** *11* (1), 160–162.

Deka, S. A Study on the Use of Indigenous Technological Knowledge (ITK) for Rice Pest Management in Assam. *Insect Environ.* **2003,** *9* (2), 91–92.

Demissie, G.; Teshome, A.; Abakemal, D.; Tadesse, A. Cooking Oils and Triplex in the Control of *S. zeamais* M. (Coleoptera : Curculionidae) in Farm Stored Maize. *J. Stored Prod. Res.* **2008,** *44,* 173–178.

Deshpande, V. K.; Sibi, V. G.; Vyakaranahal, B. S. Effect of Botanical Seed Treatment against *Callosobruchus chinensis* (L.) on Viability and Vigour of Blackgram Seeds During Storage. *Seed Res.* **2004,** *32* (2), 193–196.

Devaraj, K. C.; Srilatha, G. M. In *Antifeedant and Repellent Properties of Certain Plant Extracts against the Rice Moth, Corcyra cephalonica (Stainton),* The Symposium of Botanical Pesticides in Integrated Pest Management, Rajahmundry, India, Jan 21–24, 1990; 1993, pp 159–165.

Dhakshinamoorthy, G.; Selvanarayanan, V. Evaluation of Certain Natural Products against Pulse Beetle, *Callosobruchus maculatus* (F.) Infesting Stored Green Gram. *Insect Environ.* **2002,** *8* (1), 29–30.

Diarisso, N. Y.; Pendleton, B. B. Effectiveness of Plant Powder in Controlling Lesser Grain Borer in Stored Sorghum Grain. *Int. Sorghum Millets Newslet.* **2005,** *46,* 62–63.

Doharey, R. B.; Katiyar, R. N.; Singh, K. M. Eco-Toxicological Studies on Pulse Beetle Infesting Green Gram Effect of Edible Oil Treatments on the Germination of Green Gram, *Vigna radiata* (L.) Seeds. *Indian J. Ent.* **1983,** *45,* 414–419.

Doharey, R. B.; Katiyar, R. N.; Singh, K. M. Effect of Edible Oils in Protection of Greengram (*Vigna radiata)* Seed from Pulse Beetles (*C. chinensis* and *C. maculatus*). *Indian J. Agric. Sci.* **1988,** *58,* 151–154.

Don-Pedro, K. N. Insecticidal Activity of Some Vegetable Oils against *Dermestes maculatus*, Degeer (Coleoptera: Dermestidae) on Dried Fish. *J. Stored Product Res.* **1989,** *25* (2), 81–96.

Don-Pedro, K. N. Fumigant Toxicity of Citrus Peel Oils against Adult and Immature Stages of Storage Insect. *Pest. Sci.* **1996,** *47* (3), 213–223.

Dubey, N. K.; Kumar, R.; Simmonds, M. S. J.; Houghton, P. J. Effect of Some Essential Oils on Oviposition and Adult Mortality of Bruchids in Stored Cow Peas. *Indian J. Entomol.* **2004,** *66* (4), 374–376.

Dwivedi, S. C.; Kumari, A. Effectiveness of Citrus Clean against *Callorobruchus chenesis* (L.) Infesting Cowpea. *J. Adv. Zool.* **2000,** *21* (1), 61–64.

Dwivedi, S. C.; Kumar, R. Evaluation of *Cassia occidentalis* Leaf Extract on Development and Damage Caused by *Trogoderma granarium*, Khapra Beetle. *J. Ecotox. Environ.* **1998,** *8,* 55–58.

Dwivedi, S. C.; Sharma, Y. Evaluation of *Anethum sowa* as Seed Protectant against Larvae of *Trogoderma granarium* (E), *Baltic J. Coleopterol.* **2003,** *3* (1), 57–61.

Ecija, N. *Screening and Evaluation of Candidate Botanical Insecticides against Major Storage Pests of Legumes. Bureau of Postharvest Research and Extension;* Department of Agriculture: CLSU Compd., Munoz, Philippines, 2000; p 44.

El-Sheikh, T. A. A.; Hassanein, A. A.; Radwan, E. M. M.; Abo-Yousef, H. M. Biochemical Effects of Certain Plant Oils on the Lesser Grain Borer, *Rhyzopertha dominica* (F.). *Ann. Agric. Sci. Cairo.* **2005,** *50* (2), 729–737.

El-Ghar, G. E. S. A.; El-Sheika, A. E. Effectiveness of Some Plant Extracts as Surface Protectants of Cowpea Seeds against the Pulse Beetle, *Callosobruchus chinensis. Phytoparasitica.* **1987,** *15,* 109–113.

El-Lakwah, F. A.; Khaled, O. M.; Khattab, M. M.; Abdel-Rahman, T. A. Toxic Effects of Extracts and Powders of Certain Plants against the Rice Weevil *Sitophilus oryzae* (L.). *Ann. Agric. Sci.* **1997,** *35,* 553–566.

El-Sayed, F. M. A.; Etman, A. A. M.; Abdel-Razik, M. Effectiveness of Natural Oils in Protecting Some Stored Products from Two Stored Product Pests. *Bull. Faculty Agric. Uni. Cairo.* **1989,** *40* (20), 409–418.

Estrela, J. L.V.; Fazolin, M.; Catani, V.; Alecio, M. R.; Lima, M. S. Toxicity of Essential Oils of *Piper aduncum* and *Piper hispidinervum* against *Sitophilus zeamais. Pesq. Agropec. Bras.* **2006,** *41* (2), 217–222.

FAO 1996; The Use of Spice and Medicinal as Bioactive Proctants for Grains. Agriculture service Bulletin No. 137; Food and Agricultural Organisation of the United Nations: Rome, 1996, pp 201–213.

Fields, P. G. Effect of *Pisum sativum* Fractions on the Mortality and Progeny Production of Nine Stored-Grain Beetles. *J. Stored Prod. Res.* **2006,** *42,* 86–96.

Fukami, J.; Nakatsugaua, T.; Narahashi, J. The Relation between Chemical Structure and Toxicity in Rotenone Derivatives. *Japan J. Appl. Ent. Zool.* **1959,** *3,* 259–265.

Gakuru, S.; Foua, B. K. Compared Effect of Four Plant Essential Oils against Cowpea Weevil *Callosobruchus maculatus* F. and Rice Weevil *Sitophilus oryzae* L. *Tropicultura.* **1995,** *13,* 143–146.

Ghosal, T. K.; Senapati, S. K.; Deb, D. Efficacy of Different Edible and Non-Edible Oils in Suppressing Egg Laying, Emergence of Adults and Developmental Period of *C. chinensis. Environ. Ecol.* **2008,** *26* (2), 673–678.

Ghosal, T. K.; Senapati, S. K.; Deb, D. C. Pesticidal Effect of Edible and Non Edible Oils on Pulse Beetle, *Callosobruchus chinensis* (Coleoptera: Bruchidae). *J. Ecobiol.***2005,** *17* (4), 321–327.

Girish, G. K.; Jain, S. K. Studies on the Efficacy of Neem Seed Kernel Powder against Stored Grain Pests. *Bull. Grain Tech.* **1974,** *12,* 226–228.

Globade, A. A.; Adebayo, T. A. Fumigant Effects of Some Volatile Oils on Fecundity and Adult Emergence of *Callosobruchus maculatus* F. *Insect Sci. Appl.* **1993,** *14* (5), 631–636.

Golob, P.; Gudrups, I. *The Use of Spices and Medicinals as Bioactive Protectants for Grains;* FAO Agricultural and Science Bulletin No. 137; FAO: Rome, 1999.

Golob, P.; Webley, D. J. *The Use of Plants and Minerals as Traditional Protectants of Stored Products*; Report of Tropical Product Institute, No-G138, VI +32, 1980.

Goel, D.; Goel. R.; Singh, V.; Ali, M.; Mallavarapu, G.R.; Kumar, S. Composition of the Essential Oil from the Root of *Arlemisi annua. J. Natural Med.* **2009,** *61*(4), 458–461.

Grainge, M.; Ahmed, S. *Handbook of Plants with Pest-Control Properties;* Wiley & Sons: New York, NY, USA, 1988; p 112.

Gundurao, H. R.; Majumder, S. K. Repellency of *Kaempferia galanga* L. (Zingiberaceae) to Adults of *Tribolium castaneum* (H.) *Sci. Cult.* **1966,** *32,* 461–462.

Gu, Y. F.; Shang, F. D.; Xue, J. Z. The Repellencies of Six Plant Substances to Adult *Tribolium castanesum* (Herbst). *Acta Agric. Univ. Henanensis.* **1997,** *31,* 277–279.

Haque, M. T.; Islam, M. Z.; Husain, M. Comparative Study of Different Plant Oils in Controlling the Pulse Beetle in Mungbean. *Bangladesh J. Training Devt.* **2002,** *15* (12), 201–205.

Haryadi, Y.; Yuniarti, S. In *Study on the Insecticidal Effects of Custard Apple (Annona reticulata L.) and Mindi (Melia azedarach L.) Leaves against Sitophilus zeamais Motschulsky (Coleoptera: Curculionidae), Advances in Stored Product Protection,* Proceedings of the 8th International Working Conference on Stored Product Protection, UK, 2003; pp 600–602.

Haubruge, E.; Lognay, G.; Marlier, M.; Danhier, P.; Gilson, J. C.; Gaspar, C. A Study of the Toxicity of Five Essential Oils Extracted from *Citrus sp.* to *Sitophilus zeamais* M, *Prostephanus truncatus* (Horn) and *Tribolium castaneum* Herbst. *Madedelingen van de Faculteit Landbouwwe Tenschappen.* **1989,** *54,* 1083–1093.

Ho, S. H.; Koh, L.; Ma, Y.; Huang, Y.; Sim, K. Y. The Oil of Garlic, *Allium sativum* L. (Amaryllidaceae), as a Potential Grain Protectant against *Tribolium castaneum* (Herbst) and *Sitophilus zeamais* Motsch. *Postharvest Bio. Technol.* **1996,** *9* (1), 41–48.

Ho, S. H.; Ma, Y.; Huang, Y. Anethole, a Potent Insecticide from *Ilicium verum* Hook F., against Two Stored Product Insects. *Int. Pest Control.* **1997a,** *39,* 50–51.

Ho, S.H.; Koh, L.; Ma, Y.; Huang, Y.; Sim, K.Y. The Oil of Garlic, *Allium sativum* L. as a Potential Grain Protectant against *Tribolium castaneum* and *Sitophilus zeamais. Postharvest Bio. Tech.* **1997b,** *9,* 41–48.

Hitokoto, H.; Morozumi, S.; Wauke, S.; Kurata, H. Inhibitory Effects of Spices on Growth and Toxin Production of Toxigenic Fungi. *Appl. Environ. Microbiol.* **1980,** *39,* 818–822.

Hou, H M.; Feng, J. T.; Chen, A. L., Zhang, X. Studies on the Bioactivity of Essential Oils against Insects. *Natural Prod. Res. Devel.* **2002,** *14* (6), 27–30.

Hou, X.; Fields, P. G. Granary Trial of Protein-Enriched Pea Flour for the Control of Three Storedproduct Insects in Barley. *J. Econ. Entomol.* **2003,** *96,*1005–1015.

Huang, Y.; Ho, S. H.; Kini, R. M.; Huang, Y.; Ho, S. H. Bioactivities of Safrole and Isosafrole on *Sitophilus zeamais* and *Tribolium castaneum*. *J. Econ. Entomol.* **1999,** *92* (3), 676–683.

Huang, Y.; Chen, S. X.; Ho, S. H. Bioactivities of Methyl Allyl Disulfide and Diallyl Trisulfide from Essential Oil of Garlic to Two Species of Stored-Product Pests, *Sitophilus zeamais* (Coleoptera: Curculionidae) and *Tribolium castaneum* (Coleoptera: Tenebrionidae). *J. Econ. Entomol.* **1999,** *93* (2), 537–543.

Huang, Y., Lam, S. L.; Ho, S. H. Bioactivities of Essential Oil from *Elletaria cardamomum* (L) Maton. to *Sitophilus zeamais* and *Tribolium castaneum*. *J. Stored Prod. Res.* **2000,** *36,* 107–117.

Huang, Y.; Ho, S. H. Toxicity and Antifeedant Activities of Cinnamaldehyde against the Grain Storage Insects, *Tribolium castaneum* (Harbst) and *Sitophilus zeamais* Motsch. *J. Stored. Prod. Res.* **1998,** *34,* 11–17.

Huang, Y.; Tang, J. M. W. L.; Kini, R. M.; Ho, S. H.Toxic and Antifeedant Action of Nutmeg Oil against *Tribolium castaneum* (Herbst) and *Sitophilus zeamais* Motsch. *J. Stored Prod. Res.* **1997,** *33* (4), 289–298.

Ignatowicz, S.; Wesolowska, B. Repellency of Powderd Plant Material of the Indian Neem Tree, the Labrador tea, and the Sweet-Flag, to Some Stored Product Pests. *Pol. P. Entomol.* **1996,** *65,* 61–67.

Ilbudo, Z.; Dabire, L. C. B.; Nebie, R. C. H.; Dicko, I. O.; Dugravot, S.; Cortesero, A. M.; Sanon, A. J. Biological Activity and Persistence of Four Essential Oils towards the Main Pest of Stored Cowpeas *Calosobruchus maculatus* F. (Coleoptera : Bruchidae). *J. Stored Prod. Res.* **2010,** *46,* 235–239.

Iloba, B. N.; Ekrakene, T. Comparative Assessment of Insecticidal Effect of *Azadirachta indica, Hyptis suaveolens* and *Ocimum gratissimum* on *Sitophilus zeamais* and *Callosobruchus maculatus*. *J. Biol. Sci.* **2006,** *6* (3), 626–630.

Isikber, A. A.; Alma, M. H.; Kanat, M.; Karchi, A. Fumigant Toxicity of Essential Oils from *Laurus nobilis* and *Rosmarinus officinalis* against all Life Stages of *Tribolium confusum*. *Phytoparasitica.* **2006,** *34* (2), 167–177.

Isman, M. B. Pesticides Based on Plant Essential Oils. *Pestic. Outlook.* **1999,** *10,* 68–72.

Isman, M. B. Plant Essential Oils for Pest and Disease Management. *Crop Prot.* **2000,** *19,* 603–608.

Isman, M. B. Botanical Insecticides, Deterrents, and Repellents in Modern Agriculture and an Increasingly Regulated World. *Annu. Rev. Entomol.* **2006,** *51,* 45–66.

Ivbijaro, M. F.; Ligan, C.; Youdeowei, A. Control or Rice Weevils, *Sitophilus oryzae* (L.) in Stored Maize with Vegetable Oils. *Agric. Ecosys. Environ.* **1985,** *14,* 237–242.

Ivbijaro, M. F. The Efficacy of Seed Oils of *Azadirachta indica* A. Juss and *Piper guineenese* Schum and Thonn on the Control of *Callosobruchus maculatus* F. *Insect Sci. Appl.* **1990,** *11,* 149–152.

Ivbijaro, M. F. Toxic Effects of Groundnut Oil on the Rice Weevil *Sitophilus oryzae* (L) *Insects Sci. Appl.* **1984,** *5,* 251–252.

Jacob, S.; Sheila, M. K. Treatment of Green Gram Seeds with Oils against the Infestation of the Pulse Beetle *Callosobruchus chinensis* L. *Plant Prot. Bull.* **1990,** *42,* 9–10.

Jadhav, K. B.; Jadhav, L. D. Use of Some Vegetable Oils, Plant Extracts and Synthetic Products as Protectants from Pulse Beetle, *Callosobruchus maculatus* Fabr. in Stored Gram. *J. Food Sci. Technol.* **1984,** *21,* 110–113.

Jaipal, S.; Singh, Z.; Malik, O. P. Insecticidal Activity of Various Neem Leaf Extract against *Rhyzopertha dominica*, a Stored Grain Pests. *Neem Newslett.* **1984,** *1,* 35–36.

Jamil, K.; Rani, U.; Thyagarajan, G. Water Hyacintha-a Potential new Juvenile Hormone Mimic. *Int. Pest. Control.* **1984**, *26*, 106–108.

Jbilou, R.; Ennabili, A.; Sayah, F. Insecticidal Activity of Four Medicinal Plant Extracts against *Tribolium castaneum* (Herbst) *Afr. J. Biotechnol.* **2006**, *5* (10), 936–940.

Jembere, B.; Obengofori, D.; Hassanali, A. Products Derived from the Leaves of *Ocimum kilimandscharicum* (Labiatae) as Post-Harvest Grain Protectants against the Infestation of Three Major Stored Product Insects Pests. *Bull. Entomol. Res.* **1995**, *85*, 361–367.

Jilani, G.; Malik, M. M. Studies on Neem Plant as Repellent against Stored Grain Insects. *Pakistan J. Sci. Ind. Res.* **1973**, *16*, 251–254.

Jilani, G.; Saxena, R. C. Repellent and Feeding Deterrent Effect of Turmeric Oil, Sweet Flag Oil, Neem Oil, and Neem Based Insecticide against Lesser Grain Borer *J. Econ. Ent.* **1990**, *83*, 629–634.

Jilani, G.; Saxena, R. C.; Rueda, B. P. Repellent and Growth - Inhibiting Effects of Turmeric Oil, Sweet Flag Oil, Neem Oil and Margosan-O on Red Flour Beetle. *J. Econ. Ent.* **1988**, *81*, 1226–1230.

Jilani, G. *Use of Botanical Materials for Protection of Stored Food Grains against Insect Pest*, A Review Research Planning Workshop on Boanical Pest Control Project, IRRI: Los Banos, CA, 1984.

Jotwani, M. G.; Sircar, P. Neem Seed as a Protectant against Stored Grain Pests, Infesting Wheat Seed. *Indian J. Ent.* **1965**, *27*, 160–164.

Juneja, R. P.; Patel, J. R. Persistence of Botanical Materials as Protectant of Green Gram (*Vigna radiata* (L.) Wilczek) against Pulse Beetle (*Callosobruchus analis* F.). *Seed Res.* **2002**, *30* (2), 294–297.

Kalinovic, I.; Martincic, J.; Rozman, V.; Guberac, V. Insecticidal Activity of Substances of Plant Origin against Stored Product Insects. *Ochrana Rostlin.* **1997**, *33*, 135–142.

Karakoc, O. C.; Gokce, A.; Telci, I. Fumigant Activity of Some Plant Essential Oils against *Sitophilus oryzae* L., *Sitophilus granarius* L. (Col.: Curculionidae) and *Acanthoscelides obtectus* Say. (Col.: Bruchidae). *Turkiye Entomol. Dergisi.* **2006**, *30* (2), 123–135.

Karci, A.; Isikber, A. A. Ovicidal Activity of Various Essential Oils against Confused Flour Beetle, *Tribolium confusum* Jacquelin duVal (Coleoptera: Tenebrionidae). *Bulletin-OILB/SROP.* **2007**, *30* (?), 251–258.

Kathirvelu, C.; Ezhilkumar, S. Preliminary Screening of Botanicals against Lesser Grain Borer (*Rhizopertha dominica* F.) in Stored Paddy. *Insect Environ.* **2003**, *9* (4), 160–161.

Kavallieratos, N. G.; Athanassiou, C. G.; Saitanis, C. J.; Kontodimas, D. C. Roussos, Effect of Two Azadirachtin Formulations against Adults of *Sitophilus oryzae* and *Tribolium confusum* on Different Grain Commodities. *J. Food Prot.* **2007**, *70*, 1627–1632.

Kemabonta, K. A.; Okogbue, F. *Chenopodium ambroisoides* (Chenopodiaceae) as A Grain Protectant for the Control of the Cowpea Pest *Callosobruchus maculatus* (Coleoptera Bruchidae). *J. Fruit Ornam. Plant Res.* **2002**, *10*, 165–171.

Kestenholz, C.; Stevenson, P. C. In *Gardenia* spp. *as a Source of Botanical Pesticide against the Rice Weevil, Sitophilus oryzae L. (Coleoptera; Curculionidae), in Sri Lanka,* Proceedings of an International Congerence, Brighton, UK, 1998; pp 543–548.

Kestenholz, C.; Stevenson, P. C.; Belmain, S. R. Comparative Study of Field and Laboratory Evaluation of the Ethnobotanical *Cassia sophera* L. (Leguminosae) for Bioactivity against the Storage Pests *Callosobruchus maculatus* (F.) (Coleoptera : Bruchidae) and *Sitophilus oryzae* (L.) (Coleoptera : Curuculionidae). *J. Stored Prod. Res.* **2007**, *43* (1), 79–86.

Ketkar, C. M. In Use of Tree-Derived Non-Edible Oils as Surface Protectants for Stored Legumes against *Callosobruchus maculatus* and *C. chinensis*, Proceedings of the 3rd International Neem Conference, Nairobi, Kenya, 1987; pp 535–542.

Ketkar, C. M. *Utilization of Neem, Azadirachta indica Juss. and its By-Products*; Final Technical Report: Khadi and Village Industries Commission, Bombay, India, 1976; p 224.

Ketoh, G. K.; Glitho, A. I.; Huignard, J. Susceptibility of the Bruchid *Callosobruchus maculatus* (Coleoptera: Bruchidae) and its Parasitoid *Dinarmus basalis* (Hymenoptera: Pteromalidae) to Three Essential Oils. *J. Econ. Entomol.* **2002**, *95* (1), 174–182.

Ketoh, G. K.; Koumaglo, H. K.; Glitho, I. A.; Auger, J.; Huignard, J. Essential Oils Residual Effects on *Callosobruchus maculatus* (Col. : Bruchidae) Survival and Female Reproduction and Cowpea Seed Germination. *Int. J. Trop. Insect Sci.* **2005**, *25* (2), 129–133.

Khaire, V. M.; Kachare, B. V.; Mote, V. N. Efficacy of Different Vegetable Oils as Grain Protectants against Pulse Beetle, *Callosobruchus chinesis* L. in Increasing Storability of Pigeonpea. *J. Stored Prod. Res.* **1992**, *28*, 153–156.

Khajuria, S.; Malik, K.H. Preliminary Screening of Botanicals against *Sitophilus oryzae* (L.). *Insect Environ.* **2003**, *9* (1), 3–4.

Khan, M. I. Efficacy of *Acorus calamus* L. rhizome Powders against Pulse Beetle (*Callosobruchus chinensis* L.). *PKV Res. J.* **1986**, *10*, 72–74.

Khanuja, S. P. S.; Kumar, S.; Shasany, A. K.; Dhawan, S.; Darokar, M. P.; Tripathi, A. K.; Satapathy, S.; Kumar, T. R. S.; Gupta, V. K.; Tripathi, A. K.; Awasti, S.; Prajapati, V.; Naqvi, A. A.; Agrawal, K. K.; Bahl, J. R.; Singh, A. K.; Ahmed, A.; Bansal, R. P.; Krishna, A.; Saikia, D. Multiutility Plant 'Ganga' of *Mentha spicata* var. *viridis*. *J. Medic. & Arom. Plant Sci.* **2001**, *23* (2), 113–116.

Khare, B. P. *Insect Pests of Stored Grain and their Control in Uttar Pradesh;* Pantnagar University College of Agriculture and Technology: Pantnagar, India, 1972; p 135.

Kim, S. L.; Roh, J. Y.; Kim, D. H.; Lee, H. S.; Ahn, Y. J. Insecticidal Activites of Aromatic Plant Extracts and Essential Oils against *Sitophilus oryzae* and *Callosobruchus chinensis*. *J. Stored Prod. Res.* **2003**, *39* (3), 293–303.

Klingauf, F.; Bestmann, H. J.; Vostrowsky, O.; Michaelis, K. The Effect of Essential Oils on Insect Pests. *Mitt. Dtsch. Ges. Allge. Angew. Entomol.* **1983**, *4*, 123–126.

Kostyukovsky, M.; Ravid, U.; Shaaya, E. *InThe Potential Use of Plant Volatiles for the Control of Stored Product Insects and Quarantine Pests in Cut Flowes*, Proceedings of the International Conference on Medicinal and Aromatic Plants Possibilities and Limitations of Medicinal and Aromatic Plant Production in the 21st Century, Jul 8–11, 2001; Bernath, J., Zamborine Nemeth, E., Crakeram, L., Kock, O., Eds.; Hungary Acta Horticulturae: Budapest, Hungary, 2002;Vol. 576, pp 347–358.

Koul, O.; Isman, M. B.; Ketkar, C. M. Properties and Uses of Neem, *Azadirachta indica. Can. J. Bot.*1990, *68*, 1– 11.

Koumaglo, K. H.; Akpagana, K.; Glitho, A. I.; Garneau, F. X.; Gagnon, H.; Jean, F. I.; Moudachirou, M.; Addae, M. I. Geranial and Neral, Major Constituents of *Lippia multiflora* Moldenke Leaf Oil. *J. Essen. Oil Res.* **1996**, *8*, 237–240.

Kouninki, H.; Haubruge, E.; Noudjou, F. E.; Lognay, G.; Malaisse, F.; Ngassoum, M. B.; Goudoum, A.; Mapongmetsem, P. M.; Ngamo, L. S. T.; Hance, T. Potential use of essential oils from Cameroon Applied as Fumigant or Contact Insecticides against *Sitophilus zeamais* Motsch. (Coleoptera: Curculionidae). *Comm. Agric. Appl. Bio. Sci.* **2005**, *70* (4), 787–792.

Kouninki, H.; Ngamo, L. S. T.; Hance, T.; Ngassoum, M. B. Potential use of Essential Oils from Local Cameroonian Plants for the Control of Red Flour Weevil *Tribolium castaneum* (Herbst.) (Coleoptera: Tenebrionidae). *Afr. J. Food, Agric., Nutr. Dev.* **2007**, *7* (5), 205–212.

Krishna, A.; Prajapati, V.; Bhasney, S.; Tripathi, A. K.; Kumar, S. Potential Toxicity of New Genotypes of *Tagetes* (Asteraceae) Species against Stored Grain Insect Pests. *Int. J. Trop. Insect. Sci.* **2005**, *25* (2), 122–128.

Krishnarajah, S.R.; Ganesalingam, V. K.; Senanayake, U. M. Repellancy and Toxicity of Some Plants Oils and their Terpene Components to *Sitotroga cerealella* (Oliver) (Lep., Gelechiidae). *Tropical Sci.* **1985**, *25*, 249–252.

Kumar, R. Fumigant toxicity of Essential oils against *Sitophilus oryzae, Rhizopertha dominica* and *Tribolium castaneum* in Stored Wheat. *Indian J. Ento.* **2015**, Accepted.

Kumar, R. Effect of Essential Oils on Mortality of *Sitophilus oryzae* at Different Time Interval in Stored Wheat. *Ann. Plant Protect. Sci.* **2016**, *24* (1),182–184.

Kumar, R. Insecticidal Activity of Some Essential Oils against *Rhyzopertha dominica* on Stored Wheat. *Ann. Plant Protect. Sci.* **2016**, *24* (2), 236–239.

Kumar, R. Chemical Constituents and Fumigant Toxicity of Essential Oils against Insect Pests of Stored Rice. *Ann. Plant Protect. Sci.* **2016**, *24* (1), 18–23.

Lale, N. E. S. The Biological Effects of Three Essential Oils on *Callosobruchus maculatus* (F.) (Coleoptera: Bruchidae). *J. Afr. Zool.* **1991**, *105*, 357–362.

Lale, N. E. S. An Overview of the Use of Plant Products in the Management of Stored Product Coleoptera in the Tropics. *Post Harvest News Inf.* **1995**, *6*, 69–75.

Lale, N. E. S.; Ajayi, F. A. Suppressing Infestation of Stored Pearl Millet *Pennisetum glaucum* (L.) R. Br. by *Tribolium castaneum* (Herbst) with Insecticidal Essential Oils in Maiduguri, Nigeria. *Z. Pflanzenkr. Pflanzenschutz.* **2000**, *107*, 392–398.

Lale, N. E. S.; Kabeh, J. D. Pre-Harvest Spray of Neem (*Azadirachta indica*) Seed Products and Pirimiphos-Methyl as a Method of Reducing Field Infestation of Cowpeas by Storage Bruchids in the *Nigerian sudan savanna*. *Int. J. Agric. and Biol.* **2004**, *6* (6), 987–993.

Laudani, H.; Swank, G. R. A Laboratory Apparatus for Determining Repellency of Pyrethrum when Applied to Grain. *J. Economic Entomology.* **1954**, *47* (6), 1104–1107.

Lee, S. E.; Won, S. C.; Hoi, S. L.; Byoung, S.P. Cross –resistance of a Chlorpyrifos Methyl Resistant Strain of *Oryzaephilus surinanensis* (Coleoptera : Cucujidae) to Fumigant Toxicity of Essential Oil Extracted from *Eucalyptus globules* and its Major Monoterpene, 1,8-cineol. *J. Stored Prod. Res.* **2000**, *36*, 383–389.

Lee, B. H.; Annis, P. C.; Tumaalii, F., Lee, S. E. Fumigant toxicity of *Eucalyptus blakelyi* and *Melaleuca fulgens* Essential Oils and 1,8-cineole against Different Developmental Stages of Rice Weevil. *Phytoparasitica.* **2004**, *32* (5), 498–506.

Lee, B. H.; Lee, S. E.; Annis, P. C.; Pratt, S. J.; Park, B. S.; Tumaalii, F. Fumigant Toxicity of Essential Oils and Monoterpenes against the Red Flour Beetle, *Tribolium castaneum* Herbs. *J. Asia Pacif. Entomol.* **2002**, *5* (2), 237–240.

Lee, S. E.; Lee, B. H.; Choi, W. S.; Park, B. S.; Kim, J.G.; Campbell, B. C. Fumigant Toxicity of Volatile Natural Products from Korean Species and Medicinal Plants Towards the Rice Weevil, *Sitophilus oryzae* (L.). *Pest Mang. Sci.* **2001**, *57* (6), 548–553.

Liu, Z. L.; Ho, S. H. Bioactivity of the Essential Oil Extracted from *Evodia rutaecarpa* Hock. f. et Thomas against the Grain Storage Insects, *Sitophilus zeamais* Motsch and *Tribolium castaneum* (Herbst). *J. Stored Prod. Res.* **1999**, *35* (4), 317–328.

Liu, Z. L.; Goh, S. H.; Ho, S. H. Screening of Chinese Medicinal Herbs for Bioactivity against *Sitophilus zeamais* Motschulsky and *Tribolium castanum* (Herbst). *J. Stored Prod. Res.* **2007**, *43* (3), 290–296.

Lognay, G. C.; Verscheure, M.; Steyer, B.; Marlier, M.; Haubruge, E.; Knaepen, M. Volatile Constituents of *Agastache scrophulariaefolia* (Wild.) Kurtze leaves. *J. Essen. Oil Res.* **2002**, *14* (1), 42–43.

Lopez, M. D.; Jordan, M. J.; Pascual,V. M. J. Toxic Compounds in Essential Oils of Coriander, Caraway and Basil Active Against Stored Rice Pests. *J. Stored Prod. Res.* **2008,** *44* (3), 273–278.

Maga, R.; Broussalis, A.; Clemente, S.; Mareggiani, G.; Ferraro, G. 1,8 Cineol: Responsible for the Insecticide Activity of *Lavandula spica* Mill (lavender). *Revista Latinoamericana de Quimica.* **2000,** *28* (3), 146–149.

Mahfuz, I; Khalequzzaman, M. Contact and Fumigant Toxicity of Essential Oils Against *Callosobruchus maculates. Univ. J. Zool. Rajshahi Univ.* **2007,** *26,* 63–66.

Maina, Y. T.; Lale, N. E. S. Integrated Management of *Callosobruchus maculatus* (F.) Infesting Cowpea Seeds in Storage Using Varietal Resistance, Application of Neem (*Azadirachta indica*) Seed Oil and Solar Heat. *Int. J. Agric. and Biol.* **2004,** *6* (3), 440–446.

Malek, M. A.; Parveen, B.; Talukder, D. Insecticidal Properties of Four Indigenous Plant Extracts Against Adults of CR-1 Strain of *Tribolium castaneum* Herbst. *Bangladesh J. Ent.* **1996,** *6,* 7–11.

Malkani, A. B.; Shah, G. G.; Parihar, R.; Bhattacharyya, S. C.; Sen, N.; Sethi, K. L. In *Mentha longifolia* subsp. *himalaiensis*: A New Species as a Source of Aroma Chemicals. Proceedings of the 11th International Congress of Essential Oils, Fragrances And Flavours, New Delhi, India, 1990; pp177–180.

Mansour, M. H.; Kleeberg, H.; Zebitz, C. P. W. In *The Effectiveness of Plant Oils as Protectants of Mung Bean Vigna radiata against Callosobruchus chinensis Infestation,* Proceedings 5th Workshop Wetzlar, Germany, 1997; pp 189–200.

Matasyoh, L.G.; Matasyoh, J. C.; Wachira, F. N.; Kinyua, M.G.; Muigai, A.T.M.; Mukiama, T.K. Chemical Composition and Antimicrobial Activity of the Essential Oil of *Ocimum gratissimum* L. Growing in Eastern Kenya. *Afr. J. Biotech .* **2007,** *6,* 760–765.

Michaelraj, S.; Sharma, R. K. Toxicity of Essential Oils against Rice Moth *Corcyra cephalonica* (Stainton) in Stored Maize. *J. Entomol. Res.* **2006,** *30* (3), 251–254.

Michaelraj, S.; Sharma, K.; Sharma, R. K. Fumigation Toxicity of Some Phyto Essential Oils against Stored Insect Pests of Maize. *Pestic. Res. J.* **2007,** *19* (1), 9–14.

Michaelraj, S.; Sharma, K.; Sharma, R. K. Fumigant Toxicity of Essential Oils against Key Pests of Stored Maize. *Ann. Plant Protect. Sci.* **2008,** *16* (2), 356–359.

Mishra, D. N.; Mishra, A. K.; Tiwari, R.; Upadhyaya, P. S. Toxicity of Some Essential Oils against Pulse Beetle, *Callosobruchus maculatus* (F.). *Indian Perfumer.* **1989,** *33,* 137–141.

Mishra, A.; Dubey, N. K.; Singh, S.; Chaturvedi, C. M. Biological Activities of Essential Oil of *Chenopodium ambrosioides* Against Storage Pests and its Effect on Puberty Attainment in Japanese Quail. *Natl. Acad. Sci. Lett.* **2002,** *25* (5–6), 174–177.

Modgil, R.; Mehta, U. Effect of Oil Treatments against the Infestation of *Callosobruchus chinensis* (L.) on the Levels of B Vitamins in Stored Legumes. *Nahrung.* **1997,** *41,* 167–169.

Mohamed, M. I. E.; Abdelgaleil, S. A. M. Chemical Composition and Insecticidal Potential of Essential Oils from Egyptian Plants against *Sitophilus oryzae* (L.) (Coleoptera: Curculionidae) and *Tribolium castaneum* (Herbst) (Coleoptera: Tenebrionidae). *Appl. Entomol. Zool.* **2008,** *43* (4), 599–607.

Mohapatra, S.; Sawarkar, S. K.; Patnaik, H. P. Comparative bioassay of different solvent extracts of neem seed kernels against *Sitophilus oryzae* L. in rice. *Current Agri. Res.* **1996,** *9,* 89–93.

Mohiuddin, S.; Qureshi, R. A.; Ahmed, Z.; Qureshi, S. A.; Jamil, K.; Jyothi, K. N.; Prasuna, A. L. Laboratory Evaluation of Some Vegetable Oils as Protectants of Stored Products. *Pakistan J. Sci. Indus. Res.* **1993,** *36,* 377–379.

Mohiuddin, S.; Qureshi, R. A.; Khan, A.; Nasir, M. K. A.; Khatri, L. M.; Qureshi, S. A. Laboratory Investigations on the Repellency of Some Plant Oils to Red Flour Beetle, *Tribolium castaneum* Herbst. *Pakistan J. Sci. Indust. Res.* **1987,** *30,* 754–756.

Mondal, M.; Khalequzzaman. M. Toxicity of Essential Oils against Red Flour Beetle, *Tribolium castaneum* H. (Coleoptera : Tenebrionidae). *J. Bio. Sci.* **2006,** *14,* 43–48.

Mostafa, T. S. The Efficiency of Neem Flower and Fruit Powder against *Trogoderma granarium* Everts Adults Infesting Stored Rice Grains. *Bull. Ent. Soc. Egypt.* **1988,** *17,* 93–99.

Nakahara. K.; Alzoreky, N. S.; Yoshihashi, T.; Nguyen, H. T. T.; Hanboonsong, Y.; Trakoontivakorn, G. Insect Growth Inhibitory and Antifungal Activities of Natural Volatile Compounds from Aromatic Plants. *JIRCAS-Working Rep.* **2006,** *45,* 117–124.

Nawrot. J.; Harmatha, J. Natural Products as Antifeedants Against Stored Product Insects. *Post Harvest News Inf.* **1994,** *5,* 17–21.

Negahban, M.; Moharramipour, S. Sefidkon, F. Fumigant Toxicity of Essential Oil from *Artemisia sieberi* Besser against Three Stored Product Insects. *J. Stored Prod. Res.* **2007,** *43* (2), 123–128.

Negi, R. S.; Srivastava, M.; Saxena, M. M.; Shrivastava, M. Egg-laying and adult emergence of *Callosobruchus chinensis* on Green Gram (*Vigna radiata)* Treated with Pongam Oil. *Indian J. Ent.* **1997,** *59,* 170–172.

Nerio, L. S.; Olivero, V. O. J.; Stashenko, E. E. Repellent Activity of Essential Oils from Seven Aromatic Plants Grown in Colombia against *Sitophilus zeamais* Motschulsky (Coleoptera). *J. Stored Prod. Res.* **2009,** *45* (3), 212–214.

Ngamo, T. L. S.; Goudoum, A.; Ngassoum, M. B.; Mapongmetsem, P. M.; Kouninki, H.; Hance, T. Persistance of the Insecticidal Activity of Five Essential Oils on the Maize Weevil *Sitophilus zeamais* (motsch.) (Coleoptera: Curculionidae). *Comm. Agric. Appl. Biol. Sci.* **2004,** *69* (3), 145–147.

Ngamo, L. S. T.; Ngassoum, M. B.; Jirovetz, L.; Ousman, A.; Nukenine, E. C.; Mukala, O. E. Protection of Stored Maize against *Sitophilus zeamais* (Motsch.) by use of Essential Oils of Spices from Cameroon. Mededelingen Faculteit Landbouwkundige en Toegepaste Biologische Wetenschappen. *Universiteit Gent.* **2001,** *66* (2a), 473–478.

Niber, B. T. The Ability of Powder and Slurries from Ten Plant Species to Protect Stored Grain form Attack by *Prostephanus truncatus* Horn (Coleoptera: Bostrichidae) and *Sitophilus oryzae* (Col., Curculionidae). *J. Stored Prod. Res.* **1994,** *30,* 297–301.

Novokmet, R. K.; Kalinovic, I.; Rozman, V. Pests Control in Birds' Stored Food with Lavander Essential Oil Volatiles. *Agri. Sci. Prof. Rev.* **2002,** *8* (2), 29–36.

Nualvatna, K.; Sukprakarn, C.; Urairong, P. Use of Oil Paddy to Cover on Surface of Paddy Seed to Control Stored Paddy Insects. *Entomology and Zoology Gazette (Thailand). Warasan Kita Lae Sattawawitthaya.* **2000,** *22* (4), 299–311.

Nukenine, E. N.; Adler, C.; Reichmuth, C. Efficacy Evaluation of Plant Powders from Cameroon as Post-Harvest Grain Protectants against the Infestation of *Sitophilus zeamais* Motschulsky (Coleoptera : Curculionidae). *J. Plant Dis. Protect.* **2007,** *114* (1), 30–36.

ObengOfori, D.; Reichmuth, C. Bioactivity of Eugenol, a Major Component of Essential Oil of *Ocimum suave* (Wild.) against Four Species of Stored-Product Coleoptera. *Int. J. Pest Manage.* **1996,** *43* (1), 89–94.

Odeyemi, O. Insecticidal Properties of Certain Indigenous Plant Oils against *Sitophilus zeamais* Mots. *Appl. Entomol. Phytopathol.* **1993,** *60,* 19–27.

Ogendo, J. O.; Kostyukovsky, M.; Ravid, U.; Matasyoh, J. C.; Deng, A. L.; Omolo, E. O.; Kariuki, S. T.; Shaaya, E. Bioactivity of *Ocimum gratissimum* L. Oil and Two of its

Constituents against Five Insect Pests Attacking Stored Food Products. *J. Stored Prod. Res.* **2008,** *44* (4), 328–334.

Ohsawa, K.; Kato, S.; Honda, H.; Yamamoto, I. Pesticidal Active Substances in Tropical Plants-Insecticidal Substance from the Seeds of Annonaceae. *J. Agril. Sci. Tokyo Nogyo Daigaku.* **1990,** *34,* 253–258.

Oka, Y.; Nacar, S.; Putievsky, E.; Ravid, U.; Yaniv, Z.; Spiegel, Y. Nimaticidal Activity of Essential Oils and their Constituents against the Root –Knot Nematode. *Phytopathology.* **2000,** *90,* 710–715.

Onolemhemen, O. P.; Oigiangbe, O. N. The Biology of *Callosobruchus maculatus* Fab. on Cowpea (*Vigna unguiculata)* and Pigeonpea (*Cajanus cajan* L. Druce) Treated with Vegetable Oil and Thioral. *Samaru J. Agric. Res.* **1991,** *8,* 57–63.

Ousman, A.; Ngassoum, M. B.; Essia, N. J. J.; Ngamo, L. S. T.; Ndjouenkeu, R. Insecticidal Activity of Spicy Plant Oils against *Sitophilus zeamais* in Stored Maize in Cameroon. *Agriculture.* **2007,** *2* (2), 192–196.

Owolabi, M. S.; Oladimeji, M. O.; Lajide, L.; Singh, G.; Marimuthu, P.; Isidorov, V. A. Bioactivity of Three Plant Derived Essential Oils against the Maize Weevils *Sitophilus zeamais* (Motschulsky) and Cowpea Weevils *Callosobruchus maculatus* (Fabricius). *Electronic J. Environ. Agric. Food Chem.* **2009,** *8* (9), 828–835.

Padin, S.; Ringuelet, J. A.; Bello, D.; Cerimele, E. L.; Re, M. S.; Henning, C. P. Toxicology and Repellent Activity of Essential Oils on *Sitophilus oryzae* L. and *Tribolium castaneum* Herbst. *J. Herbs Spices Med. Plants.* **2000,** *7* (4), 67–73.

Paneru, R. B.; Duwadi, V. R.; Bhattarai, M. R. Second Year of Testing Locally Available Materials against Weevil (*Sitophilus oryzae* Linn.) in Stored Wheat. PAC Working Paper, Pakhribas Agricultural Centre: Dhankuta, Nepal. 1993; pp 44–47.

Paneru, R. B.; Patourel, G. N. J.-le; Kennedy, S. H.; Le-Potourel, G. N. J. Toxicity of *Acorus calamus* Rhizome Powder from Eastern Nepal to S*itophilus granarius* (L.) and *Sitophilus oryzae* (L.) (Col. Curculionidae). *Crop Prot.* **1997,** *16,* 759–763.

Papachristos, D. P.; Stamopoulos, D. C. Repellent, Toxic and Reproduction Inhibitory Effects of Essential Oil Vapours on *Acanthoscelides obtectus* (Say). (Coleoptera :Bruchidae). *J. Stored Prod. Res.* **2002,** *38,* 117–128.

Papachristos, D. P.; Stamopoulos, D. C. Fumigant Toxicity of Three Essential Oils on the Eggs of *Acanthoscelides obtectus* (Say). (Coleoptera :Bruchidae). *J. Stored Prod. Res.* **2004,** *40,* 517–525.

Paranagama, P. A.; Adhikari, A. A. C. K.; Abeywickrama, K. P.; Bandara, K. A. N. P. Toxicity and Repellant Activity of *Cymbopogon citratus* (D.C.) Stapf. and *Murraya koenigii* Sprang. against *Callosobruchus maculatus* (F.) (Coleoptera; Bruchidae). *Trop. Agric. Res. and Ext.* **2002,** *5* (1/2), 22–28.

Paranagama, P.; Abeysekera, T.; Nugaliyadde, L.; Abeywickrama, K. Effect of the Essential Oils of *Cymbopogon citratus, C. nardus* and *Cinnamomum zeylanicum* on Pest Incidence and Grain Quality of Rough Rice (paddy) Stored in an Enclosed Seed Box. *J. Food Agric. Envir.* **2003,** *1* (2), 134–136.

Park, I.; Le, S. G.; Choi, D. H.; Park, J. D.; Ahn, Y. J. Insecticidal Activities of Constituents Identified in the Essential Oil from Leaves of *Chamecyparis obtusa* against *Callosobruchus chinensis* (L.) and *Sitophilus oryzae* (L.). *J. Stored Prod. Res.* **2003,** *39* (4), 375–384.

Pascual, V. M. J.; Ballesta, A. M. C. Chemical Variation in an *Ocimum basilicum* Germplasm Collection and Activity of the Essential Oils on *Callosobruchus maculatus.* *Biochem. Syst. Ecol.* **2003,** *31* (7), 673–679.

PascualVillalobos, M. J.; Robledo, A. Screening for Anti-Insect Activity in Mediterranean Plants. *Indus. Crops Prod.* **1998**, *8,* 183–194.

Pascual-Villalobos, M. J. In *Volatile Activity of Plant Essential Oils against Stored-Product Beetle Pests. Advances in Stored Product Protection,* Proceedings of the 8th International Working Conference on Stored Product Protection, New York, UK, 2003; pp 648–650.

Paster, N.; Menasherov, M.; Ravid, U.; Juven, B. Antifungal Activity of Oregano and Thyme Essential Oils Applied as Fumigants Against Fungi Attacking Stored Grain. *J. Food Prot.* **1995**, *58,* 81–85.

Patel, R. A.; Patel, B. R. Evaluation of Certain Plant Products as Grain Protectants against the Rice Moth, *Corcyra cephalonica* Stainton, in Stored Rice. *Pest Mang. Eco. Zool.* **2002**, *10* (2), 121–124.

Pathak, N.; Yadav, T. D.; Jha, A. N.; Vasudevan, P. Contact and Fumigant Action of Volatile Essential Oil of *Murraya koengii* against *Callosobruchus chinensis. Indian J. Ento.* **1997**, *59* (2), 198–202.

Pathak, P. H.; Krishna, S. S. Post Embroyonic Development and Reproduction in *Corcyra cephalonica* (Staint.) (Lepidoptera:Pyralidae) on Exposure to Eucalyptus and Neem Oil Volatiles. *J. Chem. Ecol.* **1991**, *17,* 2553–2558.

Pereira, A. C. R. L.; Oliveira, J. V. de.; Gondim-Junior, M. G. C.; Camara, C. A. G. Insecticide Activity of Essential and Fixed Oils in *Callosobruchus maculatus* F (Coleoptera: Bruchidae) in Cowpea Grains *Vigna unguiculata* (L.) *Cienc. Agro. Tecnol.* **2008**, *32* (3), 717–724.

Pereira, A. C. R. L.; Oliveira, J. V. Gondim-Junior, M. G. C.; Camara, C. A. G. Influence of the Storage Period of Cowpea *Vigna unguiculata* (L.) Treated with Essential and Fixed Oils, for the Control of *Callosobruchus maculatus* F. (Coleoptera, Chrysomelidae, Bruchinae). *Cienc. Agro. Tecnol.* **2009**, *33* (1), 319–325.

Peterson, G. S.; Kandil, M. A.; Abdallah, M. D.; Farag, A. A. A. Isolation and Characterisation of Biologically Active Compounds from Some Plant Extracts. *Pest. Sci.* **1989**, *25,* 337–342.

Popovic, Z.; Kostic, M.; Popovic, S.; Skoric. S. Bioactivity of Essential Oils from Basil and Sage to *Sitophilus oryzae* L. *Biotechnol. Biotechnol.* **2006**, *20* (1), 36–40.

Prakash, A.; Rao, J. Exploitation of Newer Botanicals as Rice Grain Protectants against Angoumois Grain Moth, *Sitotroga cerealella* Oliv. *Entomon.* **2006**, *31* (1), 1–8.

Prakash, A.; Rao, J. Wild Sage, *Lippia geminata* a Paddy Grain Protectant in Storage. *Oryza.* **1984**, *21,* 209–212.

Prakash, A.; Pasalu, I. C.; Mathur, K. C. Evaluation of Plant Products as Grain Protectants in Paddy Storage. *Int. J. Ent. India.* **1982**, *1,* 75–77.

Prakash, A.; Pasalu, I. C.; Mathur, K. C. Ovicidal Activity of *Eclipta alba* Hassk. (Compositae). *Curr. Sci.* **1979**, *48,* 1090.

Prakash, A.; Pasalu, I. C.; Mathur, K. C. Use of Bel (*Aegle marmelos*) Leaf Powder as a Protectant Against Storage Pests of Paddy. *Agric. Res. J. Kerala.* **1983**, *21,* 79–80.

Prakash, A.; Rao, J.; Gupta, S. P.; Behra, J. In *Evaluation of Botanical Pesticides as Grain Protectants against Rice Weevil, Sitophilus oryzae Linn,* Proceeding of Botanical Pesticides in Integrated Pest Management, Indian Society of Tobacco Science: Rajahmundry, India, 1993; pp 360–365.

Pungitore, R. C.; Garcia, M.; Gianello, J. C.; Sosa, M. E.; Tonn, C. E. Insecticidal and Antifeedant Effects of *Junellia aspera* (Verbenaceae) Triterpenes and Derivative on *Sitophilus oryzae* (Coleoptera: Curculionidae). *J. Stored Prod. Res.* **2005**, *41,* 433–443.

Qadri, S. S. H.; Hasan, S. B. Growth Retardant Effect of Some Indigenous Plant Seeds against Rice Weevil *Sitophilus oryzae* (L.). *J. Fd. Sci. Technol.* **1978,** *15,* 121–123.

Qadri, S. S. H.; Rao, B. H. B. Effect of Combining Some Indigenous Plant Seed Extacts against Household Insects. *Pesticides.* **1977,** *11,* 21–23.

Qadri, S. S. H. Some New Indigenous Plant Repellents for Storage Pests. *Pesticides.* **1973,** *7,* 18–19.

Qureshi, S. A.; Muhiuddin, S.; Qureshi, R. A. Repellent Values of Some Common Indigenous Plants against Red Flour Beetle, *Tribolium castaneum* (Herbst.). *Pakistan J. Zool.* **1988,** *20,* 201–207.

Raghvani, B. R.; Kapadia, M. N. Efficacy of Different Vegetable Oils as Seed Protectants of Pigeonpea against *Callosobruchus maculatus* (Fab.), *Indian J. Plant Prot.* **2003,** *31* (1), 115–118.

Raguraman, S.; Singh, D. Biopotentials of *Azadirachta indica* and *Cedrus deodara* Oils on *Callosobruchus chinensis. Int. J. Pharmacog.* **1997,** *35* (5), 344–348.

Rahim, M. Biological Activity of Azadirachtin-Enriched Neem Kernel Extracts against *Rhyzopertha dominica* (F.) (Col. Bostrichidae) in Stored Wheat. *J. Stored Prod. Res.* **1998,** *34,* 123–128.

Rahman, S. M. M.; Gupta, G. P.; Sidik, M.; Rajesus, B. M.; Garcia, R. P.; Champ, B. R.; Bengston, M.; Dharmaputra, O. S.; Halid, H. In *Application of Neem Oil and Bacillus thuringiensis Preparations to Control Insects in Stored Paddy,* Proceeding of the Symposium on Pest Management for Stored Food and Feed, Bogor, Indonesia, BIOTROP Special Publication; Indonesia, 1997; Vol. 59, pp 173–197.

Raja, N.; Albert, S.; Ignacimuthu, S.; Dorn, S. Effect of Plant Volatile Oils in Protecting Stored Cowpea *Vigna unguiculata* (L.) Walpers against *Callosobruchus maculatus* (F.) (Col. : Bruchidae) Infestation. *J. Stored Prod. Res.* **2001,** *37* (2), 127–132.

Raja, N.; Ignacimuthu, S. Use of *Madhuca longifolia* Macbride Seed Oil in Controlling Pulse Beetle *Callosobruchus maculatus* F. (Coleoptera: Bruchidae). *Entomon.* **2001,** *26* (3/4), 279–283.

Rajendran, S.; Sriranjini, V. Plant Products as Fumigants for Stored-Product Insect Control. *J. Stored Prod. Res.* **2008,** *44,* 126–35.

Rajapakse, R.; Emden, H. F.V. Potential of Four Vegetable Oils and Ten Botanical Powders for Reducing Infestation of Cowpea by *Callosbruchus maculatus, C.chinensis* and *C.rhodesianus. J. Stored Prod. Res.* **1997,** *33* (1), 59–68.

Rajapakse, R.; Rajapakse, H. L. Z.; Ratnasekera, D. Effect of Botanicals on Oviposition, Hatchability and Mortality of *Callosobruchus maculatus* L. (Coleoptera: Bruchidae). *Entomon.* **2002,** *27* (1), 93–98.

Rajasekaran, B.; Kumaraswami, T. In *Control of Storage Insects with Seed Protectants of Plant Origin,* Proceeding of Behavioural and Physiological Approaches in Pest Management ‚T N A U: India, 1985; pp 15–17.

Rani, C. S.; Vijayalakhmi, K.; Rao, P. A. Vegetable Oils as Surface Protectants against Bruchid, *Callosobruchus chinensis* (L.) Infestation on Chickpea. *Indian J. Plant Prot.* **2000,** *28* (2), 184–186.

Ranpal; Verma, R. S.; Singh, S.V. Vegetable Oils as Grain Protectant against *Sitophilus oryzae* Linn. *Farm Sci. J.* **1988,** *3,* 14–20.

Rao, D. S. The Insecticidal Potential of the Corolla of Flowers. *Indian J. Ent.* **1955,** *17,* 121–124.

Rao, D. S. The Insecticidal Property of Petals of Several Common Plants of India. *Eco. Bot.* **1957,** *11,* 274–276.

Rao, J.; Prakash, A. Evaluation of New Botanicals as Paddy Grain Protectants against Lesser Grain Borer, *Rhyzopertha dominica* F. *J. Appld. Zool. Res.* **2002,** *13* (2/3), 258–259.

Reddy, A. V.; Singh, R. P. Fumigant Toxicity of Neem (*Azadiracta indica* A. juss.) Seed Oil Volatiles against Pulse Beetle, *Callosobruchus maculatus* Fab. (Col., Bruchidae). *J. Applied Ento.* **1998,** *122,* 607–611.

Reddy, V. S.; Babu, T. R.; Hussaini, S. H.; Reddy, B. M. Effect of Edible and Non-Edible Oils on the Development of Pulse Beetle, *Callosobruchus chinensis* L. and on Viability of Mungbean Seeds. *Pest Manag. Eco. Zool.* **1994,** *2* (1), 15–17.

Rees, D. P.; Dales, M. J.; Golob, P. *Alternative Methods for the Control of Stored-Product Insect Pests: A Bibliographic Database;* Natural Resources Institute: Chatham, UK, 1993; p 91.

Regnault-Roger, C. The Potential of Botanical Essential Oils for Insect Pest Control. *Int. Pest Manage. Rev.* **1997,** *2,* 25–34.

Richa, E. M.; Hashem, M. Y.; Rabie, M. Use of Some Essential Oils as Protectants against the Pulse Beetle, *Callosobruchus chinensis* (L.). *Bull. Entomol. Soci. Egypt Econo. Series.* **1993,** *20,* 151–159.

Risha, E. M.; El-Nahal, A.; K. M.; Schmidt, G. H. Toxicity of Vapours of *Acorus calamus* L. Oil to the Immature Stages of Some Stored Product Coleoptera. *J. Stored Prod. Res.* **1990,** *26,* 133–137.

Rozman, V.; Kalinovic, I.; Korunic, Z. Toxicity of Naturally Occuring Compounds of Lamiaceae and Lauvaceae to Three Stored Product Insects. *J. Stored Prod. Res.* **2007,** *43* (4), 349–355.

Sahaf, B. Z.; Moharramipour, S. Comparative Study on Deterrency of *Carum copticum* C. B. Clarke and *Vitex pseudo-negundo* (Hausskn.) Hand.-Mzt. Essential Oils on Feeding Behavior of *Tribolium castaneum* (Herbst). *Iranian J. Medi. Aro. Plants.* **2009,** *24* (4), 385–395.

Sahaf, B. Z.; Moharramipour, S.; Meshkatalsadat, M. H. Chemical Constituents and Fumigant Toxicity of Essential Oil from *Carum copticum* against Two Stored Product Beetles. *Insect Sci.* **2007,** *14* (3), 213–218.

Salama, S. L.; Mariy, F. M.; Marzouk, A. A. Efficacy of Some Plant Extracts as Seed Protectants against *Sitophilus oryzae* (L.) and *Callosobruchus muculutes* (F.). *Arab Univ. J. Agric. Sci.* **2004,** *12* (2), 771–781.

Salas, J.; Protection of Maize Seeds, *Zea mays* against Attack by *Sitophilus oryzae* through the Use of Vegetable Oils. *Agronomia Trop.* **1985,** *35,* 13–18.

Sarac, A.; Tunuc, I. Residual Toxicity and Repellency of Essential Oils to Stored Product Insects. *J. Plant Dis. Protec.* **1995,** *10* (2), 429–432.

Saradamma, K.; Dale, D.; Nair, M. R. G. K. On the Use of Neem Seed Kernal Powder as a Protectant for Stored Paddy. *Agric. Res. J.* **1977,** *15,* 102–103.

Saraswathi, L.; Rao, A. P. Repellent Effect of Citronella Oil on Certain Insects. *Pesticides.* **1987,** *21,* 23–24.

Saxena, B. P.; Koul, O. Utilization of Essential Oils for Insect Control. *Indian Perfumer.* **1978,** *22,* 139–149.

Saxena, B. P.; Koul, O.; Tikku, K.; Atal, C. K.; Suri, O. P.; Suri, K. A. Aristolochic Acid—an Insect Chemosterilant from *Aristolochia bracteata* Retz. *Indian J. Exp. Biol.* **1979,** *17,* 354–360.

Saxena, B. P.; Tikku, K.; Atal, C. K.; Koul, O. Insect Antifertility and Antifeedant Allelochemics in *Adhatoda vasica. Insect Sci. Applic.* **1986,** *7,* 489–493.

Saxena, R. C.; Dixit, O. P and Harshan, V.; Insecticidal Action of *Lantana camara* against *Collosobruchus chinensis* (Coleoptera: Bruchidae). *J. Stored Prod. Res.* **1992,** *28* (4), 279–281.

Saxena, S. C.; Yadav, R. S. In *A New Plant Extract to Suppress the Population Build-up of Tribolium castaneum (Herbst),* Proceedings of the 3rd International Working Conference on Stored Product Entomology, Manhattan, USA, 1984; pp 209–212.

Schmidt, G. H.; Risha, E. M.; El-Nahal, A. K. Reduction of Progeny of Some Stored Product Coleopteran by Vapours of *Acorus calamus* oil. *J. Stored Prod. Res.* **1991,** *27* (2), 121–127.

Schuhmacher, A.; Reiching. J; Schnitzler, P. Virucidal Effect of Peppermint Oil on the Enveloped Viruses Herpes Simplex Virus Type 1 and type 2 *in vitro*. *Phytomedicine*. **2003,** *10,* 504–510.

Seck, D.; Lognay, G.; Haubruge, E.; Wathelet, J. P.; Marlier, M.; Gaspar, M.; Severin, M. Biological Activity of the Shrub *Boscia senegalensis* (Pers.) Lam. ex Poir. (Capparaceae) on Stored Grain Insect. *J. Chem. Eco.* **1993,** *19,* 377–389.

Sendi, J. J.; Haghighian, F.; Akbar, A. R. A. Insecticidal Effects of *Artemisia annua* L. and *Sambucus ebulus* L. Extracts on *Tribolium confusum* Duv. *Iranian J. Agric. Sci.* **2003,** *34* (2), 313–319.

Shaaya, E.; Ravid, U.; Paster, N.; Juven, B.; Zisman, U.; Pissarev, V. Fumigant Toxicity of Essential Oils against Four Major Stored- Product Insects. *J. Chem. Ecol.* **1991,** *17* (3), 499–504.

Shaaya, E.; Kostjukovski, M.; Eilberg, J.; Sukprakarn, C. Plant Oils as Fumigants and Contact Insecticides for the Control of Stored Product Insects. *J. Stored Prod. Res.* **1997,** *33* (1), 7–15.

Shaaya, E.; Ravid, U.; Paster, N.; Juven, B.; Zisman, U.; Pissarev, V. Fumigant Toxicity of Essential Oils Against Four Major Stored Product Insects. *J. Chem. Eco.* **1990,** *17* (3), 499–504.

Shakarami, J.; Kamali, K.; Moharramipour, S.; Meshkatalsadat, M. Fumigant Toxicity and Repellency of the Essential Oil of Artemisia Aucheri on Four Species of Stored Products Pests. *Appl. Entomol. Phytopathol.* **2004,** *71* (2), 61–75.

Shakarami, J.; Kamali, K.; Moharramipour, S. Fumigant Toxicity and Repellency Effect of Essential Oil of *Salvia bracteata* on Four Species of Warehouse Pests. *J. Entomol. Soc. Iran.* **2005,** *24* (2), 35–50.

Sharaby, A. Some *Myrtaceae* Leaves as Protectants of Rice Against the Infestation of *Sitophilus oryzae* (L.) and *S. granarius* (L.) (Coleoptera). *Polskie Pismo Entomologiczne Bulletin Entomologique de Pologne*. **1989,** *59,* 77–382.

Sharma, M. M.; Mathur, N. M.; Srivastava, R. P. Effectiveness of Neem Kernel Powder against Lesser Grain Borer, *Rhizopertha dominica* Fab. and Rice Weevil, *Sitophilus oryzae* Linn. *Indian J. appl. Ent.* **1989,** *3,* 59–60.

Sharma, J. K.; Singh, H. B. After Effects of Some Oils on the Seedling Traits in Stored Chickpea Seeds. *Crop Res. Hisar* **1993,** *6,* 317–322.

Sharma, R. K. Efficacy of Neem Products against Storage Pests in Maize. *Ann. Agric. Res.* **1999,** *20,* 198–201.

Sharma, S.; Sachan, G. C.; Bhattacharya, A. K. A Note on the Effect of Rapeseed Mustard Oils on the Growth and Development of *Corcyra cephalonica* on Sorghum. *Indian J. Ent.* **2002,** *64* (2), 235–237.

Sharma, Y. A New Indigenous Plant Antifeedant against *Rhizopertha dominica* F. *Bull. Grain Tech.* **1983,** *21,* 223–225.

Sharma, Y. Effect of Aak Flower Extract on Different Larval Stages of the Lesser Grain Borer. *Rhizopertha dominica* Fab. *J. Adva. Zool.* **1985,** *6* (1), 8–12.

Shukla, A. C.; Shahi, S. K.; Dikshit, A. *Eucalyptus pauciflora-* a Potential Source of Sustainable, Ecofiendly Storage Pesticide. *Biotechnol. Microb. Sust. Util.* **2002,** 93–107.

Shukla, R. M.; Chand, G.; Saini, M. L. Laboratory Evaluation of Effectiveness of Edible Oils against Three Species of Stored Grain Insects. *Plant Prot. Bull.* **1992,** *44,* 14–15.

Sighamony, S.; Anees, I.; Chandrakala, T. S.; Osmani, Z. Efficacy of Certain Indigenous Plant Products as Grain Protectants against *Sitophilus oryzae* (L.) and *Rhyzopertha dominica* (F.). *J. Stored Prod. Res.* **1986,** *22,* 21–23.

Silva, G.; Lagunes, A.; Rodriguez, J. Control of *Sitophilus zeamais* (Coleoptera: Curculionidae) with Vegetable Powders Used Singly and Mixed with Calcium Carbonate in Stored Corn. *Ciencia e Investigacion Agraria.* **2003,** *30* (3), 153–160.

Singh, G.; Upadhyay, R. K. Essential Oils: A Potent Source of Natural Pesticides. *J. Sci. Ind. Res.* **1993,** *52,* 676–683.

Singh, M.; Srivastava, S.; Srivastava, R. P.; Chauhan, S. S. Effect of Japanese Mint (*Mentha arvensis*) Oil as Fumigant on Nutritional Quality of Stored Sorghum. *Plant Foods Human Nutr.* **1995,** *47* (2), 109–114.

Singh, O. P.; Singh, K. J. Effectiveness of Soybean Oil against Pulse Beetle, *Callosobruchus chinensis* (L.) on Pigeon Peas. *Farm Sci. J.* **1989,** *4,* 144–147.

Singh, R. P.; Tomar, S. S.; Attri, B. S.; Parmar, B.S.; Maheshwari, M. L.; Mukherjee, S. K. Search for New Pyrethrum Synergists in Some Botanicals. *Pyrethrum Post.* **1976,** *13,* 91–93.

Singh, S. C. Effect of Neem Leaf Powder on Infestation of the Pulse Beetle *Callosobruchus chinensis* in Stored Khesari. *Indian J. Ent.* **2003,** *65* (2), 188–192.

Singh, S.; Sharma, G. Efficacy of Different Oils as Grain Protectant against *Callosobruchus chinensis*, in Greengram and their Effect on Seed Germination. *Indian J. Ent.* **2003,** *65* (4), 500–505.

Singh, V. N.; Pandey, N. D.; Singh, Y. P. Effectiveness of Vegetable Oils on the Development of *Callosobruchus chinensis* Linn. Infesting Stored Gram. *Indian J. Ent.* **1994,** *56,* 216–219.

Singh, V.; Yadav, D. S.; Efficacy of Different Oils against Pulse Beetle, *Callosobruchus chinensis* in Greengram, *Vigna radiata* and their Effect on Germination. *Indian J. Ent.* **2003,** *65* (2), 281–286.

Srinivasan, D.; Nadarajan, L. Preliminary Studies on Plant Extract as Grain Protectant against Angoumois Grain Moth, *Sitotroga cerealella* (Olivier) (Lepidoptera: Gelechiidae). *Insect Environ.* **2004,** *10* (1), 9–11.

Srivastava, S.; Gupta, K. C.; Agarwal, A. Effect of Plant Product on *Callosobruchus chinensis* L. Infection On Red Gram. *Seed Res.* **1988,** *16* (1), 98–101.

Srivastava, S.; Gupta, K. C.; Agrawal, A. Japanese Mint Oil as Fumigant and its Effect on Insect Infestation, Nutritive Value and Germinability of Pigeonpea Seeds during Storage. *Seed Res.* **1989,** *17* (1), 96–98.

Srivastava, A. S.; Awasthi, G. P. In *An Insecticide from the Extract of a Plant Justicia adhatoda Linn. Harmless to Man,* Proceedings of the 10th International Congress on Entomology, Montreal, 1956; Vol. 2, pp 245–246.

Srivastava, A. S.; Saxena, H. P.; Singh, D. R. *Adhatoda vasica,* a Promising Insecticide against Pests of Storage. *Labdev J. Sci. Technol.* **1965,** *3,* 138–139.

Su, H. C. F.; Speirs, R. D.; Mahany, P. G. Toxic Effect of Soybean Saponin and its Calcium Salt on the Rice Weevil. *J. Econ. Entomol.* **1972,** *65,* 844–847.

Su, H. C. F. Laboratory Evaluation of Toxicity of Calamus Oil against Four Species of Stored Product Insects. *J. Entomol. Sci.* **1991,** *26,* 76–80.

Su, H. C. F. Biological Activities of Hexane Extract of *Piper cubeba* Against Rice Weevils and Cowpea Weevils (Coleoptera : Curculionidae). *J. Entomol. Sci.* **1990,** *25,* 16–20.

Su, H. C. F. Effect of *Myristica fragrans* Fruits Family: Myristicaceae) to Four Species of Stored Product Insects. *J. Entomol. Sci.* **1989,** *24,* 168–173.

Su, H. C. F. Laboratory Evaluation of Biological Activity of *Cinnamomum cassia* to Four Species of Stored-Product Insects. *J. Entomol. Sci.* **1985a,** *20,* 247–253.

Su, H. C. F. Laboratory Study on Effects of *Anethum graveolens* Seeds on Four Species of Stored-Products Insects. *J. Econ. Ent.* **1985b,** *78,* 451–453.

Su, H. C. F. Laboratory Evaluation of the Toxicity and Repellency of Coriander Seed to Four Species of Stored Product Insects. *J. Entomol. Sci.* **1986,** *21* (2), 169–174.

Subramanya, S.; Babu, C. K.; Krishnappa, C.; Murthy, K. C. K. Use of Locally Available Plant Products against *Callosobruchus chinensis* in Redgram. *Mysore J. Agric. Sci.* **1994,** *28,* 325–345.

Sunarti, C. Oils from Plants and their Toxicity to Cacao Moth, *Ephestia cautella* Walker (Lepidoptera:Pyralidae), *College, Laguna (Philippines).* **2003,** *76.*

Suss, L.; Locatelli, D. P.; Cavalieri, M. Evaluation of Repellent activity of *Azadirachta indica* A. Juss Extracts on Food Stuff Insects. *Tecnica Molitoria.* **1997,** *48,* 1105–1112.

Tadesse, A.; Basedow, T. Comparative Evaluation of SilicoSec, Mexican Tea Powder and Neem Oil against the Granary Weevil, *Sitophilus granarius* L., on Wheat in the Laboratory. *Mitteilungen der Deutschen Gesellschaft fur allgemeine und angewandte Entomologie.* **2004,** *14* (1–6), 343–346.

Tadesse, A.; Basedow, T. Laboratory and Field Studies on the Effect of Natural Control Measures against Insect Pests in Stored Maize in Ethiopia. *Zeitschfuer Pflanzenkrankheitch und Pflanzenschutz.* **2005,** *112* (2), 156–172.

Tapondjou, A. L.; Adler, C.; Bouda, H.; Fontem, D. A. Efficacy of Powder and Essential Oil from *Chenopodium ambrosioides* Leaves as Post-Harvest Grain Protectants against Six-Stored Product Beetles. *J. Stored Prod. Res.* **2002,** *38* (4), 395–402.

Tapondjou, A. L.; Adler, C.; Fontem, D. A.; Bouda, H.; Reichmuth, C. Bioactivities of Cymol and Essential Oils of *Cupressus sempervirens* and *Eucalyptus saligna* against *Sitophilus zeamais* Motschulsky and *Tribolium confusum* du Val. *J. Stored Prod. Res.* **2005,** *41* (1), 91–102.

Tapondjou, L. A.; Adler, C.; Bouda, H.; Fontem, D. A. Bioefficacy of Powders and Essential Oils from Leaves of *Chenopodium ambrosioides* and *Eucalyptus saligna* to the Cowpea Bruchid, *Callosobruchus maculatus* Fab. (Coleoptera, Bruchidae). *Cahiers Agric.* **2003,** *12* (6), 401–407.

Tewari, N.; Tiwari, S. N. Fumigant Toxicity of Lemon Grass, *Cymbopogon flexuosus* (D.C.) Stapf Oil on Progeny Production of *Rhyzopertha dominica* F., *Sitophilus oryzae* L. and *Tribolium castaneum* Herbst. *Environ. Ecol.* **2008,** *26* (4A*),* 1828–1830.

Tinkeu, L. S. N.; Goudoum, A.; Ngassoum, M. B.; Mapongmetsem, P. M.; Kouninki, H.; Hance, T. Persistence of the Insecticidal Activity of Five Essential Oils on The Maize Weevil *Sitophilus zeamais* (Motsch.) (Coleoptera: Curculionidae). *Comm. Agric. Appl. Biol. Sci.* **2004,** *69* (3), 145–147.

Tiwari, R.; Dixit, V; Tiwari, R.; Dixit, V. Insect Repellent Activity of Volatiles of Some Higher Plants against Pulse Beetle - *Callosobruchus chinensis* L. *National Aca. Sci. Lett.* **1997,** *20,* 91–93.

Toews, M. D.; Campbell, J. F.; Arthur, F. H. Temporal Dynamics and Response to Fogging or Fumigation of Stored-Product Coleoptera in a Grain Processing Facility. *J. Stored Prod. Res.* **2006,** *42,* 480–98.

Trehan, K. N.; Pingle, S. V. Moths Infesting Stored Grain and their Control. *Indian Farming.* **1947,** *8,* 404–407.

Tripathi, A. K.; Prajapati, V.; Aggarwal, K. K.; Kumar, S. Effect of Volatile Constituents of *Mentha* Species against the Stored Grain Pests, *Callosobruchus maculates* and *Tribolium castaneum. J. Med. Arom. Plant Sci.* **2000,** *22* (1B), 549–556.

Tripathi, A. K.; Prajapati, V.; Aggarwal, K. K.; Sushil, K. Toxicity, Feeding Deterrence and Effect of Activity of 1,8-cincole from *Artemisia annua* on Progeny Production of *Tribolium castanaeum* (Coleoptera : Tehebrionidae). *J. Econ. Ent.* **2001,** *94* (4), 979–983.

Tripathi, A. K.; Prajapati, V.; Verma, N.; Bahl, J. R.; Bansal, R. P.; Khanuja, S. P. S.; Kumar, S. Bioactivities of the Leaf Essential Oil of *Curcuma longa* (var. Ch-66) on Three Species of Stored Product Beetles (Coleoptera). *J. Eco. Ent.* **2002,** *95* (1), 183–189.

Tripathi, A. K.; Sharma, S.; Susil Kumar; Sharma, S.; Kumar, S.; Edison, S.; Ramana, K. V.; Sasikumar, B.; Babu K. N.; Eapen, S. J. In *Repellent and Insecticidal Properties of Piper retrofractum against Insect Pests of Crops and Stored Grain,* Proceedings of the National Seminar on Biotechnology of Spices and Aromatic Plants, Calicut, India, 1997; pp 134–138.

Trivedi, T. P. Use of Vegetable Oil Cakes and Cowdung Ash as Dust Carrier for Pyrethrin against *Rhizopertha dominica* F. and *Tribolium castaneum* Herbst. *Pl. Prot. Bull.* **1987,** *39,* 27–28.

Tunc, I.; Erler, F. Fumigant activity of Anethole, a Major Component of Essential Oil of Anise *Pimpinella anisum* L. *Bulletin OILB/SROP Turkey.* **2000,** *23* (10), 221–225.

Tunc, I.; Berger, B. M.; Erler, F.; And Dagli, F. Ovicidal Activity of Essential Oils from Five Plant against Two Stored Product Insects. *J. Stored Prod. Res.* **2000,** *36* (2), 161–168.

Umoetok, S. B. A. The Control of Damages Caused by Maize Weevils (*Sitophilus oryzae* Motsc) to Stored Maize Grains Using Cardamon and Black Pepper. *J. Food Agric. Envt.* **2004,** *2* (2), 250–252.

Upadhyay, R. K.; Jaiswal, G.; Yadav, N. Toxicity, Repellency and Oviposition Inhibitory Activity of Some Essential Oils against *Callosobruchus chinensis. J. Appl. Biosci.* **2007,** *33* (1), 21–26.

Verma, N.; Tripathi, A. K.; Prajapati, V.; Bahal, J. R.; Bansal, R. P.; Khanuja, S. P. S.; Kumar, S. Toxicity of Essential Oil from *Lippa alba* toward Stored Grain Insects. *J. Med. Agrom. Plant Sci.* **2001,** *22/23* (4A/1A), 117–119.

Verma, S. P.; Singh, B.; Singh, Y. P. Studies on the Comparative Efficacy of Certain Grain Protectants against S*itotroga cerealella* Olivier. *Bull. Grain Tech.* **1983,** *21* (1), 37–42.

Vishweshwariah, K.; Jayaram, M.; Prasad, N. K.; Majumdar, S. K. Toxicological Studies of the Seeds of *Annona squamosa. Indian J. Exp. Biol.* **1971,** *9,* 519 521.

Wang, C.; Yang, D. J.; Tan, Q. Determination of Toxicity of Plant Essential Oils to Muscum Insect Pests. *J. South West Agric. Univ.* **2000,** *22* (6), 494–495.

Wang, J.; Zhu, F.; Zhon, X. M.; Niu, C. Y.; Lei, C. L. Repellent and Fumigant Activity of Essential Oil from *Artemisia vulgaris* to *Tribolium castaneum* (Herbst) (Coleoptera : Tenebrionidae). *J. Stored Prodt. Res.* **2006,** *42* (3), 339–347.

Wanyika, H. N.; Kareru, P. G.; Keriko, J. M.; Gachanja, A. N.; Kenji, G. M.; Mukiira, N. J. Contact Toxicity of Some Fixed Plant Oils and Stabilized Natural Pyrethrum Extracts against Adult Maize Weevils (*Sitophilus zeamais* Motschulsky). *Afr. J. Pharm. Pharmacol.* **2009,** *3* (2), 66–69.

Weaver, D. K.; Subramanyam, B. Botanicals. In *Alternatives to Pesticides in Stored-Product IPM;* Subramanyam, B., Hagstrum, D.W., Eds.; Kluwer Academic Press: Massachusetts, USA, 2000; pp 303–320.

Xu, H. H.; Chiu, S. F.; Jiang, F. Y.; Huang, G. W. Experiments on the Use of Essential Oils against Stored Product Insects in a Storehouse. *J. South China Agric. Univ.* **1993,** *14,* 42–47.

Xu, H. H.; Zhao, S. H.; Xu, H. H. Studies on Insecticidal Activity of the Essential Oil from *Cinnamomum micranthum* and its Bioactive Component. *J. South China Agric. Univ.* **1996,** *17,* 10–17.

Yadav, J. P.; Bhargava, M. C. Effect of Certain Botanical Products on Biology of *Corcyra cephalonica* Stainton. *Indian J. Plant Prot.* **2002,** *30* (2), 207–209.

Yadav, R. L. Use of Essential Oil of *Acorus calamus* L. as an Insecticide against the Pulse Beetle, *Bruchus chinesis* L. *Zeitschrift fur Angewandte Entomologie.* **1971,** *68* (3), 289–294.

Yadav, T. D. Antiovipositional and Ovicidal Toxicity of Neem (*Azadirachta indica* A. Juss) Oil against Three Species of *Callosobruchus. Neem Newslett.* **1985,** *2,* 5–6.

Yalanchilli, R. P.; Punukollu, B. Bioefficacy Studies on Leaf Oil of *Curcuma domestica* Valeton: Grain Protectant Activity. *J. Med. Arom. Plant Sci.* **2000,** *22* (1B), 715–716.

Yang, F. L.; Liang, G.W.; Xu, Y.J.; Lu. Y.Y.; Zeng, L. Diatomaceous Earth Enhance the Toxicity of Garlic, *Allium sativum* Essential Oil against Stored Product Pests. *J. Stored Prod. Res.* **2010,** *46,* 118–123.

Yuya, A.I.; Tadesse, A.; Azerefenge, F.; Tefera, T. Efficacy of Combining Niger Seed Oil with Malathion 5% Dust Formulation on Maize against the Maize Weevil *Sitophilus zeamais* (Coleoptera : Curculionidae). *J. Stored Prod. Res.* **2009,** *45,* 67–70.

Zhang, H. Y.; Deng, Y. X.; Wang, J. J. Fumigant Activity of Several Kinds of Plant Essential Oils against *Cryptolestes pusillus. Plant Prot.* **2005,** *31* (2), 60–63.

Zhang, X.; Zhao, S. H. Experiments on Some Substances from Plants for the Control of Rice Weevils. *J. Grain Storage.* **1983,** *1,* 1–8.

CHAPTER 12

INTEGRATED MANAGEMENT OF STORAGE INSECTS

CONTENTS

ABSTRACT

The need of integrated pest management in storage conditions is those that have led the field since the beginning, and more immediate objectives have been given impetus by government regulations, public demands, and broader commercial storage. The main objectives are to store grain and seeds with minimum impact from insects or from chemical insecticides that may be used in pest management. Since the beginning of storage to modern era we are using only two chemical insecticides, for example, methyl bromide and aluminum phosphide but due to their bad effect methyl bromide has been banned and aluminum phosphide is restricted in use in most of the countries. Recently, research on toxic chemical free or integration of all suitable methods of pest management for stored grain insects are encouraged and supported. The term integrated pest management is used for flexible and technically sound approach which specifies the inclusion of scientific and cost-effective pest management practices and permit judicious and safe use of chemical insecticides. The most of stored grain insects are reported to developed resistance against traditional insecticides so there are several methods are used for management of storage insects. Management in the storage environment is an essential tool in stored grain pest management. It involves, primarily, the controls on in-store climate and infestation pressure which can be achieved by technically sound storage structures. Equally important, however, is the climatic control attainable by scientific management of the commodity to ensure that the stored grain is itself both dry and cool when loaded or, in ventilated stores and bins with aeration equipment, that the storage procedure achieves drying and cooling sufficiently rapidly. In a fully loaded store it is the stored grain itself which largely determines and stabilizes the temperature and humidity conditions in the store. Modern practice of pest management depends upon the concept of economic control thresholds. The economic justification of control measures, including where necessary the use of chemical pesticides, should take account of all costs and benefits. Many of these may be assessable only in subjective terms that are greatly dependent upon local attitudes and sensibilities. However, measurable losses of quantity and certain quality parameters can be objectively determined.

12.1   INTRODUCTION

Green revolution in India, has led to self-sufficiency in food and also surplus food grain and seed in storage. Post-harvest system in India is poor as compare to developed countries. In the storage of grains and seeds, extent of infestation may be lesser but the economic loss may quite high in terms of quality and quantities. The need of integrated pest management in storage conditions are those that have led the field since the beginning, and more immediate objectives have been given impetus by government regulations, public demands, and broader commercial storage. The main objectives are to store grain and seeds with minimum impact from insects or from chemical insecticides that may be used in pest management. Since the beginning of storage to modern era we are using only two chemical insecticides, for example, methyl bromide and aluminum phosphide but due to their bad effect methyl bromide has been banned and aluminum phosphide is restricted in use in most of the countries. Recently research on toxic chemical free or integration of all suitable methods of pest management for stored grain insects are encouraged and supported. The major problems of stored grains insect management in the tropics, with particular reference to developing countries, are reported by Taylor (1992) and Golob and Hodges (1992). They indicate the crucial importance of effective pest management for a broad range of storage entrepreneurs, including resource-poor farmers, commercial operators, and national marketing organizations; and a wide variety of storage situations in which simple, traditional on-farm systems and large-scale bag-storage systems predominate. Bulk storage occurs quite commonly for small bulks, at farm level; but is less common than bag-storage in large-scale operations except at large grain mills and at commercial storage system. The term integrated pest management is used for flexible and technically sound approach is also required. In defining this term it may be considered necessary (McFarlane, 1989) to specify the inclusion of scientific and cost-effective pest management practices which permit judicious and safe use of chemical insecticides. The most of stored grain insects are reported to developed resistance against traditional insecticides so there are several methods are used for management of storage insects.

Management of the storage environment is an essential tool in stored grain pest management. It involves, primarily, the controls on in-store climate and infestation pressure which can be achieved by technically sound storage structures. Equally important, however, is the climatic control attainable by scientific management of the commodity to ensure that the stored grain is itself both dry and cool when loaded or, in ventilated stores and bins with aeration equipment, that the storage procedure achieves drying and cooling sufficiently rapidly. In a fully loaded store it is the stored grain itself which largely determines and stabilizes the temperature and humidity conditions in the store.

The condition of stored products can also control, to a considerable extent, the initial insect infestation level in the stored grain. However, in tropical countries, where preharvest infestation by storage insects is hardly ever completely preventable, the ideal of loading insect-free grain into the store is not often attainable. Special facilities to completely disinfest the grain before loading may not prove cost-effective. The common alternatives, if early disinfestation is required, are to treat the grain, at intake, with a suitable admixed insecticide or to disinfest the loaded grain by in-store fumigation.

The quality of grain before storage, play vital role in minimizing infestation of store grain insects, so avoid the intake of heavily infested and damaged or uncleaned grain and seeds. Even at the small farm level it is possible to segregate the crop at harvest, especially with maize on the cob and unthreshed sorghum and millet, selecting relatively undamaged material with good storage potential and setting aside the more evidently infested or otherwise damaged material which, if there is no other option, can at least be consumed first. By such means, the rate of deterioration due to insect infestation can be reduced.

The integration of all suitable management techniques, within the framework of integrated pest management, has become a focus for research in stored products work (Evans, 1987). The importance of a multidisciplinary approach to stored grain research has also been stressed (White, 1992). This is very valid but it is useful to recall that a great deal of the research done in the past has been of this nature. It was pointed out (McFarlane, 1981), at a stored products pest management symposium held in 1978, that the need for an interdisciplinary approach is generally well known. Entomology, mycology, chemistry, engineering, and food science are commonly involved, but effective integration of technical solutions is often lacking; possibly because some of the more pragmatic disciplines, notably economics, sociology and business management, are not always sufficiently involved. It is the interface between the research team and the storage managers, whether these are individual farmers or a large storage organization, which may sometimes be the most crucial barrier to progress of pest management program.

A clear perception of the need for solutions which can be integrated into the management system, because they meet the business objectives and can be accommodated within existing management capabilities, is probably the most important requirement (Hindmarsh & McFarlane, 1983). In this context it is of some interest, although not very surprising, that a correlation has been found amongst farmers in India between "grain hoarding capacity" (which relates to the farmers' existing business objectives and capabilities) and the adoption of improved storage practices (Thakre & Bansode, 1990).

Modern practice of pest management depends upon the concept of economic control thresholds. An ECT is defined as the level of pest damage which justifies the cost benefit ratio (Hebblethwaite, 1985). It is always a variable threshold because the costs and benefits of any action will depend upon the situation and its circumstances.

The economic justification of control measures, including where necessary the use of chemical pesticides, should take account of all costs and benefits. Many of these may be assessable only in subjective terms that are greatly dependent upon local attitudes and sensibilities. However, measurable losses of quantity and certain quality parameters can be objectively determined. A manual of methods for the evaluation of postharvest losses (Harris & Lindblad, 1978) is available and provides useful information, subject to a need for modification of some methods in particular circumstances. A critical review of the methodology (Boxall, 1986) gives further guidance based upon experience gained, since 1978, from field work in many developing countries. Choices among methods, several of which have become subjects of considerable controversy, should always be made with due regard to the actual circumstances and the prevailing objective. There is no single "best method" for all circumstances and, for practical purposes, operational facility and repeatability are generally more important that fine precision. From the storekeeper's viewpoint a demonstrable loss reduction from, say, 5– 4%, however statistically significant, may be of little or no practical importance; whereas a reduction from 10 to 5% (or from 2 to 1% in a sophisticated storage system) may be of considerable interest: provided, always, that the demonstrated reduction can be achieved in routine practice.

12.2   SAMPLING AND ESTIMATION OF INSECT POPULATION

Sampling is an essential part of integrated pest management because it will provide decision-making capacity in pest management program to take protection measures only when pest populations reach levels that justify the cost of remediation. Several techniques have been tested and many are used under storage condition. The most commonly used manual commercial method for grain stored in steel bins and grain in transit vehicles is the use of a grain trier, which is a metal spear up to 4 m in length that can be inserted into grain to withdraw a sample. Once the grain sample is removed with the trier, the external-feeding pests in the grain are removed by sieving. Mechanically operated pneumatic grain triers are routinely used to sample grain at points of sale in commercial transport by truck, rail, or barge. A deep-bin probe cup can be used to take samples from deeper in a grain mass (Arbogast et al., 1997), but this is not usually done because of the difficulty in pushing the probe into the grain mass. Sieving the sample to remove adult insects but not for immature stages, which might make up a substantial proportion of all insects in a grain mass (Perez et al., 2004). The immature stages of stored grain insects can be detected by various techniques (Throne et al., 2008), but none of which is practically used at farm storage. Utilization of digital X-ray technique for detection of immature stages in stored grain is quite practical at commercial storage. The method is quick, but only a small sample can be scanned and the equipment is relatively expensive (Throne et al., 2008). Probe traps have long been used for detection of insects in stored grain (Philips et al., 2000; White et al., 1990). Conventional probe-pitfall traps can be used throughout the grain mass (Arbogast et al., 1997), but they rarely are used in this manner because of the difficulty of pushing them into the grain mass and need for regular servicing. Insects in concrete silos can be sampled throughout the grain mass using a vacuum probe sampler (Flinn et al., 2007). Several automatic grain sampling devices are used in large terminal and export grain handling facilities to collect samples from flowing grain at regular time intervals for the purpose of quality grading and insect sampling (Manis, 1992). The numbers of insects present per sample usually considered to be absolute estimates of insect population. Relative estimation of insect populations in stored grain by pheromone traps or light trap and other insect sampling methods are very popular.

12.3   COMPONENTS OF IPM UNDER STORAGE CONDITION

The following components of IPM are commonly used under storage condition.

12.3.1   PHYSICAL METHOD

12.3.1.1   TEMPERATURE

Temperature management of stored grain and seeds is a best method which effectively kills the various life stages of insects. Almost all storage insects cannot be tolerate extreme temperature, for example, heating and cooling, super heating and super cooling of grains and seeds provide extra protection without any chemical treatment. The maximum rate of growth and reproduction for most storage insects occurs between 25 and 35 °C and is reduced at temperatures above and below this range, with complete cessation of development and eventual death at ~12–40 °C (Bartholomaeles et al., 2005). Use of aeration to cool grain or seeds and reduce insect population growth rate is generally used in steel bins but is less common in concrete silos earthen pot and flat storages in the developing country. Aeration can be useful even for summer harvested crops because grain temperatures can be reduced 3–4 °C by aeration at night; however, research on summer aeration for controlling grain insects in the USA has had mixed results (Arthur & Casada, 2005; Finn et al., 2004). Summer aeration reduced insect populations in same years, but not in next year's (Arthur & Casada, 2005). This may have been due to temperatures at the grain surface, where many insects occur, being lowered from lethal levels at 40 °C in warm climates to more favorable levels for insects by summer aeration. Aeration can be effective for pest management in fall-stored grain and seeds in cool climates. Aeration is compatible with other management strategies of integrated pest management. Use of automatic aeration controllers, which turn fans on and off based on grain and ambient temperatures, can be more efficient than manual aeration for cooling grain (Arthur et al., 2008). Various forms of heating have been used to kill insects in bulk grain (Beckett et al., 2007), such as microwave or infrared radiation. However, the methods have not been widely used because of the time required to treat large amounts of grain and seeds. Recent studies show efficacy of infrared catalytic heaters for disinfestations of rice and wheat (Pan et al., 2008), but again the method is not well adopted. Heater is effective for disinfestations of empty grain bins by raising the temperature to 50 °C for 2 h (Tilley et al., 2007); however, the cost is high as compared to insecticides (Tilley et al., 2007). Heat has long been used to kill insects in mills (Beckett et al., 2007; Dean, 1913; Mills et al., 1990). With the impending loss of the fumigant methyl bromide, heat is gaining popularity as an alternative method of disinfestation. Either the whole godowns or specific areas may be heated. Generally, the goal is to raise the temperature of the mill from 50 to 60 °C for 24 h, which can be effective for insect control (Roesli et al., 2003). A challenge in heat disinfestation is uniformity of temperature throughout the treated area, which may be improved with fans. One valid concern with use of heat disinfestation, which is shared anecdotally among food industry sanitarians, is that older buildings may be structurally damaged and that some equipment is heat sensitive (Field, 2002). Although freezing can be effective for insect control, it is generally not used because of economic conditions. However, freezing is one of the few options available for disinfestation of durable organic commodities, such as grains, and seeds infested with insects. For the freezing usually 2–3 weeks of storage in commercial freezers is required for disinfestations (Johnson, 2007; Johnson & Valero, 2003).

12.3.1.2 MOISTURE

Moisture content of stored grain and seeds play a vital role in the development and infestation of insects pests. For insect development most favorable moisture range are 12–15% (How, 1965). Moisture content also affects the insect survival and reproduction inside the grain as traditional insecticides whose moisture content is below 9%. If in developing country like India farmers applied or aware to optimum moisture content of grain and seeds, they can store commodity for longer periods. Moisture content of stored commodity is directly correlated with insect growth and development. The safe moisture content of stored commodity for long-term storage is given in Table 12.1.

TABLE 12.1 Safe Moisture Content of Different Commodity for Long Term Storage Under Indian Condition.

Commodity	Moisture content
Millets	16
Cowpea	15
Beans	14
Paddy	13
Rice	13
Wheat	13
Maize	13
Coffee bean	13
Sorghum	12
Soybean	09
Groundnut pods	08
Mustard/ rapeseeds	09
Sunflower	08
Sesame	07
Groundnut kernels	07

TABLE 12.1 *(Continued)*

Commodity	Moisture content
Niger	07
Linseeds	07
Castor	07
Copra	07
Cocoa	07
Safflower	06
Palm kernel	05

12.3.1.3 SANITATION AND EXCLUSION

Sanitation of stored commodities and storage structure and the effective exclusion of stored grain insects from them are the keys of preventive management. For bulk-stored grain it is imperative that newly harvested commodities to be stored in sanitized bins and never filled grains in those bins having some older grain or products, because they may act as harbor of insects. Harvesting equipment, transportation containers, loading areas, and storage bins need to be sanitized as possible before storage of the commodities, and sometimes it is prudent to treat the surfaces of inside walls, floors, and ceilings of such structures and machinery with an insecticide to prevent previous infestation (Arthur et al., 2009; Reed, 2006). The mechanical and structural aspects of buildings and bins containing stored products must be considered during construction and maintenance of such structures. In food mills and other food-processing facilities it follows that the raw grains, which may be stored for several months before processing and might harbor growing stored-product insect populations, be physically located in bins that are separated from the processing areas and even further separated from the packaging and finished-product warehouse or loading areas. Light can attract stored grain insects (Soderstrom, 1970), so it is recommended that light fixtures not be mounted directly over outside doors but that lighting is mounted on poles away from, but directly illuminating, warehouse or any storage structures. Window screens and doors to the outside of buildings, as well as those between major processing, bulk storage, and warehouse areas in a building or complex of buildings, need to be in good service to reduce movement of storage insect. Cleaning in large processing area should employ careful sweeping and/or vacuum cleaning of food debris for complete

removal, rather than conducting blow downs of debris in order to concentrate it for removal as this can result in spreading dust and food products to
inaccessible areas such as ledges and tops of beams where insects can easily
breed without disturbance. Double-wall construction and suspended ceilings
should be avoided or removed so that hidden infestation in storage structures
can be minimized (Mill et al., 1990). Effective exclusion of stored-product
insects from storage bins, processing plants, and finished food packages
can prevent infestation. Roofs and sidewalls of storage structures should
be sealed to prevent insect entry as well as moisture damage and mold
growth following water dampness. Sealing of warehouse or any structures
is critical for effective fumigants when needed. Proper roof ventilation and
subfloor intake aeration vents are needed for proper temperature and moisture management of grain, but these must be equipped with effective insect
proof screening when in use and sealed when needed for fumigation (Casada
et al., 2001). Packaging materials for food grain and their products at both
wholesale and retail levels of marketing must be resistant to penetration by
stored grain insects (Mullen et al., 2006). Thus, food grains need to be sealed
very well to deter invaders and need to be constructed of durable materials to
resist penetrators. Technology has been developed to impregnate food packaging material with low-risk insecticides (Radwan et al., 1997), and research
has been conducted on insect repellents applied to reduce infestation (Hou et
al., 2004), but commercial adoption of insect-repellent or insecticidal food
packages has not in common use, likely owing to low cost-effectiveness and
low potential for consumer acceptance.

12.3.1.4   INERT DUSTS

The utilization of inert dusts has several industrial and agricultural uses, and
their important role in insecticide dust diluents and carriers (Ebeling, 1971).
Watkins and Norton (1947) reported a classification of the inert dusts used
as diluents and carriers. Inert dusts are used to control insect pests of field
crops, stored products, and animals (Mel'zina et al., 1982). Inert dusts used
in stored product protection can be categorized into four groups (Banks &
Fields, 1995). The first group consists of clays, sand, paddy husk ash, and
wood ash. This group also includes volcanic ash, which has been shown
to possess insecticidal activities (Edwards & Schwartz, 1981). The second
group consists of a great number of minerals such as dolomite, magnesite,
copper oxychloride, rock phosphate, ground sulfur, calcium hydroxide,
calcium carbonate, and sodium chloride (Golob, 1997). Such types of inert

dust have been tested in the past, and provide protection against stored grain insects at minimum quantities, for example, 10 g/kg of grain or seeds. Some of these minerals extensively used in developing and underdeveloped countries at small level storage (Golob & Webley, 1980). The third group consists of dusts that contain silicon dioxide which are light and hygroscopic, and are produced by drying an aqueous solution of sodium silicate (Quarles, 1992). The fourth group consists of dusts that contain natural silica, such as diatomaceous earth which are fossilized skeletons of diatoms. Haryadi et al. (1994) reported that Zeolites similar to that of inert dusts having insecticidal activities, may used for management of maize weevils, *Sitophilus zeamais* M. at 50 g/kg.

Mittal and Wightman (1989) reported that attapulgite dust was effective against rice moth, *Corcyra cephalonica* S., red flour beetle, *Tribolium castaneum* H., and Bruchid *Caryedon serratus* O., when applied at 5 g/kg. Panday and Verma (1977) tested attapulgite dust as a protectant against the cowpea weevil, *Callosobruchus maculatus* F. on black gram at the rate of 10–50 g/kg and found complete mortality after 48 h of treatment. Swamiappan et al. (1976) tested cowpeas with activated kaolin at 10 g/kg against the bruchid *Callosobruchus chinensis* L., and reported that such dust may protect cowlea up to 225 days. Permual and Patourel (1992) reported in Guyana, treated paddy rice with acid-activated kaolin at 7.5 g/kg, found effective against *Cryptolestes pusillus* S., sawtoothed grain beetle, *Oryzaephilus surinamensis* L., lesser grain borer, *Rhyzopertha dominica* F., rice weevil, *Sitophilus oryzae* L. and red flour beetle, *T. castaneum* H. and they protect paddy up to 250 days. They found kaolin to be more effective than rice hull ashes, especially against *R. dominica*. In underdeveloped countries, research is underway to replace these traditional dusts with more effective synthetic silica dusts that work at rates less than 1 g/kg of grain (Golob, 1997).

The application of mineral dusts and salts for controlling stored grain insects, stored Grain Research laboratory Australia were determining the effectiveness of calcium carbonate, magnesium oxide, rock phosphate, and zinc oxide and a rate of 10 g/kg and recommended on wheat (Gay, 1947). The efficacy of dolomite and magnesite as barriers for the control of eight stored-product insects, including *O. surinamensis, S. oryzae, R. dominica,* and *Tribolium* species, dusts were applied on surface of grain and left undisturbed. In Victorian bulk-wheat depots (Wilson, 1945) reported that mineral dusts on the grain surface as a barrier to control insect infestations, the dusts were sprinkled on the wheat surface at the rate of 187 g/m², but at this rate the infestation of *R. dominica* was not completely suppressed.

Tricalcium phosphate (TCP) is an inorganic salt that is toxic to stored grain insects (Davis & Boczek, 1987). TCP suppresses insects when blended with cereal foods at the rate of 20–30 g/kg of stored grain. TCP in blended cereals completely suppressed all species of stored grain moths, except the Mediterranean flour moth, *Ephestia kuehniella Z.*, and it was also effective on bean weevils (Davis et al., 1984) infesting navy beans and cowpeas at the rate of 1–2.5 g/kg, and at these rates progeny production was also suppressed. Stored product mites, especially the mold mite, *Tyrophagus putrescentiae* (Schrank) and the flour/grain mite, *Acarus siro* L. are susceptible to TCP at 15–60 g/kg. Typical symptoms on insects and mites included mortality, increase in developmental time, reduced fecundity, and egg viability. The use of TCP in flour was found to be unacceptable because of severe dust problems encountered when handling treated products (Highland, 1975).

The synthetic silica or silica gels are made up of 99.5% silicon dioxide, compounds such as tricalcium trisilico phosphates (Singh et al., 1984) and silica gels (Quarles, 1992a) are examples of synthetic silica. The silicophosphates have a bulk density of 140–160 g/L and are effective against stored grain insects at the rate of 10–15 g/kg of grain (Singh et al., 1984). Silica gels are dusts that contain extremely small particles (less than three micrometers), and they have a bulk density in the range of 72–450 g/L and specific surface in the range of 200–850 m^2/g and fond effective against stored grain insects as compare to DE dusts (Quarles, 1992). Silica gels used for insect control are fluffy (have low bulk density), light, and contain extremely small particles. Therefore, if applied as a dust, the silica gel particles become suspended in the air for extended periods and may not be deposited on the intended site. To overcome this problem, silica gels are applied as a liquid aerosol. The liquid aerosol formulation (Tri-Die PT) consists of 0.3% pyrethrins, 3% piperonyl butoxide, 4% silica gel, 5.7% petroleum distillate, and 87% inert ingredients. Several commercial formulations (Dri-Die, Tri-Die, and Drione) are available for insect control in homes, warehouses, and food industries. Cotton and Frankenfeld (1949) recommended different rates of a silica aerogel to control seed wheat and various types of milled cereal products from the confused flour beetle, *Tribolium confusum S. oryzae, Sitophilus granarius* L., and *E. kuehniella* larvae. The rate (in g/kg) was 0.25 for 12% wheat and feed in pellet form, 0.5 for 14% wheat, 5 for ground feed, and 10 for bird food (mixture of seeds and finely ground feed). These rates reflect the performance of the silica gel on these various forms of the food and feed products. Silica gels appear to be suitable alternatives to protect grain in countries where access to quality insecticides. Silica gels are now used as an admixture with, or to coat, the heavier DE particles. There are presently two commercial DE formulations, Dryacide and Protect-It, registered in the USA and elsewhere that contain silica gels. The presence of silica gel significantly improves DE effectiveness when compared with formulations containing DE alone.

Admixture of finely ground silica-based dusts for stored grain insect management is a traditional approach since 1930 and their commercial products "Naaki" in Germany and "Neosyl" in England were marketed for stored grain insect protection. The recommended rate was 10 g/kg of grain. In these dusts, quartz was used and Quartz is made up of crystalline silica (Goldsmith et al., 1997) but presently not in use.

DE (Korunic, 1998) is obtained from the layers of fossilized skeletons of diatoms, microscopic plants closely related to the brown algae. DE is commonly known in the USA as Celite, Diatomite, Featherstone, Filter-Cel, Super-Cel (a processed earth), Tri-O-Lite, and Sil-O-Cel. Living diatoms are microscopic organisms that occur in salt water, fresh water, stagnant water, mud, rocks, seaweed, and wherever light and water have co-existed for a sufficiently long time for their growth to accumulate. There are about 250 genera and 8000 species of diatoms. The living diatoms secrete silica, and after the diatoms die and decay, only the silica remains. All fossil diatoms are porous, and it is porosity or specific surface (m^2/g) confers their insecticidal activity, Diatoms differ in their shape and size; only 8000 species have been reported till now (Korunic, 1998). DE is predominantly made up of amorphous or shapeless silica and contains oxides of aluminum, calcium, iron, lime, and magnesium, among other elements. Stored grain insects are the most commonly tested insects with several formulation of DE dusts (Ebeling, 1971; Golob, 1997; Korunic, 1998).

Commercial DE dusts for use as insecticides, the older Insecto label recommends treating the top and bottom 10% of the grain as it is loaded into bins. Treating the entire grain mass affects flowability, test weight, and makes it difficult to unload the grain from storage structures (Jackson & Webley, 1994; Korunic et al., 1996). The mixing of DE with water (slurry application) reduces worker exposure to dust, but diminishes efficacy even at rates 2–3 times that of the dry dust application (McLaughlin, 1994). They also reported that with Dryacide, a slurry application of 5 g/m^2 resulted in less than 60% mortality of *S. granarius* and *S. oryzae* adults at 25 °C and 56% RH. The commercial formulation of DE "Protect-It" recommends application of slurries to grain as well as to surface of grain.

Cook and Armitage (1999) reported that the Dryacide dust adhering to the cuticle forms aggregations that are several micrometers thick to kill the insects. The waxes from and within the cuticle are absorbed onto these thick layers of inert dusts that form on the surface of insects and mites.

12.3.1.5   IRRADIATION

The application of irradiation for the management of stored grain insects is legally approved in most of the countries. The lower dose of irradiation completely kills or sterilizes the common stored grain insects or immature stage of insect invaded inside grains. Moreover, only single radiation exposure of store commodities is sufficient for disinfestations, and this method is suitable for grain storage at commercial level (Balasubramani et al., 2003). In this process gamma rays used to dislodge electrons from chemical bonds in molecules (Halverson et al., 2000), infrared radiation is also applied on the surface or structure as well as directly to commodities for irradiation of stored products. The dose of ionizing (Table 12.2) radiation varies from 0.250 to 14 kGy (Kilogrey) for irradiation grain safety (Hallman et al., 2008) is generally used. The sources of ionizing radiation are from radioisotopes such as cesium or cobalt, or they are generated like X-ray via an electron beam (Halverson et al., 2000).

TABLE 12.2   Food Commodity Permitted for Irradiation Under Prevention of Food Adulteration Act in India.

S. No	Food commodity	Dose of irradiation		Purpose
		Minimum	Maximum	
1	Onion	0.03	0.09	Sprout inhibition
2	Potato	0.06	0.15	Sprout inhibition
3	Rice	0.25	1.00	Insect disinfestations
4	Cereal product	0.25	1.00	Insect disinfestations
5	Pulses	0.25	1.00	Insect disinfestations
6	Dried sea food	0.25	1.00	Insect disinfcstations
7	Fresh sea food	1.00	3.00	Shelf-life extension
8	Dried fruits	0.25	0.75	Insect disinfestations
9	Mango	0.25	0.75	Quarantine treatment
10	Meat and meat product	2.50	4.00	Pathogen control and shelf-life extension
11	Spices	6.00	14.0	Microbial decontamination

12.3.2 CULTURAL METHODS

The cultural practices for the management of stored grain insects are different from other insects. The time of harvesting is very crucial and determining factor of storage durability, so always harvest the fully ripped crops. (Balasubramani et al., 2003) reported that the spraying of some systemic insecticides before harvesting is helpful in management of stored grain insects.

12.3.3 LEGISLATIVE METHODS

This method prevents the export and import or movement of infected materials. In India Destructive Insect Pests Act and Plant Quarantine Act are legally enforced for this purpose.

12.3.4 MECHANICAL METHODS

The utilization of several mechanical devices for management of stored grain insect is quite common in all over the world. The mechanical method is helpful in monitoring and mass trapping of stored grain insects. The Entoleter is used for direct controlling of insect by removing from infested stored commodities. The most extensively used direct means of mechanical control are entoleters, which use centrifugal force to impact grain or seeds containing insects. Entoleters are suitable in all storage structures as well as processing and flour industries. Entoleter cannot catch in immature stages of insects. There are several other mechanical devices, for example, probe trap, pulses beetle trap, light trap, bait trap, sticky trap, pheromone trap, and TNAU automatic insect removal bins (Balasubramani et al., 2003) are used for removal of stored grain insects and minimizing storage losses.

12.3.5 HOST PLANT RESISTANCE

The development of resistance variety against stored grain insects is not success yet. However, vertical resistance was considered as toll for management of stored grain insects in integrated program. Chanbang et al. (2008) reported that integrity of hull is best predictor of rice resistance against *R. dominica* F. In maize some phenolic compounds are responsible for hardness of kernel, treated as factor of resistance against *S. zeamais* (Arnson et al., 1992). The hull hardness and thickness of sorghum seeds considered as resistant factor against stored grain insects (Throne et al., 2000) but the mechanism of resistance against stored grain insect is not understood to till date. The transgenic avidin maize was tested in USA against all major stored grain insect and found resistant, because avidin kill the insects by sequestering the vitamin biotin (Kramer et al., 2000).

12.3.6 BIOLOGICAL METHODS

The utilization of natural enemies for the management of stored grain insects are reported by several scientists (Arbogast, 1984; Haines, 1984; Broer, 1990; Nilakhe & parker, 1990; Brower, 1991; Brower et al., 1996; Scholler, 1998; Scholler & Flinn, 2000; Scholler & Prozell, 2006; Scholler et al., 2006) throughout the world but their practical application at commercial level is very limited. The common parasitoids of major stored grain insects are listed in Table 12.3. In the USA it is legal restriction to add the parasitoids and predators in bulk storage and food too is enforced by Food and Drug Administration (FDA) and U.S. Environmental Protection Agency (EPA). The commercial production of bio control agent in European countries has setup new industries.

The stored product moths can be easily managed by *Trichogramma* sp. (Scholler et al., 2006) reported that the release of *T. evanescens* W. at the rate of 25,000 per week and *H. hebetor* at the rate of 100 per moth found highly effective against *E. kuehniella* in bakery industry. The European Cooperation in Science and Technology (COST) in European country evaluated the huge number of bio control agent against several stored product moth and beetles (Hansen & Wakefield, 2007) and their report suggested that the situation in which biological control will most promising, for example, the treatment of empty storage structure with predatory mites, parasitic wasps and entomopathogenic fungi for the management of stored product moth, beetles and mites has been found highly effective. The preventive management of stored product insects we can use parasitic wasps, predatory mites, and *Trichogramma* spp. to protect grain and processed products.

The effective release of natural enemies is very important step for the success of biological control program. Because of so many natural enemies are abundant under storage system. The *Habrobracon hebetor* is gregarious ectoparasites of larvae of pyralid moth and *Plodia interpunctella* H., *E. kuehniella* and *Cadra cautella* under storage condition. The eggs of stored

TABLE 12.3 List of Predators and Parasitoids Used Against Major Stored Grain Pests Under Indian Storage Condition.

S. no	Name of insects	Name of parasitoids	Parasitoids feed on	Predators	Predators feed on
1	*Sitophilus oryzae* L.	*Anisopteromalus caladrae*	Grub, pupae	*Acaropsellina docta*	Grub, pupae
		Lariophagus distinguendus	Grub, pupae		
		Theocolox elegans	Grub, pupae		
2	*Rhyzopertha dominica* F.	*Anisopteromalus caladrae*	Grub, pupae	*Acaropsellina docta*	Grub, pupae
		Cheyletus eruditus	Eggs	*Lyctocoris compestris*	Eggs, grub
		Pxeemotes ventricosus	Eggs	*Tenebroides mauritanicus*	Eggs
3	*Tribolium castaneum* H.	*Theocolox elegans*	Grub, pupae	*Xylocoris flavipes*	Grub, pupae, adult
		Acarophenax tribolii	Grub, pupae	*Acaropsellina docta*	Grub, pupae
		Cephalonomica tarsalis	Grub	*Lyctocoris compestris*	Eggs, pupae
				Amphiboles venator	Grub, pupae
				Blattisocius tarsalis	Grub, pupae
				Xylocoris flavipes	Grub, pupae
4	*Trogoderma granarium* E.	*Anisoptromalus calendrae*	Grub, pupae	*Acaropsellina docta*	Eggs
		Dinarmus basalis	Grub, pupae	*Amphiboles venator*	Eggs, grub, pupae, adult
		Holepyris sp.	Grub, pupae		
		Pymotes ventricosus	Grub, pupae		
		Synopeas sp.	Grub, pupae		
5	*Callosobruchus chinensis* L.	*Anisoptromalus calendrae*	Grub		
		Dinarnus basalis	Grub		
		Hetrospilus prosopidis	Grub		
		Lariophagus distinguendus	Grub		

TABLE 12.3 (Continued)

S. no	Name of insects	Name of parasitoids	Parasitoids feed on	Predators	Predators feed on
6	Sitotroga cerealella O.	Bracon hebetor	Larvae	Blattisocius tarsalis	Eggs
		Dibrachys carus	Larvae		
		Mesopolobus sp.	Larvae		
		Pteromalus cerealella	Larvae pupae, adult		
		Pymotes ventricosus	Larvae pupae, adult		
		Trichogramma sp.	Eggs		
7	Corcyra cephalonica S.	Bracon brevicormis	Larvae	Acaropsellina docta	Eggs , larvae
		Bracon hebetor	Larvae	Amphiboles venator	Larvae
		Trichogramma australicum	Larvae	Blattisocius keegani	Eggs, larvae
				Blattisocius tineivorus	Eggs, larvae
				Sycanus affinis	Larvae
8	Ephestia cautella W.	Bracon hebetor	Larvae	Aphibolus venator	Larvae, pupae
		Trichogramma chilonis	Eggs	Blattisocius tarsalis	Larvae, pupae
		Venturia canescens	Larvae	Xylocoris flavipes	Larvae, pupae
9	Plodia interpunctella L.	Bracon brevicomis	Larvae	Mus molossinus	Larvae, pupae, adult
		Bracon hebetor	Larvae	oryzaephilus surinamensis	Eggs, larvae, pupae
		Diadromus pulchellus	Larvae	Peregrinator biannulipes	Larvae, pupae
		Dibrachys covus	Larvae	Tribolium castaneum	Eggs, larvae, pupae
		Microbracon brevicomis	Larvae	Xylocoris flavipes	Larvae
		Trichogramma evanescens	Eggs		
		Trichogramma pretiosum	Eggs		
		Venturia canescens	Larvae		

product moth parasitized by *Trichogramma* sp. usually released as pupae glued to egg cards at the rate of 500 female per card and one card per linear meter of shelving (Grieshop et al., 2007). Stored product beetles *Sitophilus* sp. parasitized by *L Lariophagus distinguendus* effectively due to their penetration habit (Steidle & scholler, 2001). Recently (Niedermayer & steidle, 2007) reported that the release of *Theocolex elegans* has been found very effective against *Sitophilus granaries* under bulk storage system in Central Europe.

In the flour industry the larvae of *T. confusum* successfully managed by bio control agent *Holepyris silvanidis* (Lorenz et al., 2010), and the female of *H. silvanidis* easily developed on honey at 25–26 °C and lived up to 51 days (Frielitz, 2007). This parasitoids are highly effective against *T. confusum* in European country.

Pirate warehouse bug *Xylocoris flavipes* R. is a polyphagous predator of eggs and immature stage of several stored grain insects and this predatory bugs suppress the population of *T. confusum* in small quantity of wheat flour (Scholler & Prozell, 2010).

The commercial production of natural enemies quoted with above example is started in several countries, because they are safe, economic viable, and effective alternative to hazardous chemicals.

12.3.7 MICROBIAL METHODS

The management of stored grain insects by pathogen has been tested and reported, but very few are effective under storage condition. The laboratory evaluation of entomopathogenic fungi *Beauveria bassiana*, *Metarhizium anisopliae* and bacteria *Bacillus thuringiensis* either alone or in composition of inert dusts control the some insect of stored grain. (Abdel, 2002; Kavallieratos, 2006; Throne & Lord, 2004). *B. thuringiensis* has been registered for control of stored product Lepidoptera in several countries (Abdel, 2002). The insecticide Spinosad derived from *Saccharopolyspora spinosa* registered in the USA and used at the rate of 1.5 ppm for the management of stored grain insects (EPA, 2005).

12.3.8 PLANT PRODUCTS

The details of application of plant products for management of stored grain insects are discussed in Chapter 11.

12.3.9 BEHAVIORAL METHODS

The details of behavioral methods for management of stored grain insects are discussed in Chapter 10.

12.3.10 GENETICAL METHODS

The concepts of genetic control (Wool, 1975) for management of stored grain insects have developed some extent (Wool et al., 1992) and are not implausible. However, their efficacy against stored grain insects and cost of insect control under different storage condition remain undemonstrated (McFarlane, 1990). The resistant strains of several stored grain insects have a reduced capacity to cause grain damage, through an impaired population growth rate. There is no reason to suppose that all resistant biotypes will be less damaging than their susceptible counterparts but the monitoring of this capacity. Conversely, it would seem illogical to foster a reversion to susceptibility in cases where the resistant biotype has a substantially reduced capacity to cause damage and, therefore, a considerably reduced pest status.

12.3.11 CHEMICAL METHODS

The details of chemical control methods for management of stored gain insects are discussed in Chapter 13 under fumigation of stored products

12.4 CURRENT STATUS OF INTEGRATED PEST MANAGEMENT

The quality assurance and maintenance of stored commodities has tradition-ally been the responsibility of grain storekeepers in throughout the world. The IPM under storage condition is practical application of available knowl-edge which is ecologically, economically, socially, and technically viable for management of stored grain insects. The Stored Grain Advisor (SGA) expert system (Flinn & Hagstrum, 1990) can be used as decision making support system in farm stored grain by inputting grain abiotic and biotic conditions, then expert system will advise the future of insect infestation in any storage structure. The systems with potential relevance to grain storage in the tropics include one, announced by the Australian Centre for Inter-national Agricultural Research (ACIAR), are provide advice to optimize the use of grain protectants, and another, from Natural Research Institute (NRI), which addresses more generally the application of pest management program in grain storage. In India Directorate of Plant Protection Quarantine and Storage (DPPQS) Faridabad, National Center of Integrated Pest Management (NCIPM) New Delhi, and several Agricultural Universities are responsible for provide the information on IPM under storage condition.

KEYWORDS

- **IPM storage**
- **alternative management**
- **biotic and abiotioc**
- **entoleter**
- **irradiation**
- **component of storage IPM**

REFERENCES

Abdel-Razek, A. S. Comparative Study on the Effect of Two *Bacillus Thuringiensis* Strains of the Same Serotype on Three Coleopteran Pests of Stored Wheat. *J. Egypt. Soc. Parasitol.* **2002,** *32,* 415–424.

Arbogast, R. T.; Throne, J. E. Insect Infestation of Farm-Stored Maize in South Carolina: Towards Characterization of a Habitat. *J. Stored Prod. Res.* **1997,** *33,* 187–198.

Arbogast, R. T. Biological Control of Stored Product Insects Status and Prospect. In *Insect Management for Food Storage Band Processing;* Baur, F. J., Ed.; American Association of Cereal Chemists Inc.: Eagan, MN, 1984; 225–238.

Arnason, J. T.; Gale, J.; Conilh de Beyssa, B.; Sen, A.; Miller, S. S. Role of Phenolics in Resistance of Maize Grain to the Stored Grain Insects *Prostephanus truncates* H. and *Sitophilus zeamais* M. *J. Stored Prod. Res.* **1992,** *28,* 119–126.

Arthur, F. H.; Liu, S.; Zhou, B.; Phillips, T. W. Residual Efficacy of Pyriproxyfen and Hydroprene Applied to Wood, Metal and Concrete for Controlling Stored-Product Insects. *Pest Manage. Sci.* **2009,** *65,* 791–797.

Arthur, F. H.; Yang, Y.; Wilson, L. T.; Siebenmorgen, T. J. Feasibility of Automatic Aeration for Insect Pest Management for Rice Stored in East Texas. *Appl. Eng. Agric.* **2008,** *24,* 345–350.

Arthur, F. H.; Casada, M. E. Feasibility of Summer Aeration to Control Insects in Stored Wheat. *Appl. Eng. Agric.* **2005,** *21,* 1027–1038.

Balasubramani, V.; Mohan, S.; Ragumoorthi, K. N. *Storage Entomology an Introduction;* Tamil Nadu Agricultural University: Coimbatore, Tamil Nadu, 2003; p 100.

Banks, J. H.; Fields, P. G. Physical Methods for Insect Control in Stored-Grain Ecosystems. In *Stored-Grain Ecosystems;* Jayas, D. S., White, N. D. G., Muir, W. E., Eds.; Marcel Dekker: New York, NY, 1995; 353–409.

Bartholomaeus, A. R.; Haritos, V. S. Review of the Toxicology of Carbonyl Sulfide, a New Grain Fumigant. *J. Food. Chem. Toxicol.* **2005,** *43*(12), 1687–1701.

Beckett, S. J.; Fields, P. G.; Subramanyam, B. H. Disinfestation of Stored Products and Associated Structures Using Heat. In *Heat Treatment for Postharvest Pest Control: Theory and Practice;* CAB Int.: Wallingford, UK, 2007; pp 182–237.

Boxall, R. A. A Critical Review of the Methodology for Assessing Farm-Level Grain Losses after Harvest. *Nat. Res. Int. Report.* **1986,** *191,* 139.

Bridgeman, B. W. In *Application Technology and Usage Patterns of Diatomaceous Earth in Stored Product Protection,* 7th International Working Conference on Stored-Product Protection, Beijing, China, Oct 14–19, 1998; Zuxun, J., Quan, L., Yongsheng, L., Xianchang, T., Lianghua, G., Eds.; Sichuan Publishing House of Science & Technology: Chengdu, Sichuan Province, Peoples Republic of China, 1999, pp 785–789.

Brower, J. H. Biologicals Insect Diseases, Insect Parasites and Predators. In *Management of Grain, Bulk Commodities, and Bagged products;* Krischik, V., Cuperus, G., Galliart, D., Ed.; USDA, Cooperative Extension Service USA: Beltsville, MD, 1991; pp 195–200.

Brower, J. H.; Smith, L.; Vail, P.; Flinn, P. W. Biological Control. In *Integrated Management of Insects in Stored Products;* Subramanyam, B., Hagstrum, D. W., Ed.; Marcel Dekker: New York, 1996; pp 223–286.

Brower, J. H. Stored Product Pests. In *Classical Biological Control in the Southern United States;* Bulletin No. 355; Hebeck, D. H., Bennett, F. D., Frank, J. H., Ed.; Southern States Cooperative Series: Richmond, VA, 1990, pp 113–122.

Casada, M. E.; Noyes, R. T. In *Future Bulk Grain Bin Design Needs Related to Sealing for Optimum Pest Management: A Researcher's View,* Proceeding of the International Conference on Controlled Atmosphere and Fumigation in Stored Products. Oct 29– Nov 3, 2000; Fresno, California, Donahaye, E. J., Navarro, S., Leesch Clovis, J. G. C. A., Ed.; Exec. Print. Service: 2001, pp 457–465.

Chanbang, Y.; Arthur, F. H.; Wild, G. E.; Throne, J. E.; Subramanyam, B. H. Methodology for Assessing Rice Varieties for Resistance to Lesser Grain Borer *Rhyzopertha dominica* F. *J. Insect Sci.* **2008,** *8*(16), 1–5.

Chiu, S. F. Toxicity Studies of So-Called "Inert" Materials with the Bean Weevil, *Acanthoscelides obtectus* (Say). *J. Econ. Entomol.* **1939a,** *32,* 240–248.

Chiu, S. F.; Toxicity Studies of So-Called "Inert" Materials with the Rice Weevil and Tte Granary Weevil. *J. Econ. Entomol.* **1939b,** *32,* 810–821.

Cook, D. A.; Armitage, D. M. The Efficacy of an Inert Dust on the Mites *Glycyphagus destructor* Schrank and *Acarus siro* L. *Int. Pest Control.* **1996,** *38,* 197–199.

Cotton, R. T.; Frankenfeld, J. C. Silica Aerogel For Protecting Stored Seed or Milled Cereal Products from Insects. *J. Econ. Entomol.* **1949,** *42,* 553.

Davis, R.; Boczek, J. In *A Review of Tricalcium Phosphate as an Insect Population Suppressant: Research to Application;* Proceeding of the 4th International Working Conference on Stored-Product Protection, Tel Aviv, Israel, Sept 21–26, 1987; Donahaye, E., Navarro, S., Eds.; Caspit Press: Jerusalem, Israel, 1987, pp 555–558.

Davis, R.; Boczek, J.; Pankiewicz-Nowica, D.; Kruk, M. In *Efficacy of Tricalcium Phosphate as a Legume Grain Protectant,* Proceeding of the 3rd International Working Conference on Stored-Product Protection, Manhattan, Kansas, Oct 23–28, 1984; Mills, R. B., Wright, V. F., Pedersen, J. R., Ed.; 1984, pp 256–261.

Dean, G. Further Data on Heat as a Means of Controlling Mill Insects. *J. Econ. Entomol.* **1913,** *6,* 40–53.

Ebeling, W. Sorptive Dust for Pest Control. *Ann. Rev. Entomol.* **1971,** *16,* 123–158.

Edwards, J. S.; Schwartz, L. M. Mount St. Helens Ash: A Natural Insecticide. *Can. J. Zool.* **1981,** *59,* 714–715.

Environmental Protection Agency (EPA) 2005; Spinosad Pesticide Tolerance, Federal Register, Office of the Federal Register: National Archives and Records Administration, US, 2005; Vol. No.70, *pp* 1349–1357.

Evans, D. E. In *Integrated Pest Management;* Burner, A. J., Ed., Academic Press: Cambridge, MA, 1987; pp 425–461.

Fields, P.G.; White, N. D. G. Alternatives to Methyl Bromide Treatments for Stored-Product and Quarantine Insects. *Annu. Rev. Entomol.* **2002,** *47,* 331–359.

Flinn, P. W.; Hagstrum, D. W.; Reed, C. R.; Phillips, T. W. Stored Grain Advisor Pro: Decision Support System for Insect Management in Commercial Grain Elevators. *J. Stored Prod. Res.* **2007,** *43,* 375–83.

Flinn, P. W.; Subramanyam, B. H.; Arthur, F. H.;Comparison of Aeration and Spinosad for Suppressing Insects in Stored Wheat. *J. Econ. Entomol.* **2004,** *97,* 1465–1473.

Flinn, P. W.; Hagstrum, D. W. Stored Grain Advisor: A Knowledge-Based System for Management of Insect Pests of Stored Grain. *Appl. Nat. Res. Manage.* **1990,** *4,* 44–52.

Frielitz, C. Holepyris Sylvanidis Parasitoid zur Bekampfung der Larven des Amerikanischen Reismehlkafers. Diploma Thesis, Institute for Applied Zoology, Free University: Berlin, 2007, p 58.

Gay, F. J.; Ratcliffe, F. N.; McCulloch, R. N. Studies on the Control of Wheat Insects by Dusts. I. Field Tests of Various Mineral Dusts Against Grain Weevils. *Council Sci. Ind. Res. (Aust.) Bull.* **1947,** *182,* 7–20.

Goldsmith, D. F.; Gift, J. S.; Grant, L. D., Ed.; *Silica Risk Assessments. J. Expos. Anal. Ernviron. Epidemiol.* **1997,** *7,* 265–395.

Golob, P. Current Status and Future Perspectives for Inert Dusts for Control of Stored Product Insects. *J. Stored Prod. Res.* **1997,** *33,* 69–79.

Golob, P.; Webley, D. J. The Use of Plants and Minerals as Traditional Protectants of Stored Products. *Report Trop. Prod. Inf.* **1980,** *138,* 32,

Golob, P.; Hodges, J. Improvements in Maize Storage for The Small-Holder Farmer. *Trop. Stored Prod. Inf.* **1992,** *50,* 14–19.

Grieshop, M. J.; Flinn, P. W.; Nechols, J. R.; Scholler, M. Host Foraging Success of Three Spicices of *Trichogramma* in Simulated Retail Environments. *J. Econ. Entomol.* **2007,** *100,* 591–598.

Haines, C. P. Biological Methods for Integrated Control of Insects and Mites in Tropical Stored Products the Use of Predators and Parasites. *Trop. Stored Prod. Inf.* **1984,** *48,* 17–25.

Hallman, G.J.; Phillips, T. W. Ionizing Irradiation of Adults of Angoumois Grain Moth (Lepidoptera: Gelichiidae) and Indianmeal Moth (Lepidoptera: Pyralidae) to Prevent Reproduction, and Implications for a Generic Irradiation Treatment for Insects. *J. Econ. Entomol.* **2008,** *101,* 1051–1056.

Halverson, S. L.; Nablo, S. V. Radiation Alternatives to Pesticides.In *Stored Product IPM;* Subramanyam, B., Hagstrum D. W., Eds.; Kluwer Academic Press: Doston, MA, 2000; pp 381–400.

Hansen, L. S.; Wakefield, M. European Network on Biological Control of Pests in Stored Products. In *Integrated protection of stored products; Navaro, S., Alder, C., Riudavets, J., Stejskal, V., Eds.; IOBC WPRS Bull.* **2007,** *30,* 21–30.

Harris, K. L.; Lindblad, C. J. Postharvest Grain Loss Assessment Methods. In *AACC in Cooperation with the League for Int. Food Edu, Trop. Prod. Inst, FAO UN and the Group for Assistance on Systems Relating to Grains After-Harvest;* American Association of Cereal Chemists: Eagan, MN, 1978; 193.

Haryadi, Y.; Syarief, R.; Hubeis, M.; Herawati, I. In *Effect of Zeolite on the Development of Sitophilus zeamais M,* Proceedings of the 6th International Working Conference on Stored-Product Protection, Highley, E., Wright, E. J., Banks, H. J., Champ, B. R., Eds.; Wallingford, Oxon, United Kingdom, 1994; pp 633–634.

Hebblethwaite, M. J. The Application of Economics to Pre- and Postharvest Pest Control Programmes in Developing Countries. *Tropl. Sci.* **1985**, *25*, 215–230.

Highland, H. A. Tricalcium Phosphate as an Insect Suppressant in Flour and CSM. *J. Econ. Entomol.* **1975**, *68*, 217–219.

Hindmarsh, P. S.; McFarlane, J. A. A Programmed Approach to Food Storage Improvements. *Tropl. Stored Prod. Inf.* **1983**, *46*, 3–9.

Hou, X.; Fields, P.; Taylor, W. The Effects of Repellents on Penetration into Packaging by Stored Product Insects. *J. Stored Prod. Res.* **2004**, *40*, 47–54.

How, R. W. A Summary of Estimates of Optimal and Minimal Conditions for Population Increase of Some Stored Products Insects. *J. Stored Prod. Res.* **1965**, *1*, 177–184.

Ileleji, K. E.; Maier, D. E.; Woloshuk, C. P. Evaluation of Different Temperature Management Strategies for Suppression of *Sitophilus zeamais* M. in Stored Maize. *J. Stored Prod. Res.* **2007**.*43*, 480–488.

Jackson, K.; Webley, D. In *Effects of Dryacide® on the Physical Properties of Grains, Pulses and Oilseeds,* Proceedings of the 6th International Working Conference on Stored-Product Protection, Highley, E., Wright, E. J., Banks, H. J., Champ, B. R., EdS.; Wallingford, Oxon, United Kingdom, 1994; 635–637.

Johnson, J. A.; Valero, K. A. Use of Commercial Freezers to Control Cowpea Weevil, *Callosobruchus maculatus* (Coleoptera: Bruchidae), in Organic Garbanzo Beans. *J. Econ. Entomol.* **2003**, *96*, 1952–1957.

Johnson, J. A. Survival of Indianmeal Moth and Navel Orangeworm (Lepidoptera: Pyralidae) at Low Temperatures. *J. Econ. Entomol.* **2007**, *100*, 1482–1488.

Kavallieratos, N. G.; Athanassiou, C. G.; Michalaki, M. P.; Batta, Y. A.; Rgatos, H. A. Effect of the Combined Use of *Metarhizium anisopliae* M. Sorokin and Diatomaceous Earth for the Control of Three Stored Products Beetle Species. *Crop Prot.* **2006**, *25*, 1087–1094.

Korunic, Z. Diatomaceous Earths a Group of Natural Insecticides. *J. Stored Prod. Res.* **1998**, *34*, 87–97.

Korunic, Z.; Fields, P. G.; Kovacs, M. I. P.; Noll, J. S.; Lukow, O. M.; Demianyk, C. J.; Shibley, K. J. The Effect of Diatomaceous Earth on Grain Quality. *Postharvet Biol. Technol.* **1996**, *9*, 373–387.

Kramer, K. J.; Morgan, T. D.; Throne, J. E.; Dowell, F. E.; Bailey, M.; Howard, J. A. Transgenic Avidin Maize is Resistant to Storage Insect Pests. *Nat. Biotech.* **2000**, *18*, 670–674.

Lorenz, S.; Adler, C.; Reichmuth, C. In *Penetration Ability of Holepyris sylvanidis into the Feeding Substrate of Its Host Tribolium confusum.* Proceedings of the Tenth International Working Conference of Stored Product Protection, Estoril, Portugal, June 27 to July 02, 2010; Carvalho, O. M., Fields, P. G., Adler, C. S., Arthur, F. H., Athanassiou, C. G., Compbell, J. F., Fleurat-Lessard, F., Flinn, P.W., Hodges, R. J., Isikber, A. A., Navarro, S., Noyes, R. T., Riudavets, J., Sinha, K. K., Thorpe, G. R., Timlick, B. H., Trematerra, P., White, N. D. G., Eds.;Julius-Kuhn Archives Berlin: Germany, 2010, p 486.

Luo, H.; Wang, R.; Dai, Y.; Wu, J.; Xu, Y.; Shen, J.; Fu, J. Experimental Investigation of a Solar Adsorption Cooling System Used for Grain Low Temperature Storage of Grain. *Acta Energ. Solaris Sin.* **2006**, *27*, 588–592.

Manis, J. M. Sampling, Inspecting, and Grading. In *Storage of Cereal Grains and Their Products;* Sauer, D.B.,Ed., St. Paul: American Association of Cereal Chemists, 4th ed., 1992; pp 563–88.

McFarlane, J. A. Guidelines for Pest Management Research to Reduce Stored Food Losses Caused by Insects and Mites. *Overseas Development Nat. Res. Int Bull.* **1989**, *22*, 62.

McFarlane, J. A. In *The Integrated Approach to Pest Management. Keynote address in: Pests of Stored Products.* Proceedings of BIOTROP Symposium on Pests of Stored Products, Bogor, Indonesia, 24–26 April 1978, BIOTROP Special Publication, 1981, 9, 12–15.

McLaughlin, A. In *Laboratory Trials on Desiccant Dust Insecticides.* Proceedings of the 6th International Working Conference Stored-Product Protection CAB International, Highley, E., Wright, E. J., Banks, J. H., Champ, B. R., Eds.; Wallingford, Oxon, United Kingdom. 1994, 638–645.

Melzina, E. N.; Orekhov, S. J.; Mironov, A. N.; Ryshkov. M. M. Mineral Dusts as Possible Insecticides of Pullecidal Effect (Flea Siphonaptera Control). *Med. Parasitol. Parazit. Bolezni.* **1982,** 51, 45–49.

Mills, R.; Pederson, J. A Flour Mill Sanitation Manual. St. Paul: Eagan. 1990, 164.

Mittal, S.; Wightman, J. A. An Inert Dust Protects Stored Groundnuts from Insect Pests. ICRISAT Newsletter, **1989**, *11*, 21–22.

Mullen, M. A.; Mowery, S. V. Insect-Resistant Packaging. In *Insect Management for food Storage and Processing;* Heaps, J. W., Ed.; AACC Int: St. Paul, MN, 2006, 35–38.

Niedermayer, S.; Steidle, J. L. M. Influence of Extreme Temperatures in Grain Stores on the Parasitation Ability of Beneficial Insects in Stored Product Protection. *Html.; http// orgprints.org/9631.* **2007**, 1–9.

Nilakhe, S. S.; Parker, R. D. In *Implementation of Parasites and Predators for Control of Stored Product Pests.* Proceedings of the third National Stored Grain Pest Management Training Conference, Kansas, Missouri, Oct 20–25, 1990; pp 241–250.

Pan, Z.; Khir, R.; Godfrey, L. D.; Lewis, R.; Thompson, J. F.; Salim, A. Feasibility of Simul-taneous Rough Rice Drying and Disinfestations by Infrared Radiation Heating and Rice Milling Quality. *J. Food Eng.* **2008**, *84*, 469–479.

Pandey, G. P.;Varma, B. K. Attapulgite Dust for the Control of Pulse Beetle, *Callosobruchus maculatus* Fabricius on Black Gram (*Phaseolus mungo*). *Bull. Grain Technol.* **1977**, *15*, 188–193.

Perez-Mendoza, J.; Flinn, P. W,; Campbell, J.F.; Hagstrum, D.W.; Throne, J. E. Detection of Stored-Grain Insect Infestation in Wheat Transported in Railroad Hopper-Cars. *J. Econ. Entomol.* **2004**, *97*, 1474–1483.

Permual, D.; Patourel, G. N. J. Small Bin Trials to Determine the Effectiveness of Acid Activated Kaolin against Four Species of Beetles Infesting Paddy Under Tropical Storage Conditions. *J. Stored Prod. Res.* **1992**, 28, 193–199.

Phillips, T.W.; Cogan, P. M.; Fadamiro, H. Y. Pheromones. In *Alternative to pesticides in stored product IPM;* Subramanyam, B., Hagstrum, D. W., Eds.; Kluwer Academic: Boston, MA, 2000; 273–302.

Radwan, M. N.; Allin, G.P. U.S. Patent. 1997, 5, 509–688.

Reed, C. R. *Managing Stored Grain to Preserve Quality and Value;* AACC Int.: St. Paul, MN, 2006; p 235.

Roesli, R.; Subramanyam, B.; Fairchild, F. J.; Behnke, K. C. Trap Catches of Stored-Product Insects before and after Heat Treatment in a Pilot Feed Mill. *J. Stored Prod. Res.* **2003**, *39*, 521–540.

Scholler, M.; Flinn, P. W. Parasitoids and Predators. In *Alternative to Pesticides in Stored Product IPM;* Subramanyam, B., Hagstrum, D. W., Eds.; Kluwer Academic Publishers: Boston, MA. 2000, 229–271.

Scholler, M.; Prozell, S. Potential of *Xylocoris flavipes* (Hemiptera: Anthocoridae) to Control *Tribolium confusum* (Coleoptera: Tenebrionidae) in Central Europe.*IOBC/WPRS Bull.* **2010,** 10.

Scholler, M.; Prozell, S. Natural Enemies to Control Stored Product Pests in Grain Stores and Retail Stores. In *Implementation of Biocontrol in Practice in Temperate Regions- Present and Near Future*; Stengard, H. L., Enkegaard, A., Steenberg, T., Ravnskow, S., Larsen J., Eds.; DIAS Report: 2006; *119*, 85–106.

Scholler, M.; Prozell, S. Nutzlinge fur die Leeraumbehandlung in Vorratsschutz. *Pest. Control. News.* **2007**, *35,* 12–13.

Scholler, M.; Integration of Biological and Non-Biological Methods for Controlling Arthropods Infesting Stored Products. *Postharvest news Info.* **1998,** *9,* 15–20.

Singh, K.; Bhavnagary, H. M.; Majumder, S. K. Silicophosphate as New Insecticide. Evaluation of Silicophosphates for the Control of Stored Grain Pests in Milled Rice. *J. Food Sci. Technol.* **1984,** *21,* 302–307.

Soderstrom, E. L. Phototactic Responses of Stored-Product Insects to Various Intensities of Ultraviolet Light. *J. Stored Prod. Res.* **1970**, *6,* 275–277.

Steidle, J. L. M.; Scholler, M. Fecundity and Ability of the Parasitoid *Lariophagus distinnuendus* (Hymenoptera : Pteromalidae) to Parasitize Larvae of the Granary Weevil *Sitophilus granaries* (Coleoptera: Curculionidae) in Bulk Grain. *J. Stored Prod. Res.* **2001,** *38,* 43–53.

Subramanyam, B. H.; Swanson, C. L.; Madamanchi, N.; Norwood, S. In *Effectiveness of Insecto®, a New Diatomaceous Earth Formulation, in Suppressing Several Stored-Grain Insect Species,* Proceedings of the 6th International Working Conference on Stored-Product Protection, April, 1994; Highley, E., Wright, E. J., Banks, J. H., Champ, B. R., Eds.; CAB International: Wallingford, Oxon, U K, 1994, pp 650–659.

Swamiappan, M.; Jayaraj, S.; Chandy, K. C.; Sundaramurthy, V. T. Effect of Activated Kaolinitic Clay on Some Storage Insects. *Z. Angew. Entomol.* **1976,** *80,* 385–389.

Taylor, R. W. D.; Golob, P.; Hodges, R. J.In *Current Trends in the Protection of Stored Cereals in the Tropics by Insecticides and Fumigants,* Proceedings of the Brighton Crop Protection Conference, Pests and Diseases Section, Brighton, UK, 1992; pp 281–290.

Thakre, B. D.; Bansode, P. C. Relationship between Land Holding and Grain Hoarding Capacity and Adoption of Improved Storage Practices. *Bull. Grain Tech.* **1990,** *28*(2), 145–148.

Throne, J. E.; Pearson, T. C. Detection of Insects in Grain. In *Contribution for Integrated Management of Stored Rice Pests: Handbook;* Mancini, R., Carvalho, M. O., Timlick, B., Adler, C., Eds.; Inst. Investigacao Cientifica Tropical: Lisbon, 2008; pp 123–136.

Throne, J. E.; Lord, J.C. Control of Sawtoothed Grain Beetles (Coleoptera : Silvanidae) in Stored Oats by Using an Entomopathogenic Fungus in Conjunction with Seed Resistance. *J. Econ. Entomol.* **2004,** *97,* 1765–1771.

Throne, J. E.; Baker, J. E.; Messina, F. J.; Kramer, K. J.; Howard, J. A.; Varietal Resistance Alternatives to Pesticides. In *Stored Product IPM;* Subramanyam, B., Hagstrum D. W., Eds.; Kluwer Academic Press: Boston, MA, 2000; 165–192.

Tilley, D. R.; Casada, M. E.; Arthur, F. H.; Heat Treatment for Disinfestation of Empty Grain Storage Bins. *J. Stored Prod. Res.* **2007a**, 43, 221–28.

Tilley, D. R.; Langemeier, M. R.; Casada, M. E.; Arthur, F.H.; Cost and Risk Analysis of Heat and Chemical Treatments. *J. Econ. Entomol.* **2007b**, *100,* 604–612.

Varma, B. K.; Siddiqui, M. K. H. Control of Storage Pests through Inert Dusts. *Indian Far.* **1977,** *27, 5*, 21–25.

Watkins, T. C.; Norton, L. B. A Classification of Insecticide Dust Diluents and Carriers. *J. Econ, Entomol.* **1947,** *40,* 211–214.

White, N. D. G.; Arbogast, R. T.; Fields, P. G.; Hillman, R. C.; Loschiavo, S. R. The Development and Use of Pitfall and Probe Traps for Capturing Insects in Stored Grain. *J. Kans. Entomol. Soc.* **1990,** *63,* 506–525.

White, N. D. G. A Multidisciplinary Approach to Stored Grain Research. *J. Stored Prod. Res.* **1992,** *28*(2), 127–137.

Wilson, F. The Control of Insect Pests in Victorian Bulk Wheat Depots. *J. Council Sci. Ind. Res.* **1945,** *18,* 103–109.

Wool, D.; In Genetic Control of Insecticide Resistance in Stored Products Insects: Prospects and Preliminary Investigation, Proceedings of the 1st International Working Conference on Stored Product Entomology, Savannah, USA, Oct. 1974; 1975, 310–316.

Wool, D.; Grower, J. H.; Kamin-Belsky, N. Reduction of Malathion Resistance in Caged Almond Moth, *Cadra cautella* (Walker) (Lepidoptera: Pyralidae), Populations by the Introduction of Susceptible Males. *J. Stored Prod. Res.* **1992,** *28*(1), 59–65.

CHAPTER 13

FUMIGATION OF STORED PRODUCTS

CONTENTS

ABSTRACT

The protection of stored commodities is one of the challenging works in all over the world against stored grain insects. The admixture of chemicals with grains or seeds and sprays on structure make insect to be killed, whenever they come in contact with the insecticides; but in case of severe infestation of stored grain insects, where instant and rapid control becomes necessary, only fumigants would provide sufficient results. In storage structures, where insecticides spray cannot reach, fumigants can penetrate and kill the insects. The term fumigant is derived from fumus or smoke of chemical which at a required temperature and pressure can exist in the gaseous state in sufficient concentration to be lethal to a given pest organism. A substance that is volatile at room temperature is highly preferable as fumigant. The fumigation may also be defined as application of fumes, perfumes, smokes, or vapors especially for the purpose of disinfestations under hermitic condition. The killing action of a fumigant is influenced by its concentration in the atmosphere, the length of time it stays in the atmosphere, and the temperature and humidity of the area at the time of fumigation. Fumigants are designed to enter cracks, crevices, and other areas where insect may occur. The constraints in the use of fumigants to treat stored products include consideration of the chemical residues which they may leave in the treated stored products and affect the quality of grain, germinability, and seedling viability.

13.1   INTRODUCTION

The protection of stored commodities is one of the challenging works in all over the world against stored grain insects. The admixture of chemicals with grains or seeds and sprays on structure make insect to be killed, whenever they come in contact with the insecticides; but in case of severe infestation of stored grain insects, where instant and rapid control becomes necessary, only fumigants would provide sufficient results. In storage structures, where insecticides spray cannot reach, fumigants can penetrate and kill the insects.

The term fumigant is derived from fumus or smoke of chemical which at a required temperature and pressure can exist in the gaseous state in sufficient concentration to be lethal to a given pest organism. A substance that is volatile at room temperature is highly preferable as fumigant. The fumigation may also be defined as application of fumes, perfumes, smokes, or vapors especially for the purpose of disinfestations (Cotton, 2013). The fumes of sulfur have been utilized since twelfth century B.C. The development of fumigants for insecticidal purposes has taken place during the past 50 years. The utilization of carbon disulfide as a fumigant was reported in 1854. The application of hydrocyanic gas as a fumigant for stored grain insects started in 1986, while chloropicrin use started in 1907 and ethylene oxide and methyl bromide (MB) were first used in 1932.

The fumes fumigants reach the insect tissues through the process of respiration. If the insect inhales a toxic gas, this gas is instantly absorbed by the body cell. These gases enter into some fixed and nonvolatile combination within insect tissues. The molecules of some fumigants, for example, carbon dioxide, replace oxygen molecules in the air. The killing action of a fumigant is influenced by its concentration in the atmosphere, the length of time it stays in the atmosphere, and the temperature and humidity of the area at the time of fumigation. Fumigants are designed to enter cracks, crevices, and other areas where insect may occur. Fumigants must be applied in airtight chambers.

The constraints in the use of fumigants to treat stored products include consideration of the chemical residues which they may leave in the treated stored products and affect the quality of grain, germinability, and seedling viability. In this regard, phosphine has considerable advantages and is certainly to be preferred to MB for stored products. Problems can arise from the visible residues of the metal hydroxide which remain after the decomposition of tablet or pellet formulations. Moreover, these usually contain some undecomposed phosphide, which can also be found in stored grains.

13.2 CLASSIFICATION OF FUMIGANTS

The fumigants may by classified on the basis of their origin, the first group is hydrogenated hydrocarbon, for example, carbon tetrachloride and MB. Second group represents sulfur compounds, for example, carbon di sulphide and sulfur dioxide. Third group contains cyanides, for example, hydrocyanic acid gas (HCN) calacyanide and acrylonitrile. The fourth group consists of organic acid gases and other, for example, phosphoric, benzene, chloropicrin, and ethylene dioxide. Three forms of fumigants, gas, liquid, and solid, have been used since their invention. Gaseous fumigants are chemicals which are gases at normal temperature and pressure, for example, MB, and HCN. The liquid fumigants are generally a mixture of chemicals that are liquid, but when comes in to contact with air become gas, for example, ethylene dichloride. The solid fumigants evolve in to gas upon exposure to moisture in air, for example, aluminum phosphide. The basic properties of chemicals used as fumigants are common and all properties are influenced by each other. The molecular structure is dependent upon the formulation and molecular weight and has relationship with the components in the molecular weight. The specific gravity is the relationship of the weight of a material compared to an equal amount of water. The molecules of fumigants, being used in general, are heavier than air. Recently several essential oils have been reported to act as fumigants against almost all major stored grain insects. For pest management program under storage condition the following types of fumigants will be discussed in this chapter.

13.2.1   METHYL BROMIDE

The insecticidal properties of MB were first reported by Le Goupil 1932 in France and widely used for plant quarantine purposes. MB has been used extensively as fumigant for stored products, mills, warehouses, ships, and railway wagons. It is a colorless liquid and it may be applied by placing cylinders by releasing the gas from near the ceiling through copper pipes from vacuum fumigation. Its penetration between the bags is rapid and proved as the best fumigant for air-tight warehouses and storage bins (Majumdar & Muthu, 1960). This fumigant is non-flammable and non-explosive under ordinary circumstances and may be used without any special precautions against fire. MB is odorless at normal fumigation concentration, so it is packed with a warning gas like chloropicrin. The recommended dose of MB is 60 g per 20 cubic meters for an exposure period of 24 h.

   MB has also been used as a chemosterilant, although it has approximately one-tenth the activity of ethylene oxide against bacteria and fungi (Bruch, 1961; Richardson & Monro, 1962). MB is also used for the sterilization of space vehicles in combination with ethylene oxide (Vashkov & Prishchep, 1967). MB has a delayed effect on stored grain insects. It is commonly used for the fumigation of almost every type of cereal and their processed product, because it penetrates densely in packed materials, it is especially useful for the treatment of flours. MB reacts with the protein of wheat (Winteringham et al., 1955; Slover & Lehmann, 1972), which deteriorates the quality of wheat. The high-moisture content of wheat and its flour affects the effectiveness of MB. In West Africa MB are found most suitable for fumigation of nut under bulk storage system (Hayward, 1954; Halliday & Prevett, 1963). If the dose of MB is very high they produce foul smell in processed products and may be hazardous to human health.

13.2.2   CARBON TETRACHLORIDE

Carbon tetrachloride was reported as a fumigant in 1908. It is non-flammable and serves to reduce fire hazards of other fumigants. Its action is slow and has low toxicity to insects due to which high dose and longer duration are required (Khare, 2006). Generally, it is used with carbon disulfide, ethylene dichloride. It is less toxic to humans but their continuous use may affect some non-target organisms. Carbon tetrachloride does not affect the germination quality and vigor of seeds (Majumder et al., 1961) at the recommended dose, but higher dose may affect the quality of seeds (Roth, 1968). This fumigant is now listed by some agencies in the category of industrial substances suspected of carcinogenic potential for humans (ACGIH, 1981).

13.2.3   CARBON DISULFIDE

Carbon disulfide was one of the first fumigants utilized at large scales. Carbon disulfide was first used in France to save the grape wine industries in 1869, against the grape *Phylloxera*; at that time this chemical was injected in the soil to prevent root infestation. Carbon disulfide is formulated in mixture with non-flammable ingredients for fumigation of grains. This fumigant is highly effective when compared to carbon tetrachloride for stored grain insects (Grish, 1966). Due to their mammalian toxicity it is used in food processing industries. It is commonly used with carbon tetrachloride in the ratio of 1:4 and this mixture is widely used all over the world. It does not affect the quality of seeds (Kamel & Shabha, 1958). Carbon disulfide may react with amino acids and peptide groups of stored commodities (FAO, 1975).

13.2.4   ETHYLENE DICHLORIDE

Ethylene dichloride is flammable; so it is mixed with some non-flammable fumigants like carbon tetrachloride in the ratio of 3:1 by volume. Ethylene dichloride is soluble in fats and hence is not recommended for cereals and food having high level of oil content, and it is dangerous to human health in overdose. It has a strong, sickly chloroform-like odor, which helps in warning effect. High doses of ethylene dichloride are found to cause tumors effect rats and mice.

13.2.5   ETHYLENE OXIDE

In the bulk storage system, ethylene oxide is commonly used and accepted as substitute of carbon disulfide for fumigation of grain and seeds. Cotton and Roark)1928) discovered the insecticidal properties of ethylene oxide. It is easily handled and does not have any residues in stored grain. To reduce the flammability it is mixed with carbon dioxide (Hashiguchi et al., 1967) in the ratio of 1: 9. Ethylene oxide is a colorless, odorless liquid at low temperatures, but it can vaporize rapidly at suitable temperature. Despite a general impression to the contrary, ethylene oxide is poisonous to human beings. The acute toxic effects of ethylene oxide in humans and animals include acute respiratory and eye irritation, skin sensitization, vomiting, and diarrhea. Skin injury may result from excessive freezing following the spillage of the chemical. Continuous exposure to even low concentrations may result in a numbing of the sense of smell.

The chronic effects of ethylene oxide include respiratory irritations and secondary respiratory infection, anemia, and altered behavior. Although limited tests on mice have not revealed carcinogenic effects, the alkylating and mutagenic properties of ethylene oxide are sufficient to cause concern (Glaser, 1979). Ethylene oxide has been used for over 40 years for both insecticidal treatments and the sterilization of foodstuffs. A number of investigations have shown that it will react with food constituents and deteriorate the quality of food.

13.2.6   PHOSPHINE

The phosphine or hydrogen phosphide (PH_3) discovered from Germany contains calcium phosphide and calcium carbide. Despite the harmful effect of phosphine, that is, being highly inflammable, it is a safe and so-called convenient fumigant. Phosphine has a strong typical smell resembling garlic at concentrations below the danger level. It is one of the most toxic fumigants to stored grain and grain product insects, mites, and rodents (Khare, 2006). PH_3 is a low-molecular weight, low-boiling point compound that diffuses rapidly and penetrates deeply into stored products, such as large bulks of grain or tightly packed materials. Phosphine is very toxic to all forms of animal life; hence exposure of human beings even to small amounts must be avoided. It has also an inhibitory effect on insect respiration and is unique in that it is only toxic to insects in the presence of oxygen—in the absence of oxygen it is not absorbed and is not toxic to insects (Bond & Dumas,

1967; Bond et al., 1969). However, the action of phosphine is potentiated by carbon dioxide and the exposure time can be reduced when both gases are present (Kashi & Bond, 1975).

Some stages of insects are considerably more tolerant to phosphine than others (Bell, 1976; Hole et al., 1976; Nakakita & Winks, 1981). The eggs and pupae are usually the hardest to kill while larvae and adults are more susceptible. The tolerance in insects is developed due to long exposure period of phosphine; the laboratory tolerant strain of stored grain insects is reported by Howe (1974).

The development of resistance against phosphine can reduce their effectiveness; the *Tribolium castaneum* has developed 10 times resistance against these chemical fumigants in six generations, and the magnitude of resistance is more in immature stage than in adult stage (Winks, 1974). The resistance correlation reported by Bell et al. (1977) between adult and egg stages of *Rhyzopertha dominica*. Champ and Dyte (1976) reported the evidence of resistance to phosphine in insects from several continent of the world, particularly where inadequate techniques of fumigation were employed, and they indicated that the emergence of resistance to fumigants under practical conditions was a matter of great concern during those times. The resistance against phosphine has been reported since the start of use, and most of the countries banned this fumigant or kept under restricted use (Borah & Chalal, 1979; Taylor et al., 1975; Hole, 1981). Several scientists from all over the world reported that phosphine deteriorated the quality of seeds in terms of their germination capacity (Lindgren et al., 1968; Beratlief & Alexandrescu, 1964; Fam et al., 1974). The residues of phosphine generally occur in three forms: reaction products of formulation, unchanged phosphine absorbed in commodity, and products formed by chemical combination of phosphine with components of the commodity. The formulations of aluminum phosphide leave mainly an inert residue of the metallic hydroxide. In formulations of aluminum phosphide, a small amount of unreacted material may also remain, and hence some precautions should be taken to avoid hazards from the unspent formulation. When processed foods are fumigated, or when space fumigations are carried out, residues from the formulation should be collected and properly disposed off.

The formulation of phosphine is available in tablets and pellets from where each tablet weighs 3 g and releases 1 g of phosphine gas. For a convenience and safety point of view, pellets are sometimes supplied in predetermined quantities for specific applications. Pellets are prepared in special prepacks of 165 pellets each for treatments such as railway box-car fumigations. Similarly, sachets are joined with ropes or "blankets." In these prepared packages, the pellets or sachets are separated sufficiently to avoid a buildup of excessive heat and concentrations of the gas in small spaces. Aluminum phosphide is a solid fumigant in compressed tablet or pellets from which are round dark brown in color. It is composed of 56% W/W aluminum phosphide and 44% of inert material of which at least 26% is ammonium carbonate and 4% solid paraffin. The phosphine gas released from these tablets or pellets on coming in to contact with atmosphere along with ammonia and carbon dioxide. It is not dissolved in oils and fat and hence can be used safely on a large range of stored commodities. The bag formulation is slow in action, so it is used in fumigation program. Phosphine is used at 1–2 tablets per metric ton of stored grains.

13.2.7   HYDROGEN CYANIDE

Hydrogen cyanide was one of the important fumigants of the past. It was first used for treatment of tree against scale insects in California in 1886, but nowadays it is not in use. Hydrogen cyanide was used for fumigating dry fruits, grain, and seeds, but it is highly absorbed by many of the stored commodities and their residual effect is very high. It is a highly toxic insect fumigant and soluble in water. It affects the germination quality of sorghum as reported by Lindgren et al. (1955). The penetration of hydrogen cyanide in bulk storage of sorghum depends upon the temperature, moisture, and dockage (Kunz et al., 1964).

13.2.8   CHLOROPICRIN

The chloropicrin was first prepared by Stenhouse in 1848 by adding an aqueous solution of picric acid to an excess of bleaching powder (Roark, 1934). It has been proved to be a useful fumigant for the grain and milling industries and used extensively as a grain fumigant either alone or mixed with carbon tetrachloride. It is a colorless liquid heavier than water and non-flammable. It is also used as soil fumigant to control some fungi. It is not in use for fumigation of stored commodities till date.

13.2.9   NEW FUMIGANTS

In India, only aluminum phosphide and MB are available for fumigation of food grain. The use of aluminum phosphide is restricted by law while MB needs special infrastructures for its use. Due to injudicious use, aluminum phosphide is poisoning our environment, causing many fatal diseases in exposed persons and facilitating many fatalities. Its improper use is also resulting in the developed resistance in the insect pests of stored grain, and in many parts of world including India it has become useless for fumigation. Now a time has reached when we have to think whether it is a boon or a bane. Scientists all over the world realize that presently it is impossible to survive without its use. But they also understand that it is necessary to minimize its use at least at the farmer's level and to search for other viable alternatives. Over the last four decades, serious attempts have been made to find its alternative in plant kingdom, and hundreds of plants have been screened to search for their fumigation potential. And most interestingly, many plants have been identified which are equipped with fumigant properties against insect pests of stored grain. Now it has been proved beyond doubt that under laboratory conditions the plant components can suppress feeding and breeding of stored grain insect or even kill them completely within a few hours just like chemical fumigants. Essential oil from more than 75 plant species belonging to different families, such as *Anacardiaceae*, *Apiaceae* (Umbeliferae), *Araceae*, *Asteraceae* (Compositae), *Brassicaceae* (Cruciferae), *Chenopodiaceae*, *Cupressaceae*, *Graminaceae*, *Lamiaceae* (Labiatae), *Lauraceae, Liliaceae, Myrtaceae, Pinaceae, Rutaceae*, and *Zingiberaceae* have been studied for fumigant toxicity against insect pests of storage grain (Rajendran et al., 2008).

Plant oils have been reported to possess repellent and insecticidal (Shaaya et al., 1997; Kostyukovsky et al., 2002; Papachristos & Stamopoulos, 2002), nematicidal (Oka et al., 2000), antifungal (Paster et al., 1995), antibacterial (Matasyoh et al., 2007), virucidal (Schuhmacher et al., 2003), and antifeedant, reproduction inhibitory activity (Raja et al., 2001; Papachristos & Stamopoulos, 2002). Exploration of fumigantional potential of essential oils may lead to the development of non-synthetic, ecologically safe and easily degradable seed as well as grain protactant. Plants contain many compounds that are responsible for the ovicidal, repellent, antifeedant, chemosterilant, and toxic effects in insects (Nawrot & Harmatha, 1994; Isman, 2006). There are many reviews dealing with the use of plant products against insect pests of stored products (Lale, 1995; Golob & Gudrups, 1999; Adler et al., 2000; Weaver & Subramanyam, 2000; Isman, 2006), specifically on essential oils (Saxena & Koul, 1978; Singh & Upadhyay, 1993; Regnault-Roger, 1997) and on monoterpenoids (Coats et al., 1991). Steam distillation of aromatic plant yields essential oil, long used as fragrances and flavorings in the perfume and food industries, respectively, and more recently for aromatherapy and as herbal medicines (Buckle, 2003; Coppen, 1995; Isman, 2006).

13.3   USE OF FUMIGANTS

The fumigation of storage structures is a specialized operation and their success depends upon the care and expertise with which it is accomplished. Because of this, fumigation work is highly toxic and dangerous to handle; it is usually best to have general fumigation work done by professional fumigators who make a practice of the art and guarantee their work. Fumigants are toxic gases used for disinfestations of stored commodities in airtight storage structures. Fumigation structures should certainly be sufficiently gastight for the gas to penetrate and remain in the commodity for long enough to kill all stages of the insects present in or amongst the grains.

The purpose of fumigation is thus to obtain more-or-less immediate disinfestation of the stored commodity and the space enclosing it. Fumigation is the only chemical treatment that can achieve this effect and this relative immediacy of disinfestation, together with its completeness if done properly, is the main advantages of this particular chemical control technique. The major disadvantages are that the treatment confers no residual protection against reinfestation, once the commodity is again exposed, and the fact that the most effective fumigants are all highly toxic to humans and other non-target organisms. The precautions required to ensure the safe use of fumigants are, necessarily, much more stringent than those required to ensure the safe use of other insecticides.

The classification of fumigants as discussed above is no longer widely approved for use on stored commodities due to restrictions placed upon their use in several countries. Examples are carbon tetrachloride and ethylene dibromide, both of which are low-volatility fumigants and are recently identified as health hazardous chemicals. MB and phosphine are now the only fumigants commercially used worldwide for the protection of stored grain insects, but MB is considered as a factor for ozone depletion and phosphine developed resistance in almost all the stored grain insects throughout the world and both chemicals contaminate the environment and stored commodities. The desirable characteristics of a grain fumigant are highly efficient, good penetration ability to the stored commodities, toxicity to only the target insects, and lacking of residual effects (Taylor, 1975). In a controlled atmospheric storage system, carbon dioxide is found highly effective at lower concentration.

The solid forbmulations of metal phosphides, which are relatively easy to apply as compared to other liquid fumigants, have become the most popular and widely used fumigant all over the world. The further prolongation of recommended exposure periods for phosphine is three days in hot climate. There have been extensive investigations into the susceptibility of the developmental stages of stored grain insects (Hole et al., 1976). The pupal stage of grain weevils was found to be remarkably tolerant but other life stages were shown to be sufficiently susceptible to permit effective use of phosphine if the minimum exposure period was extended to four days, at favorable temperatures, to allow the tolerant pupae to pass into the more susceptible adult stage.

The development of resistance to phosphine in stored grain insects constitutes a problem, but does not generally invalidate the use of this fumigant which can still be expected to provide effective control of the major stored grain insects when treatments are carried out with suitable techniques (Taylor, 1975). The currently available alternative fumigant, MB, should prove effective at normal dosage rates whereas effective phosphine treatment would require an extension of the exposure period beyond the normally practicable limits for sheeted-stack fumigation. Tolerance to phosphine in the egg stage has also been observed in other insects (Hole et al., 1976; Bell, 1976), but this does not generally persist throughout egg development.

13.4   FUMIGANT APPLICATION TECHNIQUES

Fumigation techniques are responsible for the success in protecting the stored commodities in any types of storage structures.

13.4.1   STORE FUMIGATION

The free space fumigation of store rather than individual stacks of bagged grain is not a new technique; it has been practiced regularly for many years in several countries. The disinfestations of storage structures for the control of several insects invading the roof, walls, and floors are very effective. It also prevents the reinfestation of stored grain insects. Taylor (1975) reported that the repetitive dose of phosphine may induce resistance in stored grain insects. In several countries, the building is specially designed for space fumigation (Hayward, 1954). The technique involves application of fumigant in two portions, the second of these is applied 24 or 48 hours after the initial application. Using this technique, it is possible to prolong the period during which insects are exposed to a lethal concentration of fumigant, even in those storage structures where leakage of gas is taking place (Friesen, 1976).

13.4.2 STACK FUMIGATION

In the developing countries, the fumigation of stored commodities with bag stack under sheets is very common in use. This technology is relatively basic and of good standard, recommended for prevention of stored grain insect. The success of stack fumigation depends upon the sealing of stack, generally floor sealing; so for this purpose sandsnakes are recommended. Recent finding suggests that for effective sealing of heavy-duty and less flexible sheets, such as those of laminated polythene sheets, larger and heavier sandsnakes are necessary than those commonly used. The width of tubing used for the larger sandsnakes should be of the order of 150–200 mm. These, when filled, should provide a contact width on the floor of at least 100 mm. A disadvantage of this type of sandsnake is the increased weight, which is an important consideration for stored grain insect control teams with frequent operations or much travelling to do. The sandsnakes should be provided for each individual store or store complex to avoid the need for further transportation.

13.4.3 GAS CIRCULATION

An exclusive and patented technique is used for the phosphine gas circulation in storage structures. It is called "Phyto-Explo Fumigation," which enables bulk grain to be effectively treated in deep structures using phosphine. A shaft is driven into the grain, using compressed air, and is connected to a piping system which allows air circulation within the grain by means of a small pump. Fumigant is evolved from a phosphide formulation introduced into the headspace above the grain and gas is drawn down into the grain by the circulatory action of the pumping system. This technique permits effective distribution of fumigant in deep silos and in ships holds, rendering disinfestation possible without transferring the grain. The same technique can be used with MB enabling rapid treatment of silos not provided with a permanent circulatory system.

13.5   SAFETY PRECAUTIONS

The safety precaution is very important during the application of fumigants under any types of storage structure to save the persons involved in handling, application, and cleaning of fumigation work. Detailed information on safety procedures in the use of fumigation published by Food and Agriculture Organization (FAO) (Bond & Morse, 1982; Anon, 1980) is a concern in the promotion, planning, or implementation of fumigants; measures should be in sync with the recommended procedures and should ensure that all appropriate precautions are observed in the situations and circumstances for which they have responsibility. The safety information related to the use of fumigants must be read carefully before application.

The tablets and pellets of aluminum or magnesium phosphide may be dangerous if swallowed by either humans or animals. Most farmers are not aware of handling and storage of fumigants, but have quite a bit of experience in handling toxic chemicals; so the fumigants will not be ingested or swallowed by the farmers or their animals. But the handling of phosphine is different from other chemicals on the farm and does not follow the common sense rules of pesticide use. Read the label carefully for more details about the protective clothing when handling phosphine. Light cotton gloves and loose-fitting cotton clothing should be worn while fumigating with or handling phosphine so that no residues will be trapped against the skin and cause burning. There is more of a chance of having the residual dusts cling to hands, lips, hair, and clothing. As long as this dust does not become wet, it will take quite some time for the humidity in the air to cause the reaction that turns into phosphine gas. Remember that even sweat or dampness in the clothing caused by sweat can trigger this reaction. The following precautionary measures must be taken:

1. Do not eat, drink, and chew tobacco during fumigation or before completely washing up and changing clothing after fumigating any types of storage structures.
2. The protective clothing with phosphine mask must be worn before starting the fumigation work or handling phosphine.
3. The entire fumigation works should be completed in one session without any breaks.
4   If conducting fumigation of commercial storage structures or several bins, the applicator can change the protective clothing and mask at four-hour intervals.

5. Phosphide pellets and tablets produce phosphine gas when they are exposed to moisture in the air; so the applicator must leave the fumigating area before releasing the gas.

13.5.1 SYMPTOMS OF POISONING

The first symptoms of poisoning of phosphine gas are irritation in mucous membrane especially in the deep lungs and upper airways. Because phosphine gas releases highly acidic forms of phosphorus when it contacts deep lung tissues, it tends to cause pulmonary edema. Once absorbed into the body, phosphine can damage cell membranes and enzymes that are important for respiration and metabolism, while minute quantity of phosphine gas may cause mild headaches. Diarrhea, nausea, abdominal pain, vomiting, tightness of the chest, breathlessness, soreness or pain in the chest, palpitations, headache, dizziness, staggering, and skin irritation are typical symptoms of phosphine poisoning. The exposures to higher concentrations or the direct ingestion of tablets cause death to humans and other mammals such as livestock and pets. The symptoms of severe poisoning may occur within a few hours to several days after poisoning.

In case any symptom of poisoning, one must immediately call the physician for treatment. Take proper care of victim if the gas or dust from aluminum phosphide is inhaled: take the exposed person to fresh air, keep warm, and make sure that the person can breathe freely. If breathing is not proper, give artificial respiration by mouth-to-mouth or other means of resuscitation. Do not give anything by mouth to an unconscious person. Induce vomiting by touching back of the throat with finger or, if available, provide universal antidotes with lukewarm water. Change the protective clothing, keep the victim in well-ventilated area, and wash contaminated skin thoroughly with soap and water. Sometimes the dust of residue can get in the eyes; in such cases, flush the eyes with plenty of water.

The following precautions must be taken to avoid the poisoning in any non-target organisms.

1. Fumigation work must only be performed by a trained expert. For each fumigation, one person must be responsible as head of the fumigation team for tasks raning from preparing the fumigation to the release of the store for general access.
2. Read the instruction manuals carefully and follow the instructions strictly.

3. Inform all people who work in the store and all those who live in the vicinity of the store about the forthcoming fumigation.
4. Measure the length, breadth, and height of the stack and calculate the volume of the stack.
5. Attach warning signs to the stack and to the door of the store and lock the store after fumigation.
6. Make a regular check of the seals and ensure that no unauthorized persons enter the store during the entire fumigation period.
7. Only allow the most urgent work to be performed in the warehouse and take care for proper ventilation when work is taking place.
8. Measure the concentration of the gas from time to time in order to ensure that there is no danger to staff.
9. Collect the residues of the tablets, pellets, or bags and dispose them properly.

KEYWORDS

- **fumigation**
- **methyl bromide**
- **carbon tetrachloride**
- **carbon disulphide**
- **ethylene dichloride**
- **ethylene oxide**
- **phosphine**
- **hudrogine cyanide**
- **chloripicrin new fumigants**
- **Fumigation techniques**
- **symptoms of poisoning**
- **safety precaution**

REFERENCES

ACGIH. In *Threshold Limit Values for Chemical Substances in Workroom Air*, Proceedings of American Conference of Governmental Industrial Hygienists for Cincinnati Ohio, ACGIH: OH, USA. **1981**; pp 21–23.

Adler, C.; Ojimelukwe, P.; Tapondjou, A. L. In *Utilization of Phytochemicals against Stored Product Insects*, Bulletin No. 23; Proceedings of the Meeting of the OILB Working Group "Integrated Protection in Stored Products", August 22–24, 1999; Adler, C., Schoeller, M., Eds.; IOBC/WPRS: Berlin, Germany, **2000;** pp 169–175.

Anon. Fumigant Piping System for Hydrogen Phosphide in Commodities. *Pest Control.* **1980,** *48,* 54.

Bell, C. H. The Tolerance of Immature Stages of Methyl Bromide. *J. Stored Prod. Res.* **1976,** *12,* 1–10.

Bell, C. H.; Barbara, D.; Evams, P. H. The Occurrence of Resistance to Phosphine in Adult and Egg Stages of Strains of *Rhyzopertha dominica* F. (Coleoptera: Bruchidae). *J. Stored Prod. Res.* **1977,** *13,* 91–94.

Beratlief, C.; Alexandrescu, S. Effect of Phosphine Fumigation on the Germination of Wheat and Maize Seed. *Probl. Aqric.* **1964,** *16,* 45–51.

Bond, E. J.; Morse, P. M. Joint Action of Methyl Bromide and Phosphine on *Tribolium castaneum* H (Coleoptera: Tenebrionidae). *J. Stored Prod. Res.***1982,** *18,* 83– 94.

Bond, E. J.; Dumas, T. Loss of Warning Odor from Phosphine. *J. Stored Prod. Res.* **1967,** *3,* 389–392.

Bond, E. J.; Robinson, J. R.; Buckland, C. T. The Toxic Action of Phosphine. Absorption and Symptoms of Poisoning In Insects. *J. Stored Prod. Res.***1969,** *5,* 289–298.

Bruch, C. W. Gasesous Sterilization. *Annu. Rev. Microbiol.* **1961,** *15,* 245–262.

Borah, B.; Chalal, B. S. Development of Resistance in *Trogoderma granarium* E to Phosphine in the Punjab. *FAO Plant Prot. Bull.* **1979,** *27,* 77–80.

Buckle, J. *Clinical Aromatherapy: Essential Oils in Practice*; Churchill Livingstone: Edinburgh, UK, **2003,** pp 416.

Champ, B. R.; Dyte, C. E. Report of the FAO Global Survey of Pesticide Susceptibility of Stored Grain Pests. *FAO Plant Prod. Prot. Ser.* **1976,** *5,* 297.

Coats, J. R.; Kar, L. L.; Drewes, C. D. Toxicity and Neurotoxic Effects of Monoterpenoids in Insects and Earthworms. In *Naturally Occurring Pest Bioregulators. ACS Symposium Series No. 449*; Hedin, P. A., Ed.; American Chemical Society: Washington, DC., **1991;** pp 305–316.

Coppen, J. J. W. *Flavours and Fragrances of Plant Origin;* Food Agric. Org: Rome, **1995;** pp 101.

Cotton, R. T. *Insect Pests of Stored Grain and Grain Products*; Biotech Books: New Delhi, India, **2013;** pp 41.

Cotton, R. T.; Roark, R. C. Ethylene Oxide as Fumigant. *Indus. Eng. Chem.* **1928,** *20*(8), 805–807.

Fam, E. Z.; Kamel, A. H.; Mahdi, M. T.; Sheltawy, E. M. The Effect of Repeated Fumigation on the Germination of Certain Vegetable Seeds. *Bull. Entomol. Soc. Eqypt.* **1974,** *7,* 85–89.

FAO. Recommended Methods for the Detection and Measurement of Resistance of Agricultural Pests to Pesticides. Tentative Method for Adults of Some Major Pest Species of Stored Cereals with Methyl Bromide and Phosphine. *FAO Plant Prot. Bull.* **1975,** *23,* 12–25.

Friesen, O. H. Grain Drying. *Ottawa Canada Dept. Agric. Pub.* **1976,** *1497,* 44.

Glaser, Z. R. Ethylene Oxide: Toxicology Review and Field Study Results of Hospital Use. *J. Environ. Pathol. Toxicol.* **1979,** *2,* 173–207.

Golob, P.; Gudrups, I. *The Use of Spices and Medicinals as Bioactive Protectants for Grains*; Bull. No.137, FAO: Rome, Italy, **1999**.

Grish, G. K. Carbon Disulphide as a Grain Fumigant. *Bull. Grain Tech.* **1966**, *4*(1), 11–17.

Halliday, D.; Prevett, P. F. Fumigation of Pyramidal Stacks of Bagged Decorticated Ground-outs with Methyl Bromide. *J. Sci. Food Aqric.* **1963**, *14*, 586–592.

Hashiguchi, Y.; Ogahara, T.; Horiko, H. The Flammability Limit of Ethylene Oxide-Methyl Bromide-Air Mixtures. *Konyo Kayaku Kyokai-Shi.* **1967**, *28*, 128–131.

Hayward, L. A. W. Infestation Control in Stored Groundouts in Northern Nigeria. *World Crops.* **1963**, *15*(2), 63–67.

Hayward, L. A. W. The Field Fumigation of Groundnuts in Bulk. *J. Sci. Food Agr.* **1954**, *5*, 192–194.

Hole, B. D. Variation in Tolerance of Seven Species of Stored-Product Coleoptera to Methyl Bromide and Phosphine in Strains from Twenty Nine Countries. *Bull. Entomol. Res.* **1981**, *71*, 299–306.

Hole, B. D.; Bell, C. H.; Miles, K. A.; Goodship, G. The Toxicity of Phosphine to all Developmental Stages of Thirteen Species of Stored Product Beetles. *J. Stored Prod. Res.* **1976**, *12*, 235–244.

Howe, R.W. Problems in the Laboratory Investigation of the Toxicity of Phosphine to Stored Product Insects. *J. Stored Prod. Res.* **1974**, *10*, 167–81.

Isman, M. B. Botanical Insecticides, Deterrents, and Repellents in Modern Agriculture and an Increasingly Regulated World. *Annu. Rev. Entomol.* **2006**, *51*, 45–66.

Kamel, A. E. L.; Shahba, B. A. *Protection of Stored Seeds in Egypt*; Bull. No. 295, Ministry of Agric.: Cairo, Egypt, **1958**.

Kashi, K. P.; Bond, E. J. The Toxic Action of Phosphine Role of Carbon Dioxide on the Toxicity of Phosphine to *Sitophilus granarius* and *Tribolium confusum*. *J. Stored Prod. Res.* **1975**, *11*, 9–15.

Khare, B. P. *Stored Grain Pests and Their Management*; Kalyani Pub.: Ludhiana, India, **2006**; p 314.

Kostyukovsky, M.; Ravid, U.; Shaaya, E. The Potential Use of Plant Volatiles for the Control of Stored Product Insects and Quarantine Pests in Cut Flowers. In *Proceedings of the International Conference on Medicinal and Aromatic Plants Possibilities and Limitations of Medicinal and Aromatic Plant Production in the 21st Century;* Bernath, J., Zamborine Nemeth, E., Crakeram, L., Kock, O., Eds.; 8–11 July 2001, Acta Horticulturae: Budapest, Hungary, **2002**; Vol. 576, pp 347–358.

Kunz, S. E.; Morrison, E. O.; King, D. R.; The Effects of Grain Moisture Content, Grain Temperature and Dockage on the Penetration of HCN. *J. Econ. Entomol.* **1964**, *57*, 453–455.

Lale, N. E. S. An Overview of the Use of Plant Products in the Management of Stored Product Coleoptera in the Tropics. *Postharv. News Inf.* **1995**, *6*, 69–75.

Lindgren, D. L.; Sinclair, W. B.; Stupin, P. J. Tolerance of Avocado Fruit to Fumigation with Ethylene Dibromide. *Calif. Avocado Soc. Yearb.* **1955**, *39*, 202–208.

Lindgren, D. L.; Sinclar, W. B.; Vincent, L. E. Residues in Raw and Processed Foods Resulting from Post-Harvest Insecticidal Treatments. *Residue Rev.* **1968**, *21*, 1–121.

Majumdar, S. K.; Muthu, M. Recent Progress in Prophylactic Treatment for Pest Control in Stored Foodstuffs in India. *Food Sci.* **1960**, *9*(3), 89–95.

Majumder, S. K.; Muthu, M.; Srinivasan, K. S.; Natarajan, C. P.; Bhatia, D. S.; Subrahmanyan, V. Studies on the Storage of Coffee Beans. IV Control of *Araecerus fasciculatus* in Monsooned Coffee and Related Storage Experiments. *Food Sci.* **1961**, *10*, 382–388.

Matasyoh, L.G.; Matasyoh, J. C.; Wachira, F. N.; Kinyua, M. G.; Muigai, A. T. M.; Mukiama, T. K. Chemical Composition and Antimicrobial Activity of the Essential Oil of *Ocimum gratissimum* L. Growing in Eastern Kenya. *Afr. J. Biotech.* **2007,** *6,* 760–765.

Nakakita, H.; Winks, R. G. Phosphine Resistance in Immature Stages of a Laboratory Selected Strain of *Tribolium castaneum* (Coleoptera: Tenebrionidae). *J. Stored Prod. Res.* **1981,** *17,* 43–52.

Nawrot, J.; Harmatha, J. Natural Products as Antifeedants against Stored Product Insects. *Postharv. News Inf.* **1994,** *5,* 17–21.

Oka, Y.; Nacar, S.; Putievsky, E.; Ravid, U.; Yaniv, Z.; Spiegel, Y. Nimaticidal Activity of Essential Oils and Their Constituents against the Root – Knot Nematode. *Phytopathology.* **2000,** *90,* 710–715.

Papachristos, D. P.; Stamopoulos, D. C. Repellent, Toxic and Reproduction Inhibitory Effects of Essential Oil Vapours on *Acanthoscelides obtectus* (Say). (Coleoptera: Bruchidae). *J. Stored Prod. Res.* **2002,** *38,* 117–128.

Paster, N.; Menasherov, M.; Ravid, U.; Juven, B. Antifungal Activity of Oregano and Thyme Essential Oils Applied as Fumigants against Fungi Attacking Stored Grain. *J. Food Prot.* **1995,** *58,* 81–85.

Raja, N.; Albert, S.; Ignacimuthu, S.; Dorn, S. Effect of Plant Volatile Oils in Protecting Stored Cowpea *Vigna unguiculata* (L.) Walpers against *Callosobruchus maculatus* (F.) (Col.: Bruchidae) Infestation. *J. Stored Prod. Res.* **2001,** *37*(2), 127–132.

Rajendran, S.; Sriranjini, V. Plant Products as Fumigants for Stored-Product Insect Control. *J. Stored Prod. Res.* **2008,** *44,* 126–135.

Regnault-Roger, C. The Potential of Botanical Essential Oils for Insect Pest Control. *Inte. Pest Manage. Rev.* **1997,** *2,* 25–34.

Richardson, L.T.; Monro, H. A. U. Fumigation of Jute Bags with Ethylene Oxide and Methyl Bromide to Eradicate Potato Ring Rot Bacteria. *Appl. Microbiol.* **1962,** *10,* 448– 451.

Roark, R. C. Bibliography of Chloropicrin 1848–1932. *U. S. Dept. Argic. Misc. Pub.* **1934,** *88,* 176.

Roth, H.; Richardson, H. H. Permeability to Methyl Bromide of Wrappers Found on Shipments of Imported Plants. *J. Econ. Entomol.* **1968,** *61,* 771–778.

Saxena, B. P.; Koul, O. Utilization of Essential Oils for Insect Control. *Ind. Perfum.* **1978,** *22,* 139–149.

Schuhmacher, A.; Reiching, J; Schnitzler, P. Virucidal Effect of Peppermint Oil on the Enveloped Viruses Herpes Simplex Virus Type 1 and Type 2 *in Vitro. Phytomedicine.* **2003,** *10,* 504–510.

Shaaya, E.; Kostjukovski, M.; Eilberg, J.; Sukprakarn, C. Plant Oils as Fumigants and Contact Insecticides for the Control of Stored Product Insects. *J. Stored Prod. Res.* **1997,** *33*(1), 7–15.

Singh, G.; Upadhyay, R. K. Essential Oils: a Potent Source of Natural Pesticides. *J. Sci. Ind. Res.***1993,** *52,* 676–683.

Slover, H. T.; Lehmann, J. Effects of Fumigation on Wheat in Storage. *Tocopherols. Cereal Chem.* **1972,** *49,* 412–415.

Taylor, R. W. D. Fumigation of Individual Sacks of Grain Using Methallyl Chloride for Control of Maize Weevil. *Int. Pest Control.* **1975,** *17,* 48.

Vashkov, V. I.; Prishchep, A. G. Efficiency of Sterilization by Making Use of Ethylene Oxide and Methyl Bromide. *Life Sci. Spl. Res.* **1967,** *5,* 44–50.

Weaver, D. K.; Subramanyam, B. Botanicals. In *Alternatives to Pesticides in Stored-Product IPM*; Subramanyam, B., Hagstrum, D. W., Eds.; Kluwer Academic Press: MA, USA, **2000;** pp 303–320.

Winks, R. G. Characteristics of Response of Grain Pests to Phosphine; In *Commonwealth Scientific and Industrial Research Organization, Division of Entomology*; Canberra, Australia, Annual Report 1973–1974; pp 65–74.

Winteringham, F. P. W. The Fate of Labelled Insecticides in Food Products. The Possible Toxicological and Nutritional Significance of Fumigating Wheat with Methyl Bromide. *J. Sci. Food Aqric.* **1955,** *6,* 269–274.

CHAPTER 14

IMPORTANT MITES ASSOCIATED WITH STORED PRODUCE AND THEIR MANAGEMENT

CONTENTS

ABSTRACT

The storage mites form the largest group of class called Arachnids. Arachnids, being small in size, are ecologically important in most terrestrial environments. Arachnids were the first arthropods to enter terrestrial habitats as evident from the oldest fossil of *Silurian scorpions* by the end of the Paleozoic. The Acarina are minute arachnids that have adopted along a number of independent groups and have undergone convergent evolution. They are distinguished by a body that is usually divided into cephalothorax and abdomen. The mites are important to human life as they invade grain and grain products and other stored products. Stored product mites decrease the germination percentage of cereals and also infest and damage other food stuffs such as cheese, flour, seed, bulbs, tubers, and dried fruits of all kinds. Decaying materials and other such items are the main factors for the habitation of these mites. The mites cause direct and indirect damage to stored grains and their products by raising their moisture contents and generating sufficient heat for the growth of infectious bacteria and fungi.

14.1   IMPORTANT MITES ASSOCIATED WITH STORED PRODUCE

The storage mites form the largest group of class called Arachnids. Arachnids, being small in size, are ecologically important in most terrestrial environments. Arachnids were the first arthropods to enter terrestrial habitats as evident from the oldest fossil of *Silurian scorpions* by the end of the Paleozoic. The Acarina are minute arachnids that have adopted along a number of independent groups and have undergone convergent evolution. They are distinguished by a body that is usually divided into cephalothorax and abdomen. The leg consists of four pairs and there are no antennae. The anterior part of the body is known as anterior gnathosoma and posterior one is termed as ideosoma. They occupy diverse habitats, some live in plant galls, others in or on birds or mammals, while others contaminate grains and their products. Some have been found in mosses, surface litter, or upper levels of the soil, whereas some have adapted aquatic life and live in salt and fresh water. The class Arachnids and order Acarina represent distinct families; the larger one is called ticks and the smaller one is known as mites. The mites are important to human life as they invade grain and grain products and other stored products.

14.2 DAMAGE CAUSED BY MITES

They infest a wide range of hosts and commodities; for example phytophagous mites cause heavy losses to agricultural crops, vegetables, wild plants, fruits, and ornamental plants (Evans, 1992), whereas predatory mites feed on harmful mites and small soft-bodied insects like aphid, whiteflies, scales, and their eggs. On the other hand, parasitic mites parasitize animals, birds, and human beings (Srivastava, 1996). Stored grain and stored product mites decrease the germination percent of cereals (Ashfaq et al., 1996; Zdarkova, 1996) and also infest and damage other food stuffs such as cheese, flour, seed, bulbs, tubers, and dried fruits of all kinds. Decaying materials and other such items are the main factors for the habitation of these mites. The mites cause direct and indirect damage to stored grains and their products by raising their moisture contents and generating sufficient heat for the growth of infectious bacteria and fungi. Mite-infested grains undergo a series of changes in their chemical composition affecting the germination capacity; flour prepared from contaminated grains is more acidic and of fusty smell and bitter taste. They contaminate the space between the grains with their dead bodies, cast skins, and excrement and hinder the airflow in the stock. Stored fungi and mites frequently cause injury to grains and contaminate grains and agro-products by allergins and toxins

14.3 CLASSIFICATION AND IDENTIFICATION OF IMPORTANT MITES

There are several orders of mites, but only three—Astigmata, Prostigmata, and Mesostigmata—are commonly found in association with stored products. The names of these orders are derived from the position, when present of the Astigmata (equivalent to the spiracles of insects). The Prostigmata are characterized by the position of their pair of stigmata at the front of the body, but the families in this order are in other ways very variable in size and form. Many prostigmatan mites are predators or parasites of insects and of other mites. The Mesostigmata include two noticeably different suborders found commonly in stores: Gamasina and Uropodina.

14.3.1 ASTIGMATA

Astigmata mites are very small with legs of moderate length; they are whitish, creamy, or milky translucent in color .The name of the order indicates the absence of stigmata: respiration is by diffusion through the whole body surface. They are important pests of stored produce. The life cycle includes a protonymph and tritonymph, with an intervening hypopus (deutonymph) in many species under adverse conditions. Two families, Acaridae and Glycyphatidae, are common in tropical storage system.

Family Acaridae

The family Acaridae contains most of the storage mites in tropical storage, and these are usually the most abundant species in any mite-infested produce.

Acarus siro L.

Acarus siro is prominently seen in floury mass of grain and grain products. The optimum condition for the threshold of development is between 25 and 27°C and 80 and 90% relative humidity. The total life cycle is completed in 8-12 days. *Acarus siro* is able to feed cereal products (especially flour), cereal grain, cheese, and hay.

Aleuroglyphus ovatus (Troupeau)

Aleuroglyphus ovatus is a cosmopolitan species recorded from a wide variety of cereal products, cereal grains, and other products. It is a relatively large mite with a glistening whitish body and reddish brown legs. In the laboratory, it has been reared successfully on germ or on various storage fungi. No hypopal stage has been recorded.

Caloglyphus spp.

Several species of *Caloglyphus* have been recorded from stored products. They associated with damp or only partially dried produce. They produce hypopal under adverse conditions, such as decreased moisture content, and are then dispersed by insect, to which they attach themselves with suckers. The common species appear to be almost cosmopolitan, for example, Hughes (1976) reported *C. oudemansi* from the UK, Italy, the USSR, India, Indonesia, and Australia.

Lardoglyphus spp.

The three species of *Lardoglyphus* associated with stored products, *L. konoi*, *L. zasheri*, and *L. angelinae*, are pests of food that contains large amounts of protein, especially animal products, such as dried fish products, bones, hides, and skins. These mites feed directly on the produce and can increase to vast numbers, particularly in humid climates. Under adverse conditions, for example, overcrowding, hypopal are formed and these attach themselves by suckers to the adults and larvae of Dermestes beetles. *Lardoglyphus zacheri* has been reported from the UK, Germany, Mexico, South America, and Australia.

Rhizoglyphus spp.

Rhizolyphus are usually associated with perishable commodities such as bulbs and fresh root crops, but one species, *R. callae* (Hughes, 1976), has been recorded from decaying wheat spillage in the UK, and from damp cereal spillage in Ethiopia. The mites are large bloated and smooth, and possess shinning body with brown legs.

Suidasia spp.

The two species of *Suidasia* associated with stored products, *S. nesbitt* and *S. pontifica*, are the smallest acarid mites commonly found in tropical storage. The cuticle of both species has a scale-like wrinkled texture, and the form of the wrinkling is the main character for separation of the species. These mites are dull colored with tactile setae on legs and ar found damaging wheat bran and rice. *Suidasia pontifica* is common and widespread throughout the tropics of America, Africa, South Asia, and Southeast Asia, and has also been recorded from temperate countries; it infests a wide variety of produce, including milled rice, rice bran, maize, cassava, groundnuts, cowpeas, and beans.

Tyrophagus putrescentiae (Schrank)

The cosmopolitan *Tyrophagus putrescentiae* is the commonest mite pest of stored food in the tropics and it has been recorded from many stored produce. Although it is known to feed on storage fungi and to develop rapidly when preferred fungi present. Like all acarids, it develops more rapidly in higher humidity but development is very fast at 60% relative humidity and 35–40°C temperature.

Other species such as *T. tropicus* Robertson, *T. brevicrinatus* Robertson, *T. palmarum* Oudemans, and *T. perniciosus* Zachvatkin are less important mites of tropical stored products and are most usually found in small numbers in mixed populations with *Tyrophagus putrescentiae*.

Family Glycyphagidae

The family Glycyphagidae consist of a few species that are regularly found in tropical stored products, but usually in small numbers. These mites feed on linseed, tobacco, dried fruits, flour, and other grain and grain products, while it can also infest the animal products. The development period is 10–12 days at 20–28°C temperature and 80–85% relative humidity.

Glycyphagus spp.

Glycyphagus destructor is a common pest of cereals and several stored products and also occurs in house dust, but it appears to be much more abundant in temperate rather than tropical climates. It is also a fungus-feeder. It is cosmopolitan in nature.

Glycyphagus domesticus

Glycyphagus domesticus is a widespread fungus-feeding species found on many types of stored produce and frequently associated with house dust where it causes respiratory allergy in humans. It is mainly found in temperate climates.

14.3.2 PROSTIGMATA

The Prostigmata contain a stigmata at the front of the body, but their size varies from family to family. Most of the prostigmatan mites are predators or parasites of insects and of other mites.

Family Cheyletidae

Cheyletidae family is common in tropical storage condition. The typical life cycle of prostigmatans includes only two nymphal stages (protonymph and deutonymph), but in some groups these nymphal stages are missing or occur within the body of the female.

Cheyletidae spp.

Cheyletidae are predators of mites and insects; so they are considered as natural enemies of storage pests. They are large and have strong claws at the tip, and are used to catch and hold the prey, and stylet-like mouthparts.

Cheletomorpha lepidopterorum (Shaw)

Cheletomorpha lepidopterorum are orange in color and having long first pair of legs. It feeds on acarid mites, and possibly small insect larvae or nymphs. It has a cosmopolitan distribution but is not very common in tropical storage.

Cheyletus malaccensis

Cheyletus malaccensis is the dominant cheyletid in tropical stores, it preys on other mites, acarids, and Psocoptera, and it also attacks young larvae or nymphus of other insects.

Family Cuanxidae

The species of this family are predatory in nature , their body modified as long palps ending in a small claw grasping their prey.

Cuanxidae setirostris (Hermann)

Cunaxa setirostris regularly observed from stored products. Its red body is about 0.5 mm long. It has a composition in nature.

Family Pyemotidae

The mites of this family are usually parasites of insects, and two genera, *Pyemotes* and *Acarophenax*, are found in association with storage insects.

Acarophenax spp.

Some *Acarophenax* spp. is ectoparasites of storage beetles; the most commonly found species is *A. tribolii* which is a parasite of *Tribolium* spp. The female adults attach themselves by their chelicerae to the soft inter-segmental membranes of the adult beetles, especially under the elytra and feed on the beetles' body fluid.

Pyemotes spp.

Pyemotes are ecto-parasites of beetles and moth, the female *Pyemotes* feed on their host by piercing soft parts of cuticle, with their stylet-like chelic-erae and sucking the body fluid. When present in large numbers, these para-sites reduce the population growth of their hosts. The young females are 0.2–0.25 mm long and males are small and are usually found attached to the bodies of gravid females. Development from egg to young adult takes place entirely within the distended body of the female.

Four species of *Pyemotes* have been reported to parasitizing storage pests: *P. anobii*, *P. beckeri*, *P. herfsi*, and *P. tritici* (Cross & Moser,1975).

Family Tarsonemidae

The mites of this family are the smallest of the mite pests found in stored produce. They are usually 0.1–0.2 mm long, oval in shape, with a shining yellowish or colorless cuticle, and rather short legs.

Tarsonemus spp.

Most tarsonemids mites found in storage are *Tarsonemus*, and they are often infesting the cereal grain stores, especially on grain dust. Their mouth parts are piercing and sucking types and they are fungus-feeders.

Family Tydeidae

The members of Tydeidae family that occur in storage are small mites and they are predators of another mites, especially small acarids.

14.3.3   MESOSTIGMATA

The mites of this order are often larger by 0.35–1.1 mm in size than other storage mites, and are relative well-sclerotized with buff, yellowish, or brownish cuticle. The stigmata are about halfway along the body, lateral to the bases of the four pair of legs. The life cycle includes only two nymphal stages (protonymph and deutonymph) between the larva and the adult. The Mesostigmata is further divided into two sub orders: Gamasina and Uropodina.

14.3.3.1   MESOSTIGMATA GAMASINA

Have long legs and palps, and they are usually predators, through a few feed on the commodity or on fungi. In tropical storage, species of the ascid genus *Blattisocius* are the commonest gamasine mites.

Family Ameroseiidae

Kleemannia spp.

Kleemannia plumigera and *K. plumosus* are fungus-feeders and occur in moldy produce and residues in storage, but generally found in temperate climates.

Family Ascidae

Blattisocius spp.

Blattisocius mites are predators of insect eggs and larvae of other mites, in tropical storage. Large populations can occur in the presence of established populations of their prey, and in certain situations they may cause significant bio control agent. Haines (1981) reported the potential effect of *Blattisocius tarsalis* on *Ephestia cautella* populations. *Blattisocius keegani* is also often abundant in tropical stores and is known as a predator of the eggs of several storage beetles and certain mites (Hughes, 1976).

Family Dermanyssidae

Dermanyssidae family consists of predators, parasites, and fungus-feeders. Only two species occur in storage, but several species that are parasites of rats and birds are occasionally found in poor managed stores.

Androlaelaps casalis

Androlaelaps casalis is the only dermanyssid observed in tropical storage. This cosmopolitan species feeds on a mixed diet of animal and plant materials and it can survive on wheat. It also preys on another mites, and *Tribollium* larvae. It requires a high humidity for successful development.

Haemogamasus pontiger

Haemogamasus pontiger is a cosmopolitan predator on other mites and some insect larvae. Hughes (1976) reported that he found it in store residues and in cereal grains and other foods, but it may be more common in temperate stores than in tropical storage system.

Family Macrochelidae

They are more strongly sclerotized than other mites and they are usually predators of stored grain insect, but some may feed on fungi.

14.3.3.2   MESOTIGMATA: UROPODINA

The mites of this suborder are often strongly sclerotized and are dark brown in color. Their legs, which are rather shorter than other mites, can usually be folded into special cavities of the body. Because of their hardened cuticle and retractable legs, they are much less affected than other storage mites by movement of the grain.

Leiodinychus krameri

The *Leiodinychus krameri* is the common uropodine mite, and large populations are sometimes found on damp stored produce. They are fungus-feeders and transmit fungal infestation within the stored produce. They feed on several cereals, cereal products, and pulses in so many countries.

14.4 MANAGEMENT OF STORAGE MITES

There are no any specific management practices of mites under storage condition, but the following precaution must be taken to prevent the growth and development of mites.

- Maintain optimum moisture content of stored produce.
- Check the temperature and relative humidity regularly.
- Keep the store clean and dry.
- If needed, fumigate the store with sulfur at 1 kg/ 100 m³ space.

KEYWORDS

- **storage mites**
- **classification**
- **astigmata**
- **prostigmata**
- **mesostigmata**

REFERENCES

Ashfaq, M.; Ahamad, M. S.; Perveaz, A. Some Studies on Correlation between Mite Population and Germination Losses of Stored Wheat. *Pak. Entomol.* **1996,** *18*(1–2), 102–103.

Cross, E. A.; Moser, J. C. A New, Dimorphic Species of *Pyemotes* and a Key to Previously Described Forms (Acarina: Tarsonemoidea). *Ann. Entomol. Soci. Am.* **1975,** *68,* 723–732.

Evans, G. O.; *Principles of Acarology;* CAB International: Wallingford, UK, 1992; p 252 .

Haines, C. P. Insects and Arachnids from Stored Products: A Report on Specimens Received by the Tropical Stored Products Centre. *Rep. Trop. Prod. Inst.* **1981,** *54,* 73.

Hughes, A. M. *The Mites of Stored Food and Houses;* 2nd Edition, Technical Bulletin No. 9; Ministry of Agriculture, Fisheries and Food: London, 1976;Vol. 4, p 400.

Srivastava, K. P. *A Textbook of Applied Entomology;* Kalyani Publication: New Delhi, India, 1996; Vol. 2, p 507.

Zdarkova, E. The Effect of Mites on Germination of Seed. *Ochrana-rostlin.* **1996,** *32*(3), 175–179.

CHAPTER 15

IMPORTANT BIRDS ASSOCIATED WITH STORED GRAIN AND THEIR MANAGEMENT

CONTENTS

ABSTRACT

The post-harvest losses due to birds are a great matter of concern all over the world. Generally, the problem of birds starts from field when crops reach maturity stage; they also damage stored commodities in warehouse or in any storage structures. Bird population affects our economy and health. Among the vertebrate pests, only birds have feathers and they belong to class Avis of phylum chordate. Birds are highly specialized having 28,500 species and sub species. The feet and beaks of birds are highly modified to feed on different types of commodities. A number of grain-eating birds have become important pests of stored products in many countries. Several birds, for example, *Passer domesticus*, *Passer montanus*, *Acridotheres tristis*, Streptopelia species, *Corvus splendens*, and *Columba livia* are associated to crop at ripening stage and to drying and threshing floors at harvest time. Some have developed a close association with the more permanent sources of cereals and cereal products and have become a problem in warehouse and other storage structures. The contribution of birds in post-harvest losses is very few due to their geographical distribution and several types of traps and other alternative ways must be applied for the management of birds under storage condition.

15.1   INTRODUCTION

The post-harvest losses due to birds are a great matter of concern all over the world. Generally, the problem of birds starts from field when crops reach maturity stage; they also damage stored commodities in warehouse or in any storage structures. Bird population affects our economics and health. Among the vertebrate pests, only birds have feathers and they belong to class Avis of phylum Chordate. Birds are highly specialized having 28,500 species and sub species. The feet and beaks of birds are highly modified to feed on different types of commodities. A number of grain-eating birds have become important pests of stored products in many countries. Several birds, for example, *Passer domesticus*, *P. montanus*, *Acridotheres tristis*, *Streptopelia species*, *Corvus splendens*, and *Columba livia* are associated to crop at ripening stage and to drying and threshing floors at harvest time. Some have developed a close association with the more permanent sources of cereals and cereal products and have become a problem in warehouse and other storage structures. The magnitude of losses caused by birds is very less as compared to that by rodent and other vertebrate pests; they consume about

244 g grain per day per yard (Garg et al., 1966). Libay et al. (1983) reported that losses as feed to European sparrows, *P. montanus*, at farms of Philippines is about 147–177 g per day per farm.

15.2   PROBLEMS OF BIRDS

The bird population spreads at different geographic regions. The bird populations inside a store pose some major problems such as:

1. Birds may settle on top of stacks and peck holes in woven sacks in order to reach the food inside. This can cause spillage and in extreme cases, collapse of the stack.
2. Many species of birds roost inside warehouse unless access is completely restricted.
3. Birds dropping may reduce the quality of stored commodities.
4. Besides the direct losses, they also act as carrier of several mite species.
5. The nesting materials of birds act as harborage and site for breeding of some stored grain insects.
6. The grains damaged by birds are most susceptible for infection of several storage fungi, and they also carry spore of fungi.

15.3   POPULATION ESTIMATION OF BIRDS

The size of population can be estimated on the basis of their nasality, mortality, immigration, and emigration. These four factors may be put into the following formula to work out population of birds.

$$Ni + 1 + Ni + (B - D) + I - E$$

Where

Ni	+	1 Population at time initial + 1 (Estimation)
Ni		Population at time initial
B		Nasality
D		Mortality
I		Immigrants
E		Emigrants

15.3.1   TYPES OF POPULATION ESTIMATION

The estimation of bird population size or density is widely used in bird infestation, because it allows to measure changes in population and their impact on losses (Lambert, 1993). Sometimes, assessment of non-isolated population of birds is viable (Githiru & Lens, 2006).

The following two types of population estimator of birds are commonly in use:

A. Measurement of relative density
    The magnitudes of bird population can be measured on the basis of dropping, nests, and quantity of grain consumed by the birds. The relative estimation is measured in determinate units which allows comparisons in time and space, and is useful for studying the activity patterns of birds. This type of estimate is obtained by catch per unit time or effort and use of several types of bird traps. Relative population is influenced by various factors, for example, behavior of birds, level of activity, efficiency of trap, level of activity, and response of particular sex.

B. Measurements of absolute density
    The total number of birds per unit area represents the absolute population density, and such types of population estimates are helpful in preparation of life table, population dynamics. The population density also indicates the relationship between the level of bird population and resultant damage to the host, and this can be helpful in understanding any change in populations. The absolute population can be estimated by following methods.

    1. Visual counting of birds or counting by previous census of birds.
    2. Sampling of small proportion of the population to estimate the total population.
    3. Capture marking and recapture methods allow estimation of density, nasality and mortality; when distance between recaptures are known, movement and home range may be determined. The population can be estimated by using following formula.

        $P = N \times M / R$

        where,

        P = Population of birds

N = Total number of capture birds

M = Number of marked individuals released

R = Number of marked individuals recapture

The population estimation by this method has some assumptions.

i) Marked and unmarked animals are caught at the same rate (equal probability of capture).
ii) Marked and unmarked animals are subject to the same mortality rate.
iii) Marks are not lost or overlooked.

15.4 COLLECTION OF BIRDS

1. Several types of collection nets are used to capture birds. Mist nets—fine silk or nylon net with 3–4 inch mesh, 7 feet wide, and 18–38 feet long—with a taut frame of stout twine crossed by horizontal braces called "shelfatrings" are used. A bird striking the net from either side carries the net beyond the shelfatrings and hangs in the pocket of the net. A net is properly hung, with four shelves at about 6 feet height. Mist nets are effective during calm day against a dark background in the birds roosting area, of warehouse, however, this net is also effective under field condition. The captured birds are removed immediately to maintain hygiene.

2. The modified Australian crow trap is a large cage made of mesh wire with wooden frames measuring 1–2 m wide × 2 m long × 2 m height. The midsection of the truncated V-shaped cage top is provided with slots through which birds can enter. This trap is self-operating and so must contain food, water, and some of the captured birds as a decoy.

3. The funnel-shaped bird trap made of bamboo sticks or coconut midribs used to catch brooding birds called Kaliked are used in several countries. The trap is fitted into the mouth of the nest and another entrance is opened for the opposite end of the nest. The bird is trapped as it seeks its way out after incubating the eggs. A trap could only catch one bird.

4. The manually operated net trap composed of two rectangular nets each measuring 1–2 m × 2 m is called Korag. The widths are framed by light poles and the lengths by nylon stings. The nets are set flat on

the ground, parallel to each other 3 m apart such that when activated, they close in perfectly.

15.5 CLASSIFICATION OF BIRDS

Generally, the major bird pests belong to the genera *Lonchura* and family Estrildidae. The classification and marks of identification of different birds associated with stored grains are discussed in this section.

15.5.1 PIGEONS COLUMBA LIVIA GMELIN

Order—Columbiformes

Family—Columbidae

The domestic or wild pigeon developed from the rock doves of Europe and Asia was introduced into the USA in 1606, as a domestic bird. Rock doves originally nested in caves, holes, and under overhanging rocks on cliffs, so their descendants comfortably adapt to our window ledges, attics, roofs, eaves, steeples, and other components of our structures. Eight to 12 days after mating, they lay 1–2 eggs. Pigeons are used for several scientific researches on heart disease in humans. Pigeon farming for meat is a common practice in some parts of the country.

15.5.1.1 DAMAGE

Pigeons are the most serious problematic bird pest associated with stored grains, buildings, and statues.

Besides feeding on seed or grains, they also feed on garbage, fruit, spilled grains, and insects. Pigeon droppings accelerate the deterioration of grain quality, buildings, and statues.

15.5.1.2 HABITS

Pigeons are gregarious in nature, and feeding, roosting, and loafing sites are usually separate. Roosting sites are typically protected from the elements and are used for nesting, congregating at night, and shelter during rain. Pigeons prefer flat, level surfaces for landing, resting, and feeding.

Generally, pigeon feeding sites include parks, warehouse and other storage structures, squares, food loading docks, garbage areas, railroad sidings, mature crops, and wherever people eat or feed them outdoors.

The male pigeons are sexually mature after 3–4 months, while females after six months of birth. Pigeons usually mate for life unless separated by death or accident. After pairing and mating, nest construction begins. Pigeons make nest on a small twigs, straw, and debris in which they make a slight depression but sometimes nests are located in protected openings in or on warehouse and other storage structures. Eight to 10 days after matting, a single female lays one or two creamy white eggs, but sometimes three or more eggs are found in a single nest. After hatching, female pigeon nourish the young ones by providing predigested food and after 10 days, they feed on the whole grain. They started flying after 35–40 days. Pigeons nest during all seasons when conditions are unfavorable. City pigeons generally remain in one area for several years.

15.5.2   HOUSE SPARROW PASSER DOMESTICUS LINNAEUS

Order—Passeriformes

Family—Passeridae

The house sparrow introduced from England, also called as English sparrow, are nowadays common all over the world. It prefers farm buildings and houses. Due to environmental pollution, these birds virtually disappeared or their numbers fell dramatically. The house sparrow is a brown, chunky bird and their male has distinctive black bib or throat, white cheeks, a chestnut mantle around an ash-gray crown, and chestnut upper wing covers. The female and young birds have a gray breast, light eye stripe, and a streaked back. They are social birds frequently heard singing, and seen eating and flying together in flocks.

15.5.2.1   DAMAGE

House sparrows are aggressive and social birds; young birds can move out of an area to establish new territories. Sparrows are very closely associated with human activity, and will not hesitate to set up nests in high traffic areas, for example, warehouses and farms. House sparrows feed preferentially on grain. They will also feed on seedlings, buds, flowers, fruits, seeds, and garbage. During the breeding season, they feed mostly insects to their nestlings. They have been called sputzies, aggressive opportunists, animated manure machines, rank hoodlums, impudent parasites, and other unprintable terms of disrespect. Their droppings contaminate stored grain and bulk food. Droppings and feathers can make hazardous, unsanitary, and smelly wastes inside and outside of buildings, on sidewalks, and under roosting sites. Besides feeding on food grain, they also feed on pecking at rigid foam and fiberglass.

15.5.2.2   HABIT

House sparrows are prolific breeders and have 2–7 broods for an average of three broods per season with 3–8 eggs per brood or clutch. Breeding can occur throughout the year, but is most active from March to August. Eggs are hatched after two weeks of laying.

15.6   INTEGRATED MANAGEMENT OF BIRDS

The effective bird management may be achieved by holistic approaches of all suitable techniques, which do not infringe any rules and regulations related to wildlife. The primary objective of bird control in storage should be to prevent or reduce grain losses and not to merely kill animals. Effective bird-proofing of stores is strongly recommended for use in the first instance. Sanitation, elimination of harborage, proper stock and warehouse maintenance and removal of bird nests should be standard practices. Quick turnover of stocks also reduces the exposure time of grain to the pests. However, the following techniques should be applied to minimize birds' infestation under storage condition.

1. Before applying any practice of bird management, survey and inspection of storage structure must be completed for types of birds and their infestation. To constantly watch on activities of birds, survey must be conducted at early in the morning, midday, and in the evening.
2. The modification of habitat for long-term management of birds; this will helpful in nesting of birds and overcoming food scarcity.
3. The utilization of some exclusion, for example, netting of ventilator to check entry of birds in warehouse.

4. The collection and destruction of eggs or adult birds at regular interval.
5. Using the bird trap.
6. The utilization of stickers.
7. Shooting of birds is helpful to prevent the infestation.
8. Making noise by mechanical means or ultrasonic devices may keep the birds away from stored products.
9. The bigger predatory birds prey the smaller birds in nature, so allowing such birds to minimize the losses.
10. Repellents like methiocarb minimize birds from feeding on the grains.
11. The application of avicide (chemical that kills the birds) as a last resort in bird management program is very common in some parts of the country. If the infestation is very high and not manageable by any alternative method, then poison bait can be used. Poison bait should be used very carefully, for example, pigeons prefer peas, wheat, and cracked or whole maize; sparrows prefer small grains such as rice, so mix the avicide in their preferred food. Pre-baiting for 3–4 days increases the chances of success.

KEYWORDS

- post-harvest losses
- stored commodities
- bird population
- vertebrate pests
- grain-eating birds
- birds contamination

REFERENCES

Garg, S. S. L.; Singh, J.; Prakash, V. Losses of Wheat in Thrashing Yards Due to Birds and Rodents. *Bull. Grain Tech.* **1966,** *4* (2), 94–96.

Githiru, M.; Lens, L. Demography of an Afrotropical Passerine in a Highly Fragmented Landscape. *Anim. Conserv.* **2006,** *9,* 21–27.

Lambert, F. R. Trade, Status and Management of Three Parrots in the North Moluccas, Indonesia: White Cockatoo *Cacatua alba*, *Chattering Lory*, *Lorius garrulus* and *Eos squamata*. *Bird Conserv.Int.* **1993**, *3*, 145–168.

Libay, J. L.; Fiedler, L. A.; Bruggers, R. L. In *Feed Losses to European Tree Sparrows (Passer montanus) at Duck Farms in the Philippines,* Proceedings of the Ninth Bird Control Seminar, Bowling Green State University, Bowling Green, OH, Oct 4–6, 1983; pp 135–142.

CHAPTER 16

IMPORTANT RODENTS IN WAREHOUSE OR GODOWN AND THEIR MANAGEMENT

CONTENTS

Portions of this chapter have been modified from *Grain storage techniques: Evolution and trends in developing countries*, FAO AGRICULTURAL SERVICES BULLETIN No. 109, D. L. Proctor, ed. http://www.fao.org/docrep/t1838e/T1838E00.htm#Contents. Used with permission of FAO.

ABSTRACT

Rodentia is the largest order of mammals containing about 35 families and 350 genera, and more than 50% populations of this order are rodents. Rodents are the most notorious animal of stored commodities, they are not only eat the stored produce but also destroy the warehouses and godowns. The population of rodents was reported to be around 3000 million which is regularly increasing. The rats are found in every portion of land throughout the world and are known as polyestrous animals. Of all the animals that live with man, the rat was proved to be most hostile and important. The presence of rats posed great problems to mankind for food and health and in terms of money costs sizeable amount. They have been with human dwelling rather; mankind has brought rats for many thousands of years ago. The losses caused by rats varying from crop to crop, variety, location of warehouses and stores, rat species, duration and method of storage, and climate. The estimation of losses may vary from person to person, place to place, year to year depending upon the methodology opt by person and other variables that may interact. In India, there are various estimates on records but mostly, these are based on assumptions. An organized first effort was made by Indian Council of Agricultural Research in the year 1959 to determine losses in different crops at different parts of the country. The report indicate that at Kanpur only Kharif crop damage from growth to maturity was 7.1–21.5% loss to paddy tillers while in wheat and barley it was 11%. At Ludhiana, the actual loss in yield of wheat crop was 4.1%, groundnut 25.8%. In Andhra Pradesh only coconut losses were 16.0%. At Chennai, losses in paddy were 24.1%.

16.1   IMPORTANT RODENTS IN WAREHOUSES OR GODOWNS AND THEIR MANAGEMENT

Rodents are the most notorious animal of stored commodities, they are not only eat the stored produce but also destroy the warehouses and godowns. The population of rodents was reported to be around 3000 million which is regularly increasing. The rats are found in every portion of land throughout the world and are known as polyestrous animals. Of all the animals that live with man, the rat was proved to be most hostile and important. The presence of rats posed great problems to mankind for food and health and in terms of money costs sizeable amount. They have been with human dwelling rather, mankind has brought rats for many thousands of years ago. The paleo entomologist observed the bones along with relics from men of Stone Age. The detrimental habit of rats to belonging of mankind is known since long as in Vedas. The rat menace has been observed since times immemorial, and consequent efforts to combat by legendary pied piper and medieval rat catchers with all their skill and secret formulae. Deoras (1963) tried to estimate the occurrence of rat population by making catch release and catch of rats in special traps. These rats were captured at Javki and sex ratio varied from 0.45 to 1.2 and Javki with 42 times catches sex ratio was 0.46, where, in Butcher Island it was highest 1.2. The rats are omnivorous but preference or aversion may be varied from species to species, for example annual mortality being given as 95/m³ of the home range being 200 feet in diameter of the diet being grainivorous.

16.2   ECONOMIC IMPORTANCE OF RATS

The losses caused by rats vary from crop to crop, variety, location of warehouses and stores, rat species, duration and method of storage, and climate (Gratz, 1990). A report of Jackson and Meehan (Jackson, 1977; Meehan, 1984), some examples based on surveys are given below which indicate the huge economical losses that have been found and can generally be expected. Samples conducted in small warehouses in the Philippines indicated losses of 40–210 kg of grain in each (Rubio, 1971; Agnon, 1981 as cited in Benigno & Sanchez, 1984); at that time this was equivalent to about US $ 80 for each unit. Interviewing farmers in Bangladesh on rodent damage inside houses provided an estimated loss equivalent to US $ 29.50 for a six-month period (Bruggers, 1983). Mian (Mian et al., 1984) also reported that, on average, households were each infested by about eight mice and two rats. At 10.5 million households the annual losses are estimated at US $ 620 million for the entire country in houses only. Higher estimates were found by Krishnamurthy et al. (Krishnamurthy et al., 1967) in a similar study in India. In large grain stores the situation may be even worse. For example, Frantz (Frantz, 1975) estimated that each godown in Kolkata had, on average, a population of about 200 *Bandicoot* rats. At an estimated 50 g of produce one rat can destroy in one night. To these food losses, costs for cleaning produce, the losses due to damaged packaging (Meehan, 1984) as well as structural damage have to be added. It is impossible to put an exact estimate on these losses, but it is obvious that the damage caused by rodents is enormous. In the USA the annual loses by rodents is estimated at US $ 900 million (Clinton, 1969 as quoted in Meehan, 1984), while the annual cost of rodent control is estimated at US $ 100 million (Brooks, 1973 as quoted in Meehan, 1984). Sumangil (Sumangil, 1990) reported losses of rice in the Philippines were reduced from US $ 36 million to US $ 3.5 million with the advent of organized rat control programs.

In Bangladesh two national strategic multi-media rodent control campaigns were organized and analyzed in detail, Net profits were calculated at US $ 800,000 for each campaign, based on a single crop and season (Adhikarya & Posamentier, 1987). Calculated benefits would be a multiple of this amount, if all crops could have been surveyed and the reduction in structural damage and human suffering could be quantified. Further field studies in the same country have shown clearly that losses can be reduced by 40–60% at farmers level also (Posamentier, 1989).

The estimation of losses may varying from person to person, place to place, year to year depending upon the methodology opt by person and other variables that may interact. In India, there are various estimates on records but mostly, these are based on assumptions. An organized first efforts was made by Indian Council of Agricultural Research in the year 1959 to determine losses in different crops at different parts of the country. The report indicate that at Kanpur only Kharif crop damage from growth to maturity was 7.1–21.5% loss to paddy tillars while in wheat and barley it was 11%. At Ludhiana, the actual loss in yield of wheat crop was 4.1%, groundnut 25.8%. In Andhra Pradesh only Coconut losses were 16.0%. At Chennai, losses in paddy were 24.1%. Gupta (Gupta et al., 1960) reported that incidence of rats and loss to sugarcane yield and sugar recovery was on an average 7 million in all Uttar Pradesh annually. Krishnamurthy et al. (Krishnamurthy et al., 1967) made estimates of Hapur area and reported that losses due to rats ranged from 1.36 to 3.59 tons annually in four villages only. A committee on storage losses of foodgrains during post-harvest handling was appointed by government of India, their report gave an estimated loss of 9.33% out of which rodents take a toll of 2.5%. Rats also serve as a reservoir of Typhus, in Bombay 70 cases were reported to be due to endemic typhus (Savoor et al., 1948) a committee on rat control of Bombay in 1958 reported 20,000 cases of rat bites admitted in hospital in Bombay town. The tapeworm of genus *Hymenolepis* are common intestinal parasites of rat, mice, and human, it is reported that if feces of rat containing eggs of these parasites reach the stomach of man, not only infection may be caused, but hepatic cysticercosis may also develop (Srivastava, 1968). Dubock and Richards (Dubock, 1984; Richards, 1988) reported rodent control in urban and rural situations, including warehouses, in various Asian and Central American countries.

The cost-benefit ratios ranged from 1:2 to 1:30. Hernandez & Drummond (Hernandez & Drummond, 1984) found that in Cuban warehouses the loss of 1% of the amount available to human consumption could be readily preventable by standard control techniques. Rodents consume and damage human foods in the field as well as in warehouse. Besides their direct damage to the produce they also contaminate the produce by urine and droppings and degrade the quality of produce. Due to their gnawing and burrowing habit they also destroy many articles (packaging, clothes, and furniture) and storage structures (floors, buildings, bridges, etc.). They are responsible for transmitting several dangerous human diseases.

16.3 HABIT AND HABITATE

The habit and habitate of rats are varying from species to species but as mammals, the characters exhibit similarity with that of man. Man is responsive to their environment and rat population to tend to increase to the carrying capacity of their locations. The rat populations flourish when plenty of food, water, and nesting sites are present and the natural parasites and predators are limited. As per their habit they named as field rats, commensal rats, domestic rats, and wild rats. The field rats are those which feed on crops in field and grazing pastures, their damage may occurs at any crop season but mostly before the harvest. The field rat population density varies with the soil condition and rat species, the maximum density spreading on the entire field has been observed in semi-arid zone. The rats are seen confined to raized bunds only such rats with close association with man are termed as commensal rats. Domestic rats live with man and feed man's belongings, and these rats are most commonly found in houses granaries and are abundant in almost all the part of the country. The wild rats are heavier than these two types of rats and they prefer open area for surviving.

Rodentia is the largest order of mammals containing about 35 families and 350 genera, and more than 50% populations of this order are rodents. Anderson and Jones (Anderson & Jones, 1967) reported the more than 4000 species of mammals, of which about 1700 are rodents. Of the rodents the Family Muridae contains the most species, and of the genera the genus *Rattus*, Out of the 1700 rodent species about 150 species have been consider as a pest at some locality to some crop at some time or another.

16.4   SIGNS AND SYMPTOMS OF RODENT INFESTATION

The following characteristics may indicate in signs or presence rodent

Live animals

Rodents are mainly active at night. If animals are nonetheless seen during the daytime, this is a sign of an already advanced stage of infestation.

Droppings

The shape, size, and appearance of droppings can provide information as to the species of rodent and the degree of infestation. The droppings of Norway rats are around 20 mm in length and are found along their runs. The droppings of black rats are around 15 mm long and are shaped like a banana. Mouse droppings are between 3 and 8 mm in length and irregular in shape.

Runs and tracks

Runs, such as those of Norway rats, are to be found along the foot of walls, fences, or across rubble. They virtually never cross open areas of land, but always pass through overgrown territory, often being concealed by long grass. Runs inside buildings can be recognized by the fact that they are free of dust. The animal's fur coming into contact with the wall leaves dark, greasy stains. Even black rats, which do not have any fixed runs, can leave similar greasy stains at points which they pass regularly, for example when climbing over roof beams.

Footprints and tail marks

Rats and mice leave footprints and tail marks in the dust. If you suspect there might be rodent infestation, scatter some sort of powder (talcum powder, flour) on the floor at several places in the store and later check for traces. The size of the back feet serves as an indication of the species of rodent, back feet larger than 30 mm: black rat, Norway rat, bandicoot rat, back feet smaller than 30 mm: house mouse, multi-mammate rat, Pacific rat.

Tell-tale damage

Rats leave relatively large fragments of grain they have nibbled at (gnaw marks). They generally only eat the embryo of maize. Sharp and small left-overs are typical for mice. Rodent attack can further be detected by damaged sacks where grain is spilled and scattered. Small heaps of grain beneath bag stacks are a clear sign. These should be checked for using a torch on regular controls. Attention should be paid to damaged doors, cables, and other material.

Burrows and nests

Depending on their habits, rodents either build nests inside the store in corners as well as in the roof area or build burrows outside the store. Rat holes have a diameter of between 6 and 8 cm, whereas mice holes are around 2 cm in diameter. These holes can be found particularly in overgrown areas or close to the foundations of a store.

Urine

Urine traces are fluorescent in ultraviolet light. Where available, ultraviolet lamps can be used to look for traces of urine.

16.5   IMPORTANT RAT SPECIES ASSOCIATED WITH STORED PRODUCT

Posamentier (Posamentier, 1989) reported the following rat species associated with store and most prevalent in Asian country as well as the world and other species may enter buildings occasionally, but they are of locally important.

16.5.1   BROWN RAT, RATTUS NORVEGICUS

This rat is also known as grey, house, sewer, Norway, or wharf rat, this species is cosmopolitan in nature, but thought to have originated in Asia. It has spread gradually around the entire world during the last two centuries through international trade and human settlement (Meehan, 1984). Its range is limited to coastal areas especially in ports. In many Asian countries it is displaced by *Bandicota bengalensis* (Deoras & Pradhan, 1975), and it is probable that populations of *Rattus norvegicus* in these areas are replenished only by new arrivals from outside. Generally it is brown-gray dorsally and light-gray ventrally, the tail is bi-colored, and the feet are white. The length is 180 250 mm and a fully-grown adult may weigh up to 400 g, although heavier individuals have been recorded (Niethammer, 1981). The tail is shorter than the body length. The ears are thick, opaque, and short with fine hairs, while the snout is characteristically blunt. It is the most important species in Europe, because it lives in close proximity to man and has often been responsible for transmitting diseases to man.

16.5.2   HOUSE RAT, RATTUS RATTUS

This rat also known as black, roof, fruit, rice field, or Alexandrine rat and cosmopolitan in nature found in all over the world, this species originated in South East Asia (Meehan, 1984). In the same country the coloration may range from almost black to red brown dorsally and dark grey to white ventrally. The body length is 150 to 220 mm and the fully-grown adult weight is 150–250 g (Niethammer, 1981). The tail is longer than the body length. The ears are thin, translucent, relatively large, and hairless, while the snout is comparatively pointed. Beside damage in store *Rattus rattus* has become a field pest in many countries and, because of its good climbing ability, infests fruit orchards besides entering buildings. This species was responsible for carrying the fleas which spread the plague in the human.

16.5.3   HOUSE MOUSE, MUS MUSCULUS

This mouse is also cosmopolitan in nature, and originated from Central Asia on the Iranian–Russian border (Schwartz & Schwartz, 1943). It is now the most widespread rat species in the world (Meehan, 1984). The color variations are very common in *M. musculus*, the fur dorsally is usually brown to brownish grey, and grey ventrally. The body length is 70–110 mm, and a well-developed adult weighs 15–30 g. The tail is about as long as the body length. The ears are quite large in relation to the rest of the body, while the feet are comparatively small and the snout pointed. The *M. musculus* is a good climber and lives in social groups. This is a serious pest of stored grain but also infesting the fields' crops and natural vegetation.

16.5.4   PACIFIC RAT, RATTUS EXULANS

Rattus exulans is also known as the Polynesian rat and relatively small in size. It is colored gray-brown dorsally and light grey ventrally. The body length is 110–130 mm, and a well-developed adult may weigh up to 45 g. The tail is longer than the body length (Niethammer, 1981). It is common throughout the Pacific islands ranging westward to western Bangladesh (Poche, 1980). Due to its excellent climbing ability it is also infest the coconut trees.

16.5.5   BANDICOOT RAT, BANDICOTA BENGALENSIS

This is a common species in Asian country, which ranges from Pakistan eastwards. It seems to be replacing *R. rattus* and *R. norvegicus* in India (Prakash, 1975) and probably other Asian mainland countries also. The fur is dark to light brown dorsally, occasionally blackish, and light to dark grey ventrally. The body length is around 250 mm, and the uniformly dark tail is shorter than the body length. In addition to consuming or spoiling much stored produce, the lesser bandicoot rat is a very active burrower and is responsible for much structural damage to the storage buildings as well. It is also a very good swimmer able to live in deep water rice fields, where it can cause much damage to the crop. In Bangladesh and Burma it is the most important rodent species in both store and field. *B. bengalensis* is very aggressive even against individuals of the same species (Posamentier, 1989). This species is very susceptible to most rodenticides (Brooks et al., 1980; Poche et al., 1979).

16.5.6   INDIAN GERBIL, TATERA INDICA

The body color is light brownish or sandy on the upper surface and darker with light haired in the lower side. The tail is longer than body length, eyes are comparatively large. It mainly feed on grains, succulent stems, leaves, roots, fruits, and grasses. More than one rat is usually found in one burrow (Meehan, 1984). It is very poor in swimming; it makes burrow in the dry land near the field, river bands.

16.5.7   MULTIMAMMATE RAT, MASTOMYS NATALENSIS

Mastomys natalensis is economically the most important rodent pest in Africa, and a true indigenous commensal (Fiedler, 1988). In many areas, it may be replaced by the much larger *R. rattus*. The fur is soft, brownish on the back and grayish ventrally. The body length is up to 150 mm, and the well-developed adult weight is 50–100 g. The tail, which is uniformly dark, is about the same length of the body. Most distinctively, the female has up to 24 nipples on her belly while other rat species rarely have more than 10 nipples and the reproductive potential is very high, particularly since this species lives in large social groups. Consequently, very large population explosions occur from time to time, and they causing huge losses to stored produces.

16.5.8   EGYPTIAN SPINY MOUSE, ACOMYS CAHIRINUS

Acomys cahirinus occurs from Mauritania to Pakistan and is usually found in semi-desert, rocky country, dry woodland, thorn scrub, and savannah (Greaves, 1989). However it has become commensal in some places, replacing *M. musculus*, causing damage to stored grain and domestic premises. The commensal form is nearly black with a grey belly. The body length is 60–120 mm, and the tail much shorter than body. Apparently *A. cahirinus* has an unusual resistance to anticoagulant rodenticides.

16.6   BIONOMICS AND BEHAVIOR OF RATS

All rat species are polyoestrous and polytocous, a female can produce a regular succession of litters at short intervals for several months under favorable conditions. The minimum gestation period of most of species is normally around three weeks. The rats multiply fast as pair of rats multiplies into 60 in a year. A female rat may have up to five litters in her lifetime, *R. norvegicus* and *R. rattus* averaging seven or eight young in each litter. The *M. natalensis* rat can have up to 20 young in a litter. A female bandicoot rat may share a burrow with a weaned litter, have a litter suckling and be pregnant all at the same time. The house mouse can have a new litter every four weeks (Meehan, 1984).

Rats have well developed senses of smell and touch, but poorly developed eyesight. They have excellent light sensitivity but poor acuity and are color blind (Meehan, 1984). This allows poison baits to be colored, for safety reasons, without modifying their acceptability by the target species. They possess a good sense of hearing including frequencies in the ultrasonic range up to 100 Khz. This has led to the development of ultrasonic deterrent devices, of variable effectiveness. *R. rattus* and *R. exulans* are very good climbers, *R. norvegicus* is less climbers. However, all are able to use very small openings for their size or move up cracks and pipes to gain access to buildings. They are also good swimmers and readily take to the water. They are also good jumpers: *R. norvegicus* can jump vertically 77 cm and horizontally more than 120 cm, house mice can jump to a height of 24 cm (Meehan, 1984).

The burrowing behavior of rats (especially the bandicoot rats, *R. norvegicus* and the multimammate rat) is a particular nuisance to store owners in tropical countries. Floors subside, easing the entry of other individuals, providing hiding places, causing a loss of stored produce and even leading to a partial collapse of buildings. Rodents make burrows to breed in, for storing large amounts of food, and for protection against predation and extreme climatic conditions. In the case of bandicoot rats these burrow systems may be 100 cm deep and very extensive. Rats are omnivorous, an additional reason why they are successful mammals. In spite of this there may be some preferences in the field if a choice is available. Overall, rats and mice in the wild will take a balanced diet. The amount of food taken may also vary. Under laboratory conditions, rodents have been observed to consume about 10% of their body weight per day (Chitty, 1954; Meehan, 1984; Spillet, 1968; Brooks et al., 1981; Posamentier & Alam, 1981). Enclosure studies indicate that under near field conditions the amount consumed or destroyed is about five times the amount eaten in the laboratory (Haque et al., 1980), although the proportion actually consumed is uncertain. What is certain, however, is that the actual losses caused are a multiple of their dietary requirements. Most rats return to a fixed place of feeding. House mice on the other hand are haphazard, inquisitive feeders (Crowcroft, 1966).

Most of the rats start damage during the hours of darkness, which is also when they do most of their feeding. There are two peaks of activity, the major peak occurring just after sunset and a minor peak just before sunrise. This has been observed for *M. musculus* (Dewsbury, 1980), *R. rattus* (Barnett et al., 1975), *R. norvegicus* (Calhoun, 1962), and *B. bengalensis* (Parrack, 1966). When they are hungry, or under crowded conditions, they may also be active during daylight hours.

It is probably their ability to rapidly adapt their behavior to new or changing situations, above all else, that has caused some rodent species to become major pests. This is most apparent in their reaction to "new objects" placed in their environment by man. *R. norvegicus* is naturally very suspicious and tends to avoid any object that is new to it. It may take several days before an individual will enter a trap or take bait. Even then, if the new object appears to be food, only a small amount is taken. If the food contains an acute poison causing symptoms after a short while, rats may not touch the bait again. *R. rattus* behaves similarly but not to the same extent, while *M. musculus* tends to explore rather than avoid new objects.

Several rats are characteristically mobile and able to disperse rapidly. This allows them to move quickly into and take advantage of new areas with favorable conditions (Fiedler, 1988; Meehan, 1984). However once individuals have established a burrow, they will not move very far, as long as conditions remain favorable. The bandicoot rats, and others, will move from surrounding fields into villages at harvest time that is when fields suddenly no longer provide enough food (Posamentier, 1989). In built up areas containing food stores *B. bengalensis* moves within an area 30–146 m in diameter (Spillet, 1968; Chakraborty, 1975; Frantz, 1984), depending on the location of the warehouses, when they are emptied, structural conditions, and the availability of water.

16.7   MANAGEMENT OF RODENTS

The management of rodent population, the environmental condition of inhabiting area, availability of food material and habitate play pivotal role. The migration of rats from one place to other is also important to be born in mind while planning to manage rat population. Sustained efforts are made to control rats in the field but the again appeared as migrated from other sources. A considerable portion of rat population from the operational field appeared to have been migrated to the bunds at the little distance from the farm. These rats continue to revisit the farm and cause damage whenever they found in the farm field and same pattern is observed in food grain storing sits. Thus, to manage rat population in field and warehouse is only the cooperative efforts that would provide some benefits and by adopting an individual practice. The cooperative efforts made by farmers, grain handlers, and the Government agencies will not provide protection unless a scheduled programming is done with a system approach. It should thus occupy a key position in operational plan for agricultural production and protection. The first step in rodent management program is to have knowledge of the biology and behavior of their population which is known as the basic technique that must be match to the characteristic of each species and peak period of activity. The method of calculating peak period varying place to place, However the population fluctuation data is important for the success and scheduling of management program. There are several techniques and methods of management of rodent populations depending upon the situation of the infestation are described here.

16.7.1   NON-CHEMICAL MANAGEMENT OF RODENTS

Co-operation

The area is made rat-free due to good management measures, rats from near-by areas will migrate into it. It is therefore more efficient if management program are conducted in several adjacent areas simultaneously. In the case of a village all households should be motivated and organized to manage rats at the same time. In the case of stores, large and small, surrounding areas including other stores should also be disinfested. This means that all the storekeepers or managers involved should coordinate and synchronize their rodent management activities for maximum effect.

Monitoring

Important step of rodent management program is monitoring. Usually it means surveillance for the presence of rodents. However, it should also mean looking for features in the environment which would encourage rodents to migrate into it. Monitoring should be organized regularly or may be once a week to analyze the situation. The monitoring should include the following aspects:

- What species of rodent are causing damage to the produce?
- What is the approximate degree of infestation (loss estimation)?
- What is the extent of the infestation? If necessary, work must be performed in conjunction with neighbors.
- Where exactly are the rodents particularly active?
- Where are the runs, burrows, and nests?
- In what condition are the store and the surroundings?
- Conditions immediately outside the building with respect to potential infestation points.
- Recommendations for improvement, such as repairs to structures, or further action required.

Management of a rodent infestation is need for continuous monitoring even after a successful management program (Kaukeinen, 1984).

Sanitation

Rodents require food and shelter. Therefore, it is most important to reduce the availability of food and shelter, which determine of successes of management program. In the case of warehouse the most effective method of rodent prevention is the improvement of sanitation inside or outside them. Primarily this means sweeping the store and keeping both it and the surrounding area free from any objects such as empty containers, idle equipment, or discarded building materials, which could provide nesting for rodents. In a tidy store any infestation will be noticed at a very early stage, so the management practice will more effective. With reduced access to food and no places to hide, rats will not survive, that is live and breeds, inside the building. Management practice should take the life history and behavior of species present around stores (Colvin, 1990). Rats avoid clear spaces; therefore, by keeping a strip of two or more meters around a building clear of vegetation will reduce the chance of rats' infestation.

The above suggestions are enough to eliminate serious problems with rats and mice in stores where large quantities of food grains are stored. Rats feel uneasy if their "paths" and "markings" are removed or cleaned daily by sweeping. They will not feel secure enough to remain in a warehouse and damage stored product in the search of food. If they do, the damage is minimal and immediately noticeable.

Rat Proofing

The rat proofing of stores solving at least 50% of problem along with sanitation. A very intensive and careful drive has to be made to educate the objects and make them appreciate the magnitude of the position and follow the cooperative management program. Proofing is necessary to restrict access by rats; this is accomplished by proofing buildings or keeping food in rat proof containers. Hard metal strips should be fitted to the bottom edges of all wooden doors and their frames, and vulnerable windows should be protected with tight wire netting screens in hard metal frames. Steel rat guards fitted to drainpipes and other attachments to the building should be at least 1 m above ground level. Door hinges and similar fittings should be so placed or protected that rats cannot use them for climbing. Floors and walls should be kept in good repair. New holes dug by rats should be filled in immediately, with Reinforce Concrete Cement. The important point is that repairs should be carried out as soon as the damage is noticed, which should be within a few hours of it being done if the building is inspected daily. Although rats are active mainly after dark, they will move about during day as well when there is no human activity. Therefore doors of stores should stay tightly shut during the day as well, when the store is not in use (Jenson, 1965).

Cats

Cats can make a contribution toward rodent control. If the population of rats and mice is low cats directly control by feeding on them. It is their presence, which keeps most rats and mice away. It should be mentioned, however, that cats themselves may become a hygiene problem in stores if care is not taken, for example providing sand-boxes as cat cloak-room (Prakash, 1990; Wood, 1984).

Trap

The use of traps is only worthwhile if the degree of infestation is low. Often local traps are available and in some cultures people are very good at using them. They should be placed where rats move regularly. If placed along a wall, the trap should be perpendicular to it and the treadle with the bait should face the wall. There are different kinds of traps. Sticky or glue traps are another way of catching rats and mice (Prakash, 1990; Meehan, 1984). They are boards made of wood, hard, or cardboard covered with very sticky material. There are different types of glue available and they should be checked for suitability (stickiness, and usability in humid or dusty conditions).

16.7.2   CHEMICAL MANAGEMENT OF RODENTS

The use of rodenticides is only effective under good storage conditions and in particular good storage hygiene. Before the application of rodenticides, all preventive measures must be taken to ensure that no reinfestation takes place.

There are two groups of rodenticides:

1.   Acute poisons
2.   Chronic poisons

Acute poisons are used only in the case of high rodent population with the aim of reducing the degree of infestation to a low level within a short period. Subsequently, chronic poisons or other methods must be used for further control.

1. Acute Poisons

Zinc phosphide is the most common acute poison in use all over the world. It is comparatively cheap and has a good and fast effect if applied correctly. Zinc phosphide is mixed in bait in a concentration of 2.5%.

Application of acute poisons

When applying Zinc phosphide, follow all safety measures and proceed as follows:

-   Draw up a sketch of the area and of the store and mark the settings of the bait.

- Make sure you have an adequate amount of receptacles (bait boxes).
- Make sufficient amounts of untainted bait for prebaiting.
- Fill the receptacles or bait boxes with untainted bait and set them out at the planned points. Offer it until it is fully accepted.
- Control the bait daily and refill if necessary.
- If the bait has not been accepted after a number of days, change the food base or the bait positions.
- Replace all untainted baits with poisoned ones at the same time.
- For preparing poisoned bait mix zinc phosphide with freshly broken grain or meal at a ratio of 1:39, that is, each kilogram of poisoned bait will consist of: 975 g best quality grain + 25 g zinc phosphide before mixing, add ~1% edible oil in order to prevent dust developing (Never mix water with zinc phosphide).
- Attach warning signs to the doors of the store and at the entrance to the property drawing attention to the control campaign in progress, the poison used and the dangers involved, and lock the stores.

Control the baits daily.
- Note on the control sheet how much bait has been eaten.
- Refill any bait which has been eaten and stop the campaign after five days at the latest, as bait aversion will occur.
- Collect all receptacles (bait boxes).
- Thoroughly clean all materials which have come into contact with the bait and store them in a safe place.
- Burn or bury any dead bodies of rodents found.
- Further measures continue the control campaign using chronic poisons or by putting out traps.

Chronic Poisons

Chronic rodenticides have a delayed action. The rodents will die without feeling pain. They will thus not become suspicious of the poisoned bait and no bait aversion will ensue. Therefore, prebaiting is not necessary. Poisoned animals normally die in their nests or hiding places. The bodies of dead rodents are therefore not usually found during the course of treatment.

Most of the chronic poisons are anticoagulants which prevent any clotting of the blood. Animals that have been poisoned will die from internal bleeding. There are two different groups of these poisons:

"First generation" anticoagulants

These are rodenticides which only lead to death after repeated ingestion (up to seven times). They are referred to as "first generation" anticoagulants because they were the first to come on the market. They include the following products:

Active ingredient	Most common trade names
Warfarin	Warfarin
Difacinon	Ramik, Difacin
Chlorfacinon	Caid, Raviac, Quick
Coumatetralyl	Racumin
Coumachlor	Tomarin
Coumafuryl	Fumarin

"Second generation" anticoagulants

These are rodenticides which kill the animals after a single ingestion. These products are thus also categorized as "acute poisons with delayed effect."
They include the following products:

Active ingredient	Most common brand names
Brodifacoum	Talon, Klerat, Ratak Super
Difenacoum	Ratak
Bromadiolon*	Rodine, Mak
Flocoumafen	Storm

*Bromadiolon is particularly used against mice.

Application of chronic poisons

When applying chronic poisons attention has to be paid to the safety measures proceed as follows:

- Draw up a sketch of the store and its surrounding area and mark the settings of the bait.
- Make sure you have an adequate amount of bait stations.

- Make a sufficient amount of poisoned bait if there is no ready-to-use baits available.

  Example for the preparation of bait:
  Eighteen parts (900 g) of broken grain (premium quality)
  One part (50 g) of poison
  One part (50 g) of salt or sugar

- Set out the bait stations at the predetermined points.
- Fill the bait stations with the required amount of bait.
- Attach warning signs to the doors of the store and at the entrance for drawing attention to the control campaign in progress, the poison used and the dangers involved.
- Control the bait stations every 2–3 days.
- Note on the control sheet how much bait has been eaten at each bait station at every inspection.
- Refill any bait which has been eaten.
- If the bait is not accepted change the food base or the bait positions.
- Stop the campaign if it is seen on 2–3 inspections in succession that the bait is no longer eaten.
- Collect all bait stations or prefabricated baits which have been set out.
- Thoroughly clean all materials which have come into contact with the bait and store them in a safe place.

Fumigation

The control of rodents by fumigation can be very effective, but it may be expensive and dangerous. It should be remembered though that the gas must have access to burrows, if these are present in the warehouse. That is the burrows should be open and the fumigant used must be heavier than air. If the species concerned makes burrows which are easy to spot (e.g. *R. norvegicus*, *B. bengalensis* they can be fumigated directly. The simplest method is to use a powder which releases hydrogen cyanide, or aluminum phosphide tablets which release phosphine when placed into the burrows. The gases are generated when the powder or tablets come into contact with moisture in the soil. Alternatively, methyl bromide gas may be pumped into the burrow system. As soon as the fumigant has been applied all burrows must be closed, by filling the entrance holes with soil. However, fumigants cannot be used in loose or sandy soils as too much gas escapes, and the treatment may not be effective. Occasionally, rats have been known to block tunnels and prevent complete distribution of the gas, so that some individuals survive. Fumigation gases used for rodent control are also dangerous to man and other animals. Therefore, strict safety precautions must be observed (Meehan, 1984).

16.7.3   SAFETY MEASURES

Rodenticides, whether chronic or acute, are poisons and should be treated as such and at all times. Some may be more toxic to humans or non-target animals than others; some non-targets may be less affected by certain rodenticides than others. Nevertheless, it is important that safety procedures are rigidly enforced wherever they are used (Meehan, 1984).

Standard safety measures should be taken when handling rodenticides:

- Ensure that children and pets cannot come into contact with any bait that has been set out.
- Warn all people working on and living around the treated area.
- Attach warning signs to the doors of the stores and at the entrance to the property in order to draw attention to the rodent control campaigns.
- Always wear rubber gloves when working with rodenticides.
- Clearly mark bait boxes and stations with the words: "Danger" "Poison."
- Inform a doctor about the active ingredients used and provide him with a label or information sheet from the product to enable assistance in the case of poisoning.
- The following applies when using zinc phosphide:
- Always wear a breathing mask with a P_3 particle filter.
- Ensure that zinc phosphide does not come into contact with any moisture, as a
- Poisonous gas (phosphine) will be produced.
- Wearing protective clothing during operations.
- Not eating, drinking, or smoking during operations.
- Not breathing in dust during operations (wear dust mask).
- Keeping baits out of reach of others, especially children and domestic animals.
- Thoroughly washing the skin, clothing, and equipment after operations.

16.7.4 FIRST-AID MEASURES IN CASE OF POISONING

The first-aid measures listed for insecticides also apply for rodenticides. Attention should be paid to the following:

Chronic rodenticides: This group of poisons is regarded as having a relatively low toxicity. No symptoms or damage will normally result from a single ingestion. Nevertheless, always consult a doctor on suspicion of poisoning. Anemia and shock may occur with repeated ingestion of chronic poison within short time. Vitamin K1 (5–10 mg) can be administered as an antidote. A blood transfusion is necessary in serious cases of poisoning.

Zinc phosphide: Anyone suffering from zinc phosphide poisoning must be taken to the nearest hospital immediately. Symptoms of poisoning are catarrh of the throat, bronchitis and possibly pneumonoedema, and with serious poisoning sickness, vomiting (smell of carbide), diarrhea, and disturbance of consciousness and cramps. The person affected should be made to vomit immediately by sticking your fingers deep into his mouth. Potassium permanganate solution (0.1%) as well as activated carbon should then be administered.

KEYWORDS

- **rodents**
- **warehouse**
- **rats and mics infestation**
- **rat species**
- **mammalian pests**

REFERENCES

Adhikarya, R.; Posamentier, H. *Motivating Farmers for Action;* Deutsche Gesellschaft für Technische Zusammenarbeit: Eschborn, Germany, 1987; GTZ Schriftenreihe Nr. 185, pp 209.

Agnon, T. M. Surveys of Rat Damage in Nueva Ecija Ricemill-Warehouses. M.Sc. Thesis, *University of the Philippines Los Banos,* 1981; pp 40.

Anderson, S.; Jones, J. K. *Recent Mammals of the World - A Synopsis of Families;* Ronal Press: New York, 1967; pp 224.

Barnett, S. A.; Cowan, P. E.; Prakash, I. Circadian Rhythm of Movements of the House Rat, *Rattus rattus* L. *Indian J. Exp. Biol.* **1975,** *13,* 153–155.

Benigno, E. A.; Sanchez, E. F. In *Rodent Problems in the Association of Southeast Asian Nations,* Proceedings of a Conference on the Organization and Practice of Vertebrate Pest Control, Elvetham Hall, Hampshire, England, 1984; pp 37–48.

Brooks, J. E. A Review of Commensal Rodents and Their Control. *CRC Crit. Rev. Environ. Control.* **1973,** *3*(4), 405–453.

Brooks, J. E.; Htun, P. T.; Naing, H. The Susceptibility of *Bandicota bengalensis* from Rangoon, Burma, to Several Anticoagulant Rodenticides. *J. Hyg. Camb.* **1980,** *84,* 127–135.

Brooks, J. E.; Htun, P. T.; Naing, H.; Walton, D. W. Food Requirements of the Lesser Bandicoot Rat. *Acta Theriol.* **1981,** *26,* 373–379.

Bruggers, R. L. Stored Commodities Survey in Bangladesh. In *'Vertebrate Damage Control Research in Agriculture';* Annual Report, Denver Wildlife Research Center: Denver, CO, 1983; pp 101.

Calhoun, J. B. *The Ecology and Sociology of the Norway Rat;* US Department of Health, Education and Welfare: Bethesda, MD, 1962; Public Health Service Publication No. 1008, pp 24.

Chakraborty, S. In *Field Observation on the Biology and Ecology of the Lesser Bandicoot Rat Bandicota bengalensis (Gray) in West Bengal,* Proceedings of the All India Rodent Seminar, Sidhpur, Ahmedabad, India, Sep 23–26, 1975; pp 102–111.

Chitty, D. *Control of Rats and Mice;* Clarendon Press: Oxford, 1954; Vols. I and II., pp 304.

Clinton, J. M. Rats in Urban America. *Publ. Health. Rep.* **1969,** *84*(1), 1–7.

Colvin, B. A. Habitat Manipulation for Rodent Control. In *'Rodents and Rice';* Quick, R., Ed.; IRRI: Manila, Philippines, 1990; pp 61–64.

Crowcroft, P. *Mice All Over;* Foulis & Co. Ltd.: London, 1966; pp 16–21.

Deoras, P. J.; Pradhan, M. S. In *Observations on the Bandicota Rats from Goregaon Malad in Bombay,* Proceedings of the All India Rodent Seminar, Sidhpur, Ahmedabad, India, 1975; pp 61–69.

Deoras, P. J. Studies on Bombay Rats. *Curr. Sci.* **1963,** *33,* 457.

Dewsbury, D. A. Wheel-running Behaviour in 12 Species of Muroid Rodent. *Behav. Process.* **1980,** *5,* 271–280.

Dubock, A. C. In *Economic Benefits of Vertebrate Pest Control,* Proceedings of a Conference on the Organization and Practice of Vertebrate Pest Control, Elvetham Hall, Hampshire, England, Aug 30–Sept 3, 1984; *ICI* Plant Protection *Division: Fernhurst, Haslemere, Surrey, England; pp* 315–326.

Fiedler, L. A. Rodent Problems in Africa. In *'Rodent Pest Management';* Prakash, I., Ed.; CRC Press: Boca Raton, Florida, 1988; pp 35–66.

Frantz, S. C. Home Range of the Lesser Bandicoot Rat, *Bandicota bengalensis* (Gray), in Calcutta, India. *Acta Zool. Fennici.* **1984,** *171,* 297–299.

Frantz, S. C. In *The Behavioural/Ecological Milieu of Godown Bandicoot Rats: Implications for Environmental Manipulation,* Proceedings of the All India Rodent Seminar, Sidhpur, Ahmedabad, India, 1975; pp 95–101.

Gratz, N. G. Societal Impact of Rodents in Rice Agriculture. In *'Rodents and Rice';* Quick, R., Ed.; IRRI: Manila, Philippines, 1990; pp 17–26.

Greaves, J. H. *Rodent Pests and Their Control in the Near East;* Plant Production and Protection Paper, FAO of the UN: Geneva, 1989; Issue 95, pp 112.

Gupta, K. M.; Singh, R. A.; Misra, S. C. In *Economic Loss Due to Rat Attack on Sugarcane in Uttar Pradesh,* Proceedings of the International Symposium on Bionomics and Control of Rodents, Kanpur, India, 1960; pp 17–19.

Haque, E.; Sultana, P.; Mian, Y.; Poché, R. M.; Siddique, A. *Yield Reduction in Wheat by Simulated and Actual Rat Damage;* Technical Report No. 9, Bangladesh Agricultural Research Institute: Joydebpur, Bangladesh, 1980; pp 18.

Hernandez, A.; Drummond, D. C. A Study of Rodent Damage to Food in Some Cuban Warehouses and the Cost of Preventing It. *J. Stored Prod. Res.* **1984,** *20*(2), 83–86.

Jackson, W. B. Evaluation of Rodent Depredations to Crops and Stored Produce. *EPPO Bull.* **1977,** *7*(2), 439–458.

Jenson, A. G. *Proofing of Buildings against Rats and Mice;* Technical bulletin, Ministry of Agriculture, Fisheries and Food: Great Britain, 1965; Issue No. 12, pp 102.

Kaukeinen, D. E. In *Potential Non-target Effects from the Use of Vertebrate Toxicants,* Proceedings of a Conference on the Organization and Practice of Vertebrate Pest Control, Elvetham Hall, Hampshire, England, Aug 30–Sept 3, 1984; pp 11.

Krishnamurthy, K., Uniyal, V.; Singh, J.; Pingale, S. V. Studies on Rodents and Their Control. Part I. Studies on Rat Population and Losses of Food Grains. *Bull. Grain. Tech.* **1967,** *5* (3), 147–153.

Meehan, A. P. *Rats and Mice. Their Biology and Control;* Rentokil Limited: East Grinstead, UK, 1984; pp 383.

Mian, Y.; Ahmed, S.; Brooks, J. E. *Post-harvest Stored Food Losses at Farm and Village Level; Small Mammal Composition and Population Estimates;* Technical Report, Bangladesh Agricultural Research Institute: Joydebpur, Bangladesh, 1984; Issue No. 25, mimeo., pp 11.

Niethammer, J. Characteristics of Destructive Rodent Species. In *'Rodent pests and their control';* Weis, N., Ed.; Deutsche Gesellschaft für Technische Zusammenarbeit (GTZ) GmbH: Eschborn, Germany, 1981; pp 1–22.

Parrack, D. W. The Activity Cycle of the Lesser Bandicoot Rat, *Bandicota bengalensis. Curr. Sci.* **1966,** *35,* 544–545.

Poché, R. M. Range Extension of *Rattus exulans* in South Asia. *Mammalia.* **1980,** *44*(2), 272.

Poche, R. M.; Sultana, P.; Mian, Y.; Haque, E. Studies with Zinc Phosphide on *Bandicota bengalensis* (Gray) and *Rattus rattus* (Linnaeus) in Bangladesh. *Bangladesh J. Zool.* **1979,** *7*(2), 117–123.

Posamentier, H.; Alam, S. Food Choice with Pre-conditioning between Wheat and Rice of *Bandicota bengalensis* and *Rattus rattus. Bangladesh J. Zool.* **1981,** *8*(2), 99–101.

Posamentier, H. *Rodents in Agriculture, a Review of Findings in Bangladesh;* Deutsche Gesellschaft fur Technische Zusammenarbeit (GTZ) GmbH: Eschborn, Germany, 1989; Sonderpublik. No. 176, pp 107.

Prakash, I. In *Replacement of Sympatric Rodent Species in the Indian Desert,* Proceedings of All India Rodent Seminar, Sidhpur, Ahmedabad, India, 1975; pp 29–31.

Prakash, I. Rodent Control: The Need for Research. In *'Rodents and Rice';* Quick, R., Ed.; IRRI: Manila, Philippines, 1990; pp 1–8.

Richards, C. G. J. Large-scale Evaluation of Rodent Control Technologies. In *'Rodent Pest Management';* Prakash, I., Ed.; CRC Press: Boca Raton, Florida, 1988; pp 269–284.

Rubio, R. R. Survey of Rat Damage in Laguna Ricemill-Warehouses. Undergraduate Thesis, University of the Philippines Los Banos, 1971; pp 35.

Savoor, S. R.; Soman, D. W.; Vahia, N. S. *Indian med. Gaz.* **1948,** *83,* 20.

Schwartz, E.; Schwartz, H. K. The Wild and Commensal Stocks of the House Mouse *Mus musculus* Linnaeus. *J. Mammal.* **1943,** *24,* 59–72.

Spillet, J. J. The Ecology of Lesser Bandicoot Rat in Calcutta, Bombay. *Nat. His. Soc.* **1968,** *126,* 124–125.

Srivastava, R. S. In *Hymenoplis Infection in Man Caused by Rat,* Proceedings of the International Symposium on Bionomics and Control of Rats, 1968; pp 91–93.

Sumangil, J. P. Control of Ricefield Rats in the Philippines. In *'Rodents and Rice';* Quick, R., Ed.; IRRI: Manila, Philippines, 1990; pp 35–48.

Wood, B. J. Rat Pests in Tropical Crops - A Review of Practical Aspects. Proceedings of a Conference on the Organization and Practice of Vertebrate Pest Control, Elvetham Hall, Hampshire, England, Aug 30–Sept 3, 1984; *ICI* Plant Protection *Division: Fernhurst, Haslemere, Surrey, England; pp* 263–287.

CHAPTER 17

IMPORTANT STORAGE FUNGI AND THEIR MANAGEMENT

CONTENTS

ABSTRACT

The storage fungi include above all species of *Aspergillus* and *Penicillium*. All storage fungi have the ability to grow in equilibrium with 70–90 % relative humidity, but some fungi require higher osmotic pressure to grow like *Aspergillus balophilicus*. Most of the infecting bodies of fungi occur in or on the surface of grain, seed, and godowns. Among the 70,000 described species of fungi, very few are associated with stored grains and seeds. Once the grain is placed into a storage facility, a succession of new microbial species begins to grow, without intervention, microbial respiration will increase temperature and moisture, providing optimum growth conditions for several fungal species. Storage duration may vary from six (usually for maize, soybean, and sunflower) up to 20 months or longer if the seeds are to be carried over. Longevity of seed in storage is influenced by the stored seed quality as well as stored conditions. Irrespective of initial seed quality, unfavorable storage conditions, particularly air temperature and air relative humidity, contribute to accelerating seed deterioration in storage. Hence, it is difficult to assess the effective storage period because the storability of the seed is a function of initial seed quality and the storage conditions. Damage caused by fungi is often detected at advanced stage. They are major causes of spoilage in humid climate where it ranks second to insects. Fungi do not only cause direct losses but also threaten the health of both man and animals by producing toxins, called mycotoxins, which are contaminating food and feed and not fit for human consumptions. Storage fungi require a relative humidity of at least 65% which is equivalent to an equilibrium moisture content of 13% in cereal grain. They grow at temperatures of between 10 and 40 °C. Major group of fungi is devoid of the green pigment chlorophyll, including in it are the edible fungi, molds, rust, and smut yeast. They seem to have evolved at a very early stage of existence from algae which lost their pigment and came to obtain food in peculiar ways.

17.1   IMPORTANT STORAGE FUNGI AND THEIR MANAGEMENT

The storage structures are limited in microbial species because of good human efforts in order to maintain grain quality. Among the 70,000 described species of fungi, very few are associated with stored grains and seeds. Once the grain is placed into a storage facility a succession of new microbial species begins to grow, without intervention, microbial respiration will increase temperature and moisture, providing optimum growth conditions for several fungal species. Storage duration may vary from 6 (usually for maize, soybean, and sunflower) up to 20 months or longer if the seeds are to be carried over. Longevity of seed in storage is influenced by the stored seed quality as well as stored conditions. Irrespective of initial seed quality, unfavorable storage conditions, particularly air temperature and air relative humidity, contribute to accelerating seed deterioration in storage. Hence, it is difficult to assess the effective storage period because the storability of the seed is a function of initial seed quality and the storage conditions (Heatherly & Elmore, 2004). During storage, seed quality can remain at the initial level or decline to a level that may make the seed unacceptable for planting purpose what is related to many determinants: environmental conditions during seed production, pests, diseases, seed oil content, seed moisture content, mechanical damages of seed in processing, storage longevity, package, pesticides, air temperature and relative air humidity in storage, and biochemical injury of seed tissue (Simic et al., 2004; Guberac et al., 2003; Heatherly & Elmore, 2004).

Damage caused by fungi is often detected at advanced stage. They are major causes of spoilage in humid climate where it ranks second to insects. Fungi do not only cause direct losses but also threaten the health of both man and animals by producing toxins, called mycotoxins, which are contaminating food and feed and not fit for human consumptions. Storage fungi require a relative humidity of at least 65% which is equivalent to an equilibrium moisture content of 13% in cereal grain. They grow at temperatures of between 10 and 40 °C. Major group of fungi is devoid of the green pigment chlorophyll, including in it are the edible fungi, molds, rust, and smut yeast. They seem to have evolved at a very early stage of existence from algae which lost their pigment and came to obtain food in peculiar ways. Stored grain and seeds are subject to infection by several fungi. These fungi are classified into three groups:

17.1.1   STORAGE FUNGI

The storage fungi include above all species of *Aspergillus* and *Penicillium*. All storage fungi have the ability to grow in equilibrium with 70–90 % relative humidity, but some fungi require higher osmotic pressure to grow like *Aspergillus balophilicus*. Most of the infecting bodies of fungi occur in or on the surface of grain, seed, and godowns. The molds have been accepted in the tropics just as one of the routines. Tuite and Christensen (1957) reported that inoculum of storage fungi was uncommon in the air in country elevators and in the air in a terminal elevator. There is no mystery as to where these molds come from the primary source of inoculums is moldy material within the store itself. Majumder et al. (1966) recognized that the parasitic fungi may be carried internally in the tissues of the kernel.

17.1.1.1   MAJOR ASPERGILLUS SPECIES

Out of 80 species reported by Raper and Fennell (1965) only 26 species are most prominent under storage condition (Table 17.1). The two species, *A. restrictus* and *A. glauceus,* are responsible to causing serious losses. In grains where moisture content is in equilibrium or less than 78 to 80%, these fungi can grow in lots with moisture content higher than this.

TABLE 17.1   List of Major *Aspergillus* Species Associated with Stored Grains.

S. no	*Aspergillus* species
1	A. glausus
2	A. amsetelodami
3	A. chevalieri
4	A. echinulatus
5	A. repens
6	A. ruber
7	A. candidus
8	A. clavatus
9	A. giganteus
10	A. flavus
11	A. oryzae
12	A. tamari
13	A. flavipes
14	A. fumigates
15	A. nidulans
16	A. ochraceous
17	A. alliaccus
18	A. ochraceous
19	A. sulphureus
20	A. resrictus

TABLE 17.1 *(Continued)*

S. no	*Aspergillus* species
21	A. conicus
22	A. terreus
23	A. versicolor
24	A. sydowi
25	A. wentii
26	A. terricola

A. restrictus causes grain discoloration in wheat which is termed as sick wheat and blue eye in corn. It develops at 14–15% moisture but does not heat the grains. It also indicates its presence by resulting musty odor and another mold.

A. glaucus is the most dominant groups of mold that invades corn, wheat, sorghum, and soybean at 12–15 % moisture content. In the early stage these fungi are not easily detectable by the eye. It was observed that grain got wetted due to rains, get invaded by this group of fungi (Khare, 1962). It kills and discolors the grain and seeds at high moisture level in store. It does not result increase in temperature but may cause increase in moisture in grain in which it is growing.

A. candidus causes the discoloration of grains very fast, resulting heating and decay of the grains and seeds. After infection, grains become unfit for animal consumptions.

A. flavus causes rapid heating in the grain resulting decay and discoloration of seeds and grains.

A. ochraceous is black-greenish, black-brownish, black, purplish black, and radiate.

A. niger is varying in color from dark to light brown, conidiophores are smooth and causes extensive losses to stored grain or seeds.

A. ustus has conidial heads appearing radiate when young but becoming broadly to irregularly columnar at maturity, persistently white or assuming avillaneous to vinaceous buff shades.

17.1.2 FIELD FUNGI

The field fungi require high moisture content to grow, a moisture content in equilibrium with relative humidity of 90–100 %, which in cereal grains means a moisture content of 22–33%, wet weight basis or 30–33% on a dry weight basis. These fungi do not continue to grow after harvest. The major species of field fungi are *Helminthosporium*, *Alternaria*, *Cladosporium*, and *Fussarium*. The field fungi may survive for many years in grain (Christensen, 1963), but die very fast in grains held at moisture contents in equilibrium with relative humidity of 70–75%.

Helminthosporium species found in 5–10% of wheat and oats, although 100% infection by this pathogen is easily, like barley, is among highest susceptibility ranges.

Cladosporium species occurs in 5–20% of wheat and barley.

Epicoccum, *Nigrospora*, and *Popularia* also infect seed or grains near to maturity. Most of the field fungi belong to deutromycotina. Mostly field fungi produce dark color spores and these are seed-born pathogens.

17.1.3   ADVANCE DECAY FUNGI

Every species of fungus has its own optimum climatic requirements. Infestation with certain species of fungi may already occur in the grain which leads to a considerable reduction of the grains' storage periods. After infection of storage fungi or field fungi, grains or seeds may be infected by *Mucor*, *Absidia*, and *Rhizopus* species in advance stage. They require a temperature range of 3–60 °C and 80–90% relative humidity; they are therefore, not the initiators of cereals deterioration in storage. Pelhate (1968) isolated *Neurospora sitaphila* and *Sordaria fimicola* from the stored grains.

17.2   FAVORABLE CONDITION FOR INFECTION OF FUNGI UNDER STORAGE CONDITION

The major factors that determine the infection of fungi under storage condition and cause the damage to grain and seeds are:

Moisture content of grain

Moisture content below 13.5% in cereal seeds such as wheat, barley, rice, corn, and sorghum and below 12.5% in soybean prevents invasion by storage fungi regardless of how long the grains are stored. As the moisture content rises above these levels, invasion by storage fungi increases with temperature and time. It is also important to be aware that there is variation in moisture content through a grain mass. Storage fungi will grow where moisture is suitable and not according to the average moisture content of the grain mass. These moisture content limits for safe storage imply that nowhere in the bulk of grain is the moisture content higher than that specified. The development of all types of microbes is retarded when the moisture content drops below 13%. The limiting moisture contents of seeds may differ based on host plant species and fungal species.

Temperature

If the temperature is between 40 and 50 °C, storage fungi grow very slowly (Table 17.2). At 80 to 90 °C, they grow much more rapidly. The fungal pathogens such as *Botrytis* spp., *Penicillium* spp., and *Rhizopus* spp. could develop only very slowly at low temperatures. The spore germination and growth of *R. stolonifer* are entirely inhibited at temperatures below 7.5 °C (Dasgupta & Mandal, 1989). The temperatures optimal for *Monilinia fructigena* infecting apple and *A. niger* infecting mango were 23 to 25 °C and 30 °C, respectively (Xu et al., 2001).

TABLE 17.2 Temperature Required for Growth of Important Storage Fungi.

Fungus	Temperature °C		
	Optimum	**Minimum**	**Maximum**
A. resrictus	30–35	5–10	40–50
A. glausus	30–35	0–5	40–45
A. candidus	45–50	10–15	50–55
A. flavus	40–45	10–15	45–50
Penicillium spp.	20–25	−5 to 9	−35 to 40
Absidia, mucor and rhhizosphere	−	−3	60

Damage grain and foreign material

Damage grains are more likely to be contaminated with storage fungi going into storage and more likely to be invaded once they are in storage than sound kernels. Foreign material may restrict air movement through the grain mass leading to temperature and moisture problems which may favor storage mold development.

Storage site

Grain invaded by storage fungi, even if not detected in ordinary inspection, is partly deteriorated and is a much poorer storage risk than grain free of storage fungi and otherwise sound. Grain moderately invaded by storage fungi develops damage at lower moisture content, at a lower temperature, and in a shorter time than does grain free or almost free of storage fungi.

Storage period

Grain that is to be stored for only a few weeks before it is processed can be stored safely with a higher moisture content and more extensive invasion by storage fungi and can be kept at a higher temperature than grain that is to be stored for months or years.

Population of insect and mite infestation

Insects and mites may carry fungal spores on their bodies thus introducing storage fungi into the grain mass. Insect and mite activity in a grain mass tends to lead to an increase in both temperature and moisture content of the grain surrounding the insect infestation. In these "hot spots" conditions may be favorable for mold growth.

Sites of Storage Fungi in the Seeds and grains

The longevity of storage fungi is markedly affected by the location of the fungi in or on the seeds. The organisms that are deep-seated are protected by the seed tissues for a longer time, whereas the organisms present on the seed surface are eliminated soon. There are very few seed-borne biotrophic fungal pathogens, such as downy mildew pathogens present on mature seeds (Sackston, 1981). Differential survival of the two forms of inoculum, that are superficial conidia and internal mycelium, for short and longer periods has been demonstrated in the case of *Alternaria brassicola*. A linear relationship between the population of conidia produced on the seed surface of cabbage and the internal inoculum was observed by Maude and Humpherson (1980).

Previous Infestation/Infection

Seeds already infected under field conditions or prior to storage may deteriorate faster, since the storage fungi can continue to invade the seed tissues for longer periods when such seeds are stored at lower moisture and temperature conditions, favoring rapid development of storage fungi. Seed spoilage under such conditions may be at a faster rate as in corn (maize) seeds infected by *A. flavus* (Qasem & Christensen, 1960).

17.3   LOSSES CAUSED BY FUNGI IN STORAGE

The following types of losses are caused by fungi under storage environment

Discoloration

Both storage and field fungi may cause discoloration of whole seed or a portion of it, particularly the embryo. Christensen et al. (1964) reported the kernels from more than hundred lots of 100% damaged corn obtained from USA. Grain inspection office found that the embryos of all damaged kernels were decayed to some extent by fungi. Wheat in storage develops dark brown to black color after infection of fungi; these discolored lines are high in fatty acids and also very friable.

Heating

There are two types of heating observed in storage, first is dry grain heating or heating caused by infestation of insects. Second is wet grain heating caused by fungi and other microorganisms which occur in grain with moisture content more than 15% and grain temperature as high as 60–70 °C.

Decay

Decay is the last stage of spoilage caused by the fungi, when the organism involved is detectable by the unaided eye and sense. Decayed grains are not fit for human as well as animal consumptions.

Loss in germination

Storage fungi may affect the quality and percentage of germination. Several times seeds are completely losses their vigor and not fit for next cropping. The extent of reduction is influenced by moisture content, temperature, and period of storage, long term storage gives an opportunity to invade seed embryos preferentially, resulting in significant or even total loss of germination. Fungal proliferation may lead to a rapid decrease in seed viability. The germination of safflower seeds significantly decreased with increasing moisture content and length of storage when infected by *Penicillium chrysogenum* (Hasan, 2000).

Production of Mycotoxin

Several fungi produce mycotoxins in the grains and stored products (Table 17.3).

TABLE 17.3   Mycotoxin Produced by Important Storage Fungi.

Fungi	Mycotoxin	Stored product
A. flavus	Aflatoxin	Wheat, rice, maize, sorghum, oats, rye, barley
Cladyopus purpurea *Cladyopus microcephala*	Erogotamine	Rye, gram, bajra
Fusarium species	Tearalenose	Maize
A. ochraceous *P. viridicatum*	Ochratoxin	Maize, wheat, oats
P. viridicatum *A. glausus*	Citrinin	Barley, oats
A. versicolor *A. nidulans*	Sterigmatocystin	Wheat
A. glausus *P. viridicatum* *P. urticae*	Patulin	Barley, wheat
A. glausus *P. viridicatum*	Penicillic acid	Maize
P. islandicum	Luteoskyrin, stadrybotryotoxin	Rice
P. toxicarium	Citreoviridin	Rice

Lipid degradation

Lipids used as energy reserves during seed germination. They may be degraded endogenously and also through pathogen activity, by both oxidation and hydrolysis, involving lipoxygenases and lipases. It is indicated by increased fatty acid contents. High levels of activities of amylase, cellulase, lipase, and protease detected in these fungi might have had a role in seed deterioration (Table 17.4).

TABLE 17.4   Lipid Degradation by Important Storage Fungi.

Crop	Pathogen	Changes
Safflower	*P. chrysogenum*	Increase in free fatty acid, storage at 10°C
	A. flavus and *A. niger.*	Increase in free fatty acid contents at 25°C
Soybean	*Diaporthe phaseolorum var. sojae*	Higher oil and free fatty acid contents

17.4 PENICILLIUM

Penicillium are generally most dominant in temperate climate, they produce more spores as compare to *Aspergillus* species. Very low moisture required for growth of *Penicillium* like 17–19% in maize and wheat 17–20% in sorghum and 16–18% in soybean. *Penicillium* kills and discolors germs and whole seeds, they cause mustiness and caking in grain and seeds. It causes blue eye diseases in maize at moisture content 18.5% and low temperature. *Penicillium* causes hemorrhagic syndrome in duck and an occasionally serve disease of turkey poultry involving liver lesions, are suspected to be due to consumption of feed heavily involved by certain species. (Raper & Thom, 1949) reported 137 species of *Penicillium* out of which 66 species has been found in stored grain product (Table 17.5).

TABLE 17.5   List of Major *Penicillium* Species Associated with Stored Grains.

S. no	*Penicillium* species
1	P. adametzi
2	P. citresvivide
3	P. decumbens
4	P. frequencies
5	P. multicolour
6	P. purpurrescens
7	P. spinulosum
8	P. thomii
9	P. capsulantum
10	P. charlesii
11	P. cyarlesii
12	P. cyaneum
13	P. velutinum
14	P. waksmani
15	P. albidum
16	P. canescens
17	P. coralligerum
18	P. janthinellum
19	P. jenseni
20	P. kapuscinskii
21	P. nigricans

TABLE 17.5   *(Continued)*

S. no	*Penicillium* species
22	*P. pulvillorum*
23	*P. raciborskii*
24	*P. ralfsii*
25	*P. soppi*
26	*P. atramentosum*
27	*P. chrysogenum*
28	*P. citrinum*
29	*P. corylophilum*
30	*P. notatum*
31	*P. oxalicum*
32	*P. roguefotri*
33	*P. steckii*
34	*P. stoloniferum*
35	*P. comeberti*
36	*P. corybiferum*
37	*P. crustosium*
38	*P. expansum*
39	*P. granilatum*
40	*P. orchraceum*
41	*P. visidicatum*
42	*P. avellaneum*
43	*P. funiculosum*
44	*P. piceum*
45	*P. rubrum*
46	*P. variabile*
47	*P. vennriculatum*
48	*P. wortmamai*

17.5   MYCOTOXIN

Mycotoxin is derived from the Greek word *Mycos*-meaning fungus and the Latin word *toxicum*, which means poison. Mycotoxins are secondary metabolites of moulds, contaminating a wide range of crop plants and fruits before or after harvest and are toxic on consumption by animals, including human beings. The acute and chronic impact of mycotoxins on human and animal health is proven scientifically. Mycotoxin contamination is recognized as an unavoidable risk because the formation of fungal toxins is weather dependent and effective prevention is impossible.

According to the FAO, more than 25% of the world's agricultural production is contaminated with mycotoxins. This equates to economic losses estimated at $923 million annually in the US grain industry alone. Most countries have adopted regulations to limit exposure to mycotoxins, having strong impact on food and animal crop trade. The presence of mycotoxins is unavoidable and therefore testing of raw materials and products is required to keep our food and feed safe.

Mycotoxins are metabolic substances which are produced by various fungi under certain conditions and remain in the stored produce as residues. They are highly poisonous to both humans and animals, if eaten; they lead to illness known as mycotoxicoses. The best-known mycotoxins are Aflatoxin, which are produced by *A.s flavus*, Ochratoxin, Patulin and Citrin. Among the fungi colonizing seeds, species of *Aspergillus, Fusarium, Alternaria* and *Penicillium* have been shown to produce mycotoxins, affecting the quality of foods and feeds. *A. flavus* and *A. parasiticus* produce aflatoxins in seeds of many crops such as: corn (maize) (Cortes et al., 2000); peanut (groundnut) (Diener et al., 1987); mustard (Bilgrami et al., 1992); and rice and wheat (Tsai &Yu, 1999). Another mycotoxin, fumonisins, produced by *Fusarium* in corn has been detected. Several fungi (*Fusarium* spp., *Cephalosporium* sp., *Trichothecium* sp., *Phomopsis* sp.) produce a group of toxic compounds designated and corn grains invaded by *F. graminearum* contain DON, NIV, and their ester zearalenone (ZEA) (Ngoko et al., 2001). High concentrations of DON, 15 acetyl DON, and zearalenone were detected in barley seeds inoculated with *F. graminearum*, whereas barley seeds inoculated with *F. poae* had low concentrations of DON. Mycotoxins are highly stable and cannot be destroyed by boiling, pressing, or processing. This means that infested produce has to be destroyed. The problem cannot be dealt with by mixing contaminated produce with healthy grain or by feeding it to animals, as the toxins will be accumulated in their body and later consumed by people in form of milk or meat.

This danger is reflected in the laws of a number of countries concerning the maximum residue limits (MRL) of dangerous substances in foodstuffs. As an example, the maximum residue limits of Malathion and Aflatoxin B1 are shown in mg per kg of grain: Malathion 10 mg/kg and Aflatoxin B1 0.005 mg/kg This means that the maximum residue limit of Aflatoxin B1 is 2000 times less than that of Malathion.

The optimum climatic conditions for the growth of fungi and the formation of mycotoxins are often not identical and dependent on various unidentified factors. Therefore, mycotoxin contamination can only be stated with certainty by means of laboratory examinations.

17.6 MANAGEMENT OF STORAGE FUNGI

17.6.1 PREVENTIVE MANAGEMENT

To ensuring a safe final product is prevention. Some seeds are contaminated with mycotoxins in the field. If the infection occurs in the field, as in the case of wheat, barley, and corn, the fungal pathogens (*Fusarium* spp.) will continue to develop during postharvest stages and storage. Mycotoxins, such as fumonisin B1, are invariably produced. Drying the seeds to a safe water activity level is one of the most effective measures that can be applied. By reducing moisture levels to 14% for maize and 9.5% for groundnuts at 20 °C, it is possible to reduce the growth of *A. flavus* (Wareing, 1999). Insect infestation of seeds results in greater level of damaged kernels, favoring higher incidence of *A. flavus* and *A. parasiticus*. Hence, management of insect and mite infestation may help to prevent proliferation of *Aspergillus* spp. and aflatoxin production.

17.6.2 CURATIVE MANAGEMENT

Timeliness, clean-up, and drying to maintain safe moisture levels are important during harvesting. As crops left on the field for longer periods show higher levels of mycotoxin contamination, it is essential to harvest the crops at the right time, followed by adequate drying. For mycotoxin decontamination, biological methods have been explored. The possibility of degrading aflatoxins using certain fungi which produce peroxidases was reported by (Lopez-Garcia & Park, 1998). Among the several chemicals evaluated for their ability to inactivate and reduce the hazard of selected mycotoxins, ammoniation has been shown to be the most effective process. Aflatoxin contamination in maize, peanuts, and cotton could be significantly reduced by an ammoniation process (Lopez-Garcia & Park, 1998). Because of the unpredictable and heterogeneous nature of mycotoxins production and contamination, it may not be possible to achieve 100% destruction of all mycotoxins in all food systems. However, it is considered that the use of a HACCP-based hurdle system, in which contamination is monitored and controlled throughout production and postproduction operations, may be effective. The development of suitable integrated mycotoxin management systems may control at various points from the field to the consumer (Lopez-Garcia et al., 2004).

17.6.3 IMPORTANT TIPS FOR MANAGEMENT

Harvest as soon as the moisture content allows for minimum grain damage.

Adjust the harvesting equipment for minimum kernel or seed damage and maximum cleaning.

Clean all grain harvesting and handling equipment thoroughly before beginning to harvest. Clean bins or storage facilities thoroughly to remove dirt, dust and other foreign material, crop debris, chaff and grain debris.

Clean grain going into storage to remove light weight and broken kernels or seeds as well as foreign material and fines.

Moisture content is by far the most important factor affecting the growth of fungi in stored grain. After harvest, grain should be dried to safe moisture contents as quickly as possible.

Aerate grain to safe and equalized temperatures through the grain mass.

Protect grain from insect and mite damage.

Check stored grain on a regular basis and aerate as needed to maintain low moisture and proper temperature.

High moisture corn can be protected from storage molds with propionic acid or other organic acids sold under various trade names. It is important to follow label directions on rate and application methods. Grain treated with propionic acid can only be used for animal feed and it is not permitted in commercial grain channels.

KEYWORDS

- **storage fungi**
- **field fungi**
- **advance decay fungi**
- ***Penicillium***
- **mycotoxin**
- **fungal infection in stored products**

REFERENCES

Bilgrami, K. S.; Choudhary, A. K.; Ranjan, K. S. Aflatoxin Contamination in Field Mustard (*Brassica juncea*) Cultivars. *Mycotoxin Res.* **1992,** *8,* 21.

Christensen, J. J. Corn Smut Caused by *Ustilago Maydis*. Monograph No. 2. *Am. Phytopath. Soc.* **1963,** *2,* 41.

Christensen, C. M. Effect of Moisture Content and Length of Storage Period on Germination Percentage of Seed Corn, Wheat and Barley Free of Storage Fungi. *Phytopathol,* **1964,** *54,* 1464–1466.

Cortes, N. de A.; Cassetari Neto, D.; Correa, B. Occurrence of Aflatoxins in Maize Produced by Traditional Cultivation Methods in Agricultural Communities in the State of Marto Grosso. *Higiene Alimentar.* **2000,** *14,* 16–26.

Dasgupta, M. K.; Mandal, N. C. *Postharvest Pathology of Perishables;* Oxford and IBH Publishing Co. Pvt. Ltd.: New Delhi, India, 1989; p 156.

Diener, U. L.; Cole, R. J.; Sanders, T. H.; Payne, G. A.; Lee, L. S.; Klich, M. A. Epidemiology of Aflatoxin Formation by *Asperigllus flavus*. *Ann. Rev. Phytopathol.* **1987,** *25,* 249.

Guberac, V.; Maric, S.; Lalic, A.; Drezner, G.; Zdunic, Z. Hermetically Sealed Storage of Cereal Seeds and its Influence on Vigour and Germination. *J. Agron. Crop Sci.* **2003,** *189,* 54–56.

Hasan, H. A. H. Fungal Association and Deterioration of Oil-Type Safflower (*Carthamus tinctorius*) Seed During Storage. *Czech Mycol.* **2000,** *52,* 125–137.

Heatherly, L. G.; Elmore, R. W. Managing Inputs for Peak Production. In *Soybeans: Improvement, Production and Uses;* Boerma H. R., Specht, J. E., Eds.; 3rd Edition, Agronomy N-16, ASA, CSSA, SSSA: Madison, Wisconsin, USA, 2004; pp 451–536.

Khare, B. P. Ecological Studies of Some Stored Grain Insect Pests *Sitophilus oryzae* L. and *Rhyzopertha dominica* F. Ph.D. Thesis, Agra University: UP, India, 1962; p 214.

Lopez, G. R.; Park, D. L. Effectiveness of Postharvest Procedures in Management of Mycotoxin Hazards. In *Mycotoxins in Agriculture and Food Safety;* Bhatnagar, D., Sinha, S., Eds.; Marcel Dekker: New York, NY, 1998; pp 407–433.

Lopez, G. R.; Park, D. L.; Phillips, T. D. Integrated Mycotoxin Management Systems. Internet Resources: File: 11A: Mycotoxin Management Systems Food, Nutrition And Agriculture. Available at www.fao.org/documents/show_cdr.asp. 2004.

Majumder, S. K.; Bano. A.; Vinugopal, J. S. *Processing in Food Grains to Induce Immunity to Insect Attack;* Central Food Technological Research Institute: Mysore, India, 1966; pp18–19.

Maude, R. B.; Humpherson , J. M. Studies on the Seed Borne Phases of Dark Leaf Spot (*Alternaria brassicola*) and Grey Leaf Spot (*Alternaria brassicae*) of Brassicas. *Ann. Appl. Biol..* **1980,** *95,* 311–319.

Ngoko, Z.; Marasas, W. F. O.; Rheeder, J. P.; Shephard, G. S.; Wingfield, M. J.; Cardwell, K. F. Fungal Infection and Mycotoxin Contamination of Maize in the Humid Forest and Western Highlands of Cameroon. *Phytoparasitica.* **2001,** *29,* 352–360.

Pelhate, J. Besoinsen cau Chez Moissisures des Grains. *Mycopath Mycol. Appl.* **1968,** *36,* 117–128.

Qasem, S. A.; Christensen, C. M. Effect of Moisture Content and Temperature on Invasion in Stored Corn by *Aspergillus flavus*. *Phytopathology.* **1960,** *50,* 703.

Raper, K. B.; Fennel, D. *The Genus Aspergillus Baltimore*. Williams & Wilkins Co: Baltimore, MD, 1965; p 686.

Raper, K. B.; Thom, C. *A Manual of Penicillia;* Williams & Wilkins Co.: Baltimore, MD, 1949; p 326.

Sackston, W. E. Downy Mildew of Sunflower. In *The Downy Mildews;* Spencer, D. M., Ed.; Academic Press: London, UK, 1981; pp 545–575.

Simic, B.; Popovic, S.; Tucak, M. Influence of Corn (*Zea mays* L.) Inbred Lines Seed Processing on their Damage. *Plant, Soil Environ.* **2004,** *50,* 157–161.

Tsai, G. J.; Yu, S. C. Detecting *Aspergillus parasiticus* in Cereals by an Enzyme Linked Immunosorbent Assay. *Int. J. Food Microbiol.* **1999,** *50,* 181–189.

Tuite, J. F.; Christensen, C. M. Grain Storage Studies 23rd Time of Invasion of Wheat Seed by Various Species of *Aspergillus* Responsible for Deterioration of Stored Grain and Source of Innoculum of these Fungi. *Phytopathol.* **1957,** *47,* 265–268.

Wareing, P. *The Application of the Hazard Analysis Critical Control Point (HACCP) Approach to the Control of Mycotoxins in Foods and Feeds;* Postharvest convention, 1999; Cranfield University: Bedford, UK, 1999.

Xu, X. M.; Guerin, L.; Robinson, J. D. Effects of Temperature and Relative Humidity on Conidial Germination and Viability. Colonization and Sporulation of *Monilinia fructigena. Plant Pathol.* **2001,** *50,* 561–568.

CHAPTER 18

STORED GRAIN INSECTS OF QUARANTINE IMPORTANCE

CONTENTS

ABSTRACT

India enforced Destructive Insect Pests Act in 1914 to check the introduction and spread of exotic pests. This regulatory measure strictly prohibits the spread of infected or infested materials from other country or within the country. To significantly enforce this law, International Plant Protection Convention has been established for phytosanitary measures. The Government of India notified the Plant Quarantine Order 2003 for phytosanitary measures, and this order is based on pest risk analysis. *Sitophilus granaries, Sitophilus zeamais, Trogoderma sp., Prostephanus truncates, Acanthoscelides obtectus,* and *Hypothenemus hampei* are introduced due to export and import of stored commodities like rice, wheat, maize, gram, barley, berry, and vegetative propagative materials from several countries.

18.1   INTRODUCTION

India enforced Destructive Insect Pests Act in 1914 to check the introduction and spread of exotic pests. This regulatory measure strictly prohibits the spread of infected or infested materials from other country or within the country. To significantly enforce this law, International Plant Protection Convention has been established for phytosanitary measures. The Government of India notified the Plant Quarantine Order 2003 for phytosanitary measures, and this order is based on pest risk analysis. More than 700 plant pests have been identified including 190 insects and other species like bacteria, fungi, nematodes, viruses, and weeds (Narayansamy, 2009).

The flowing are the stored grain insects of quarantine importance in India.

18.2   GRANARY WEEVIL *SITOPHILUS GRANARIES*

The export and import of stored commodities like rice, wheat, maize, and barley must be free from this insect. In India, monitoring, supervision, and testing of paddy seeds against this insect are conducted by National Bureau of Plant Genetic Resources, New Delhi, and Indian Institute of Rice Research, Hyderabad, who follow the hot water treatment of paddy seeds. The consignment of wheat and rice for the purpose of consumption should be free from this insect as per phytosanitary certificate issued by the component authorities (PQR, 2003).

18.3 MAIZE WEEVIL *SITOPHILUS ZEAMAIS*

For the export and import of maize, the seeds must be free from *Sitophilus zeamais* and fumigated proper fumigants as per the protocol of quarantine department and same rules follow for consumption purpose of maize (CPC, 2003).

18.4 KHAPRA BEETLE *TROGODERMA* SP

The import of oat grain or seeds from Australia must contain phytosanitary certificate and should be free from *Trogoderma* sp. In special circumstances, the consignment must be fumigated with appropriate fumigants (PQR, 2003).

18.5 LARGER GRAIN BORER *PROSTEPHANUS TRUNCATES*

The export and import of maize must be free from infestation of *Prostephanus truncates* and the same should be followed for the purpose of their consumption with special declaration of phytosanitary measures (PQR, 2003).

18.6 PULSES BEETLE *ACANTHOSCELIDES OBTECTUS*

The import of beans either for seed or for consumption purpose must be free from infestation of *Acanthoscelides obtectus*, with required certificate of phytosanitary measures (CPC, 2003).

18.7 COFFEE BERRY BORER *HYPOTHENEMUS HAMPEI*

The coffee seeds or vegetative propagating materials are not imported from Africa and South America due to *Hypothenemus hampei* in India (PQR, 2003).

KEYWORDS

- quarantine
- stored grain insects
- exotic insects
- granary weevil
- maize weevil
- khapra beetle
- larger grain borer
- pulses beetle
- coffee berry borer

REFERENCES

CPC. *Crop Protection Compendium*. Common Wealth Agriculture Bureau International: London, 2003.

Narayansamy, P.; Mohan, S.; and Awaknavar, J. S. In *Pest Management in Stored Grains*, Proceedings National Symposium Entitled Towards Pest Free Grains and Seeds in Storage, Mar 26–27, 2007; Department of Entomology, Annamalai University: Tamil Nadu, India, Satish Serial Publishing House: New Delhi, India, 2009, p 272.

PQR. Plant Quarantine (Regulation of import in India). Ministry of Agriculture Government of India: New Delhi, India, 2003.

CALCULATION OF DOSES OF INSECTICIDES IN STORED GRAIN PEST MANAGEMENT

Calculating the dosage for surface treatment using dust formulations

Recommended application rates are given in g/m^2 (= g commercial product/ m^2 surface area).

Details required for calculation:

a. Surface area to be treated in m^2
b. Recommended application rate of insecticide in g/m^2

Amount of dust formulation required = recommended application rate × surface area.

Calculating the dosage for surface treatment using EC and WP formulations

Details required for calculation:
a. Amount of spray mixture for surface treatment
b. Amount of insecticide required for the spray mixture

a. Amount of spray mixture for surface treatment

Details required for calculation:

i. Surface area to be treated (in m^2)
ii. Recommended application rate of spray mixture (in $l/100 \ m^2$)

The following amounts are recommended for surface treatment;

Smooth walls   : 3–5 l/100 m^2

Rough walls   : 6–8 l/100 m^2

Jute bag   : 8–10 l/100 m^2

Plastic bag   : 3–5 l/100 m^2

Amount of spray mixture = recommended application rate × actual surface area to be treated.

b.   Amount of insecticide required for the spray mixture

1. Recommended application rate is given in ml/l (EC) or g/l (WP) (= ml or g of commercial product/l of spray mixture)

Details required for calculation:

i.   Amount of spray mixture (in l)
ii. Recommended application rate of insecticide (in ml/l for EC or g/l for WP formulation)

Amount of insecticide required = recommended application rate of insecticide × amount of spray mixture

2. Recommended application rate is given in % (= % of active ingredient in spray mixture)

Details required for calculation:

i.   Amount of spray mixture required (in l)
ii. Active ingredient content of insecticide (in %) (% a.i. in formulation)
iii. Recommended application rate of the insecticide (in %) (a.i. % required)

Amount of formulation required =(a.i. % required × volume of spray required)/% a.i. in formulation

Calculating the dosage for fogging

The dosage of a fogging concentrate depends on the volume of the free space in the store which may be calculated by deducting the volume of the stack from the total volume of the store. The recommended application rate of ready-to-use fog formulation is expressed in ml/100 m³.

RECORDS KEEPING IN WAREHOUSES

1. Store journal

The store journal contains a record of all procedures carried out in the store, such as incoming and outgoing produce, results of inspections, treatments, etc. Entries should be made daily and after any activities have been performed. The store journal consists of two tables: balance sheet and control sheet.

Balance sheet

The balance sheet contains all information on movements of the stored produce, the place of origin or destination, the stack number (or lot number in the case of seed), and reference to the relevant documents like invoices or receipts. These must be filed chronologically. The storekeeper confirms every procedure with his signature.

Control sheet

The control sheet contains information on all activities in the store, such as inspections and their results, treatments, cleaning and ventilation, any repairs, and weather data. An additional quality control book is required for seed stores in which the results of the laboratory tests which form a part of the essential internal quality maintenance program are recorded.

The store journal should be firmly bound and the pages numbered. The first part should consist of the balance sheets and the thicker rear part should consist of the control sheets. A separate journal should be kept for each store and should remain in the store.

2. Stock sheets

In storage centers and in seed stores, store book-keeping is composed of journals from the individual stores or lots, making it very time-consuming to calculate the actual amounts of produce present in the stores. In such cases it is practical to keep stock sheets. A stock sheet shows the current overall stock of the storage center at any one time on a single page. The stock sheet is divided up according to the type of produce and, in the case of seed, according to type, category, and state of processing. Entries should consist of the date of any movement, the new overall total stock, and the reference to the store where the movement has taken place. This enables the details of the procedure to be checked in the balance sheets of the relevant store. Stock sheets are also kept in the form of a firmly bound book.

3. Stack cards

Every stack is given a stack card placed where it is clearly visible. This serves to identify the stack and the produce and contains details on inspections and pest control measures performed.

4. Monthly report

The storekeeper's monthly report serves to inform superiors on amounts of produce and its state, on the storage conditions as well as on activities and any problems in the store. These reports should also be referred to on the inspections of the store regularly done by the superior.

SAMPLE OF STORE JOURNAL
BALANCE SHEET

Commodity... Warehouse..

Date	In (t)	Out (t)	Balance (t)	No. of bags	In/out stack no.	Origin destination	Document no.	Signature

CONTROL SHEET

	Controls					Treatment				Aeration	Climate		Sign.
Date	Commodity	Stack no.	MC %	Insect infesta-tion	Other observa-tion	Kind of Treat-ment	Chemical used/rate	Result	Cleaning/ repair	Vents manipulated time	Temp.	rh	

SAMPLE OF STOCK SHEET
Commodity..........................

Date	Store no.	Balance (t)

SAMPLE OF STACK CARD
Food Corporation of India

Warehouse no./name...

Stack no.. Commodity...

Lot no.. Variety..

Origin..

Date	In (ton)	Out (ton)	Balance (ton)	No. of bags	Signature

SAMPLE OF STACK CARD
Seed Corporation of India

Warehouse no./name...

Lot no.. Raw seed..

Commodity.. Pre-cleaned seed..

Origin.. Processed seed

Variety..Treated with..

Date	In (ton)	Out (ton)	Balance (ton)	No. of Bags	Signature

Back of Stack Card

Stack dimension

Length......................m   Amount of water needed for surface treatment.....................l

Width......................m   Amount of chemical (EC/WP) /liter water.......................ml

Height......................m   Number of tablets needed for fumigation..........................

Surface area................m²

Volume ....m³

Date of inspection	Moisture content	Insect infestation	Date of treatment	Kind of treatment	Chemical used	Rate of application	Remark	Signature

WAREHOUSE CHECKLIST

Location: Name of store: Capacity in ton

Products stored: Amount stored in ton

1. Condition of warehouse surrounding   YES   NO

Is the surrounding of the warehouse free from

   a)   Accumulation of grains, old bags, junk, and trash?
   b)   Weeds, tall grass, and bushes?
   c)   Evidence of rodents?
   d)   Standing water?

Condition of warehouse exterior

   a)   Is the roof intact?
   b)   Is the water drainage intact?
   c)   Are the walls without holes or cracks?
   d)   Do the doors close hermetiqually?
   e)   Are the ventilation openings protected against the penetration of insects, rodents, and birds?

Condition of warehouse interior

   a)   Are the walls, the floor, and the roof undamaged?
   b)   Is the floor and the roof clean?

    c)   Is the floor free of spilled grain, dirt, and trash?
    d)   Do the ventilation openings function properly?
    e)   Is the store free of residues of former treatments (empty phosphine tubes, phosphine residues, rodent baits, etc.)?

Storage practices

    a)   Are all empty bags stored on pallets?
    b)   Are all stacks at least 1 m apart?
    c)   Are insecticides, fertilizer, and other products stored separately from the grain?
    d)   Are all bags in the stacks without holes?
    e)   Are all stacks built in a safe way?
    f)   Are stack cards in use for all stacks?
    g)   Are the stock journals kept up to date?

Presence of pests

    a)   Is the store free of flying insects?
    b)   Are the walls free of crawling insects, larvae, and pupae?
    c)   Are the bags free of crawling insects, larvae, and pupae?
    d)   Is the store free of evidence of rodents?
    e)   Is the store free of evidence of birds?

Pest control

    a)   Has any pest control treatment been done shortly before or during the inspection?
    b)   If so, what kind of treatment?
    c)   Which pesticide has been applied?
    d)   In case, bait stations against rodents are in use, are they furnished with fresh baits?

GLOSSARY

Active ingredient Chemical in a pesticide formulation that controls the pest(s).

Ambient Surrounding air and its properties, temperature and humidity.

Anticoagulant Substances that kill by preventing normal blood clotting and causing internal hemorrhaging.

Asci Sac-like cell of hyphae in which meiosis occurs and which contains ascospores.

Ascospore Specialized haploid cell produced during meiosis.

Asexual Not involving union of sex cells. Usually vegetative reproduction in plants and asexual spore production in fungi.

Atomized To reduce to tiny particles or a fine spray.

Avicides Pesticide used to control or repel birds.

Basidiospores A sexual fungal spore of fungi in the division Basidiomycetes which includes mushrooms, puffballs, smuts, and rusts. Basidiospores are usually borne on a microscopic **basidium** or stalk.

Bunt Fungal disease of small grains. Bunt affects plants by destroying the contents of the infected seed kernels and replaces them with spores of the fungus.

Bushel A unit of volume that is 2152.42 cubic inches. Also denotes 60 pounds of wheat.

Bushel 1.2444 cubic feet or 149 eight-ounce cups.

Cadaver Dead body.

Cereal Crops A grass plant such as wheat, oats, or corn, the starchy grains of which are used as food.

cfm/bu Cubic feet per minute per bushel.

Chlamydiosis A rare infectious disease that causes pneumonia in humans. The illness is caused by a chlamydia, which is a type of intracellular parasite closely related to bacteria. Also known as Parrot fever.

Chlorotic Lacking chlorophyll. Yellowing of plant leaves.

Codex Maximum Residue Limits Following the application of a pesticide product, trace residues may be present at harvest on agricultural commodities, such as fruits, vegetables, or cereal grains or in processed commodities, such as fruit juices, vegetable oils, and flours. Thus, government regulatory authorities have established maximum residue limits (MRLs) or tolerances as a check for compliance with national good agricultural practices (GAP) and to facilitate international trade. MRLs are referred to as "tolerances", and in general the two terms are synonymous. On the international level, the Codex Alimentarius Commission has sought to establish a globally applicable listing of harmonized MRLs to support international trade. Thus, many countries refer to Codex MRLs when considering regulatory and trade aspects of pesticide residues.

Coleoptera An order of insects that includes the beetles. Beetles constitute the largest and most diverse order of insects on earth, making up about 30% of all animals.

Coleoptile Sheath covering the growing shoot of a young plant seedling. As growth continues, the coleoptile ruptures to expose the first leaf.

Commodity A product that can be used for commerce and includes agricultural products such as grain, livestock, etc.

Conidiophore A specialized hypha on which one or more conidia are produced.

Conidium (Conidia pl.) An asexual spore.

Detritus Parts of dead organisms and cast-off fragments and wastes of living organisms.

Diatomaceous Earth (D.E.) Fossilized skeletons of one-celled organisms called diatoms. When soft-bodied insects come in contact, D.E. causes massive loss of body fluids and death. When the dust is eaten by insects, the D.E. inhibits breathing, digestion, and reproduction. Because it kills by mechanical action rather than poison, insects do not usually develop resistance.

Diluent Substance that is added to a pesticide formulation to dilute its concentration. Usually water.

Encephalitis An inflammation (irritation and swelling with presence of extra immune cells) of the brain, usually caused by infections.

Exuviae Cast-off skins or coverings of various organisms such as insects.

Filamentous Thread-like.

Formulation The manner in which a pesticide is prepared for practical use and usually contains the active ingredients (a.i.) and inert or other ingredients.

Frass Debris and fecal matter produced by insects.

Gastroenteritis Inflammation or irritation of the stomach and/or intestine.

Genus A grouping of organisms that usually contains a group of closely related species.

Glume Small leaf-like structure found at the base of grass spikelets.

Grain Protectant Usually a liquid insecticide applied directly to grain.

Hantavirus A type of virus carried by rodents causing severe respiratory infections in humans and, in some cases, hemorrhaging, kidney disease, and death.

Haploid Having one set of chromosomes.

Harborage A place of refuge or shelter.

Hardware Cloth A metal fabric also known as wire cloth; the mesh openings are larger than window screening, but smaller than fencing.

HEPA High efficiency particulate air filters that have been tested to assure removal of 99.9% of particles that are 0.3 microns (µm) in size.

Histoplasmosis A disease caused by the fungus *Histoplasma capsulatum*; may infect lungs, skin, mucous membranes, bones, skin, and eyes.

Host A plant that is invaded by a pathogen and from which the pathogen obtains its nutrients.

Hundredweight (cwt) One hundred pounds, usually of seed. C is Roman for 100. Wt means weight.

Hydathode Gland occurring on the leaf edges of many plants and secretes water.

Inert Ingredients That part of a pesticide formulation that has non-pesticidal qualities. Inert ingredients usually aid in storage and application.

Infection peg A very fine hypha that is thrust through the cuticle or epidermis of a plant host cell.

Inoculant The pathogen or its parts that causes infection. Also refers to beneficial nitrogen-fixing bacterium that is applied to legume seeds to produce nodulation.

Inoculation Arrival or transfer of a pathogen onto a host.

Insectivorous Eats insects.

LD 50 A single dose of a material expected to kill 50% of a group of test animals. The LD50 dose is usually expressed as milligrams or grams of material per kilogram of animal body weight (mg/kg or g/kg). The material may be administered by mouth or applied to the skin. The lower the LD50, the more toxic a compound.

Lepidoptera An order of insects that includes moths and butterflies.

Lesions A localized area of discolored, diseased tissue.

Meisosis A special process of cell division during which spores are produced. Meiosis involves the reduction by half in the amount of genetic material.

Meningitis An infection which causes inflammation of the membranes covering the brain and spinal cord.

mm Millimeter. One millimeter is 1,000th of a meter or 1/25th of an inch.

Mode-of-action The way in which a pesticide affects a pest at the cellular level.

Montreal Protocol The Montreal Protocol on Substances That Deplete the Ozone Layer_is an international agreement designed to protect the stratospheric ozone layer. The treaty was originally signed in 1987 and substantially amended in 1990 and 1992. The Montreal Protocol stipulates that the production and consumption of compounds that deplete ozone in the stratosphere are to be phased out by 2005.

Mosaics Symptoms of certain viral diseases characterized by intermingled patches of normal and light green or yellowish color.

Mycelium The combined mass of microscopic thread-like strands (hyphae) that makes up the body of fungi and produces vegetative spores.

Mycotoxins Toxic substances produced by fungi or molds on agricultural crops that may cause sickness in animals or humans that eat feed or food made from contaminated crops.

Necrosis Describes the condition of being dead or discolored.

Necrotic Dead and discolored.

New Castle Disease A highly contagious virus disease of domestic poultry and wild birds characterized by gastro-intestinal, respiratory, and nervous signs.

NIOSH National Institute of Safety and Health.

Nodulation Creation of nodes (swelling or bumps) on legume roots by nitrogen-fixing bacteria.

Non-systemic Does not move within a plant or animal. Usually affects on the part that it touches.

Ovule A structure found in plants that contains an egg cell and develops into a seed after fertilization.

Pericarp Seed wall which develops from the mature ovary wall.

Pesticide A chemical substance (e.g., an insecticide or fungicide) that kills, controls, or repels harmful organisms and is used to control pests, such as insects, weeds, microorganisms, rodents, birds, etc.

Pheromone A chemical released by an insect or other animal through which it communicates with another individual of the same species through a sense of smell.

Phytoplasmas Smallest bacteria-like microorganisms. Formerly called mycoplasma-like organisms. They differ from bacteria by lack of a solid wall.

Pistil The flask-shaped female reproductive unit of a flower that is composed of the ovary, style, and stigma.

Postemergent Usually refers to herbicides applied after weeds have emerged from the soil and generally act through the foliage of the plant.

Preemergent Usually refers to herbicides applied to the soil and are absorbed by the seed or by the roots or stems of tiny seedlings before the plants emerge from the soil.

ppm Parts per million.

Pycnidia Asexual, flask-shaped fruiting bodies lined inside with conidiophores that produce conidia.

Race A genetically and often geographically distinct mating group with a species. Also a group of pathogens that infect a given set of plant varieties.

Rachis The central axis of a grain spike, flower cluster, or compound leaf.

Restricted-Use-Pesticide A pesticide that can be sold to or used by only certified (licensed) applicators.

Rodenticides Pesticides used to control rodents.

Root Rots Softening, discoloration, and often disintegration of succulent root tissue as the result of fungal or bacterial infection.

Rusts A disease giving a "rusty" appearance to a plant and caused by one of the rust fungi.

Salmonellosis An infection caused by Salmonella bacteria and commonly manifested by diarrhea, abdominal pain, nausea, and sometimes vomiting.

Saprophyte An organism that uses dead organic matter for food.

Sclerotium A compact mass of hyphae with or without host tissue, usually with a darkened outer shell, and capable of surviving under unfavorable environmental conditions.

Seed Rots Softening, discoloration, and often disintegration of seed tissue as the result of fungal or bacterial infection.

Slurry A suspension formed when a quantity of powder is mixed into a liquid in which the solid is only slightly soluble (or not soluble). Fungicides and some insecticides are applied to seeds as slurries to produce thick coating and reduce dusts.

Smuts A disease caused by the smut fungi (Ustilaginales). Smuts are characterized by masses of dark, powdery spores.

Snap Trap Usually a common mousetrap. A piece of equipment used to quickly trap or catch mice and is usually composed of three parts: base, trigger, and the spring-loaded snap.

Spores The reproductive unit of fungi consisting of one or more cells.

Sporulating Producing spores.

Stigma The upper part of the pistil of a flower on which pollen is deposited.

Stomata A microscopic pore in a leaf or stem through which gases and water vapor can pass. The exchange of air and transpiration of water are regulated by the two cells (called guard cells) on either side of a stomate.

Systemic Absorbed and translocated throughout a plant or animal. Systemic fungicides are absorbed by the plant, then translocated throughout its tissues. They can be applied after an infection occurs and still have a benefical impact.

Tampico A tough vegetable fiber used as a substitute for bristles in making brushes.

Teliospore A thick-walled spore characteristic of the rust and smut fungi.

Thermal Inertia The tendency of an object with large quantities of heavy materials to remain at the same temperature or to fluctuate only very slowly.

Thermocouple A thermoelectric device used to measure temperatures accurately, especially one consisting of two dissimilar metals joined so that a potential difference generated between the points of contact is a measure of the temperature difference between the points.

Tillering Producing a shoot, especially one that sprouts from the base of a grass.

Toxoplasmosis An infection with the protozoan intracellular parasite *Toxoplasma gondii.*

Translocated Movement of a substance within a plant or animal.

Vector Animal capable of transmitting a pathogen.

Volatile The term given to a substance that is easily converted to the gas state.

Volunteer Plants Plants outside of a defined row and are usually produced from "leftover" seed from the previous growing season. Plant arising from seed dispersed from a previous crop.

Webbing A network of threads spun by certain insect larvae.

West Nile virus (WNV) West Nile virus is transmitted by mosquitoes and causes an illness that ranges from mild to severe. Mild, flu-like illness is often called West Nile fever. More severe forms of disease, which can be life-threatening, may be called West Nile encephalitis or West Nile meningitis, depending on where it spreads.

INDEX